Handbook of Power and Energy Engineering Volume I

Edited by **Linda Morand**

NY RESEARCH PRESS

New York

Published by NY Research Press,
23 West, 55th Street, Suite 816,
New York, NY 10019, USA
www.nyresearchpress.com

Handbook of Power and Energy Engineering: Volume I
Edited by Linda Morand

International Standard Book Number: 978-1-63238-269-6 (Hardback)

Printed in China.

Contents

Preface

Energy and power are inter-related concepts. Everything derives energy from power. Since the use of fire human beings have been converting one form of energy into another. The efficiency of this transfer of energy gave birth to the field known as Energy and Power Engineering. Today, there are numerous transformation processes such as photo-voltaic, electromechanical and chemical for energy conversion.

Energy and power engineering covers a network of interconnected components such as electric generators, transformers, power electronics and electric motors, which convert different forms of energy to electrical energy and vice versa. From utilization to distribution, transmission and generation of energy, all these steps are crucial aspects of energy and power engineering.

Energy engineering examines the numerous uses of energy. This is done in order to suggest new ways to improve the existing systems. The ultimate aim is to utilize all forms of energy in the most eco-friendly way. From renewable energy sources like solar, hydro, geothermal and biomass to fossil fuels such as oil, energy academicians are concerned with all these sources.

The various researches in this book have detailed statistical models and research methodologies for upcoming students and researchers interested in this field. I would like to thank our researchers and writers from all the parts of the world for sharing their valuable research with us. I would personally like to thank the publishing house for giving me this immense opportunity and believing in me.

Editor

1

Communication Networking Schemes for Wide Area Electric Vehicle Energy Service Network

Dequan Gao[1,2], Jinping Cao[3], Yiying Zhang[1, 2], Xiuli Wang[4]
[1]Beijing University of Posts and Telecommunications, Beijing 100876, China
[2]State Grid Information & Telecommunication Company Ltd., Beijing 100761, China
[3]China Electric Power Research Institute, Beijing 100192, China
[4]Information Communication Branch of Beijing Electric Power Company, Beijing 100031, China

ABSTRACT

Electric vehicles (EVs) are an emerging type of mobile intelligent power consumption devices in Smart Grid as new green transport tools. In order to provide a powerful automation and intelligence support for wide area electric vehicles energy service network, we analyze the network infrastructure and communications demands of various terminals, devices and monitoring systems distributed in wide area electric vehicle energy service network. According to interactive user services scenarios and energy operations intelligent monitoring, we propose multimode communication integration architecture for wide area electric vehicle energy service network by means of the fusion of the Internet of Things (IoT) technology. Then, we design different networking schemes in access networks and backbone transmission networks meeting multi-scene and multi-operation interaction requirements. The networking schemes will provide efficient technical support to implement intelligent, cross-regional, interactive energy services for electric vehicle users.

Keywords: Electric Vehicle; Wide-area Energy Service Network; Communication Networking; Internet of Things

1. Introduction

Many developed countries and megacities are undertaking serious environmental pollution and great oil shortage challenges. So world automobile industries are turning to develop clean energy vehicles with huge investment. Now, many people have realized that future trends of automobile industry are green, energy-saving and intelligent. Electric vehicles (EVs) are seen as one of the most promising means to improve the near-term sustainability of the transportation and stationary energy sectors [1]. Electric vehicles have improved their performance and made suitable for commercial and domestic use during the last decades [2]. Electric vehicles will be prominently taken as the mobile infrastructures of electrified transportation mode. Electric vehicles development has been as one of new energy national strategies in China. More importantly, electric vehicle is an emerging type of mobile intelligent power consumption device and energy storage terminal for Smart Grid. With the large-scale population of electric vehicles, electric vehicles can be conventionally charged to play the role of the valley filling in the power grid while its load is low. Thus, they will improve the comprehensive utilization efficiency of power generation equipment, and achieve the effect of energy saving and emission reduction. Inevitably, to construct wide-area energy service network for electric vehicles is the core premise of their large promotion. According to the 12th Five-Year special plan on electric vehicle technology development in China, a networked power supply system, which includes 400,000 charging piles and 2000 charging-swap stations, will be built in above 20 demonstration cities and their surrounding areas to meet the energy supply requirements of large-scale commercial demonstration of electric vehicles until about 2015.

The domain of electric vehicles opens new business perspectives and opportunities [3]. Electric vehicle energy services businesses mainly cover battery charging, battery swap, battery discharging, battery distribution, electric energy metering and billing, fund settlement, scheduling monitoring, etc. the basic operations functions are devices running IntelliSense, real-time monitoring and warning of battery status, vehicle status information acquisition, charging-swap path intelligent navigation, battery life-cycle management, centralized monitoring of equipment, optimal allocation of resources, etc. The distributed smart charging-swap information in energy service network should be measured, interacted, shared and controlled, which necessarily relies on rapid and reliable information communication system.

Next generation electric vehicles will radically change the design paradigms in the automotive network domain [4]. However, information communications network supporting wide-area electric vehicle information acquisition and key charging-swap service operations and monitoring, has not met the practical, interactive requirements. These problems have made intelligent interaction more difficult between the power grid and users, and led to be unable to provide conveniently, efficient and safe services for electric vehicles users. Therefore, reliable information communication technology (ICT) should be developed to support information perception, aggregation, interaction, and highly automated, intelligent charging-swap services. This paper focuses on the objectives of providing efficient and interactive electric vehicles smart energy services. Some exploratory researches involve wide-area electric vehicles energy services network communications networking architecture. In this paper, we synthetically apply the Internet of Things (IoT) technology and power communication to improve intelligent perception of electric vehicles energy information and services integration level between electric vehicles and power grid.

2. EV Energy Service Network Facilities

Electric vehicle energy service network is indispensable infrastructure for large-scale commercialization of electric vehicles. Many Chinese cities have released preferential policies and invested massive funds to promote the construction of electric vehicle energy service facilities. Generally, electric vehicle energy service network facilities are composed of centralized charging stations, battery swap stations, battery distribution stations, AC charging pile, operation monitoring center, etc. They will guarantee reliable energy supply for electric vehicles in a wide area.

Centralized charging stations can provide electric vehicle battery charging and battery distribution stations. Typically, a centralized charging station has the capabilities of charging, discharging, maintenance and detection of large amount of standardized battery packs.

Battery swap stations are power battery supply spots by exchanging batteries for electric vehicle users. In a battery swap station; there are power supply area, charging area, battery replacement area, battery inspection and maintenance area, control room, parking lot, etc. Battery distribution stations provide battery replacement and logistics services for battery packs distribution without charging function. AC charging piles are slow charging devices for electric vehicles. The functions of a charging pile involve timing, power energy measurement, costing calculation. Operation service centers are responsible for monitoring and management of regional electric vehicle energy service network operators.

3. Communication System Architecture of Wide Area EV Energy Service Network

Shifting bulk loads and demand response programs require communication between consumers and producers of energy and will be more widely enabled with the expansion of grid communication networks across the country [5]. The communication system of electric vehicle energy service network will achieve the interconnection and interoperability of all kinds of terminal, electric vehicle and operations management system. The communication system can support orderly, seamless cross-regional communication services in the cities, provinces, inter-city and inter-provincial areas.

3.1. Communication System Demands

The characteristics of the communication system are flexible networking, multi-mode access, and massive access points. We analyze its demands from the view of security, real-time and bandwidth. As an important part of intelligent distribution and utilization networks, it is linked with the power grid enterprise businesses such as energy metering, tariff settlement, electricity management. So power grid must be safe and stable when a large number of wide-area distributed charge-swap devices are linked into it. Therefore, security requirements of wide-area communication network are rather higher for operation service center.

All business surveillance, operations monitoring, video monitoring, battery monitoring, power supply monitoring, logistics distribution information of centralized charging stations, battery-swap stations, battery distribution station, and service management center must be real-time uploaded, so the real-time requirements of communication performance are highly efficient. Moreover, electric vehicles (including logistics vehicles) will upload vehicle conditions, battery status information and receive value-added services and scheduling information, relatively, their real-time communication requirements are also higher.

Electric power information acquisition system, operation management system, station-level monitoring system need be deployed in centralized charging stations and battery swap stations. Their communication bandwidth is expected to reach MB level. The communication bandwidth of charge piles and electric vehicles is approximate 30 kbps.

3.2. Communication System Architecture

At present, there are some technical problems restricting the large-scale commercial development of electric vehicles in China. For example, the existing communication networks cannot cover both urban and rural areas, cross area to support complex electric vehicles services. Thus, we must integrate power communication network, the public network (GPRS/3G/4G), cable network, wireless sensor networks (WSN), and other communications technology to establish top-level communication system architecture of the electric vehicle energy services network. The communication system will smart, effective interaction between electric vehicles, battery packs, charging-swap facilities and power grid, and seamless mobile information services.

According to the above demand analysis of electric vehicle energy network communication system, we propose three-layer communication network architecture supporting dynamic access and switch, as shown in **Figure 1**. The network architecture will establish cross-regional information transmission covering headquarters operation service center, provincial operation service centers, municipal operation service centers and station monitoring center.

First-level network is the backbone telecommunication network linking operation service headquarters and subordinate operation service centers. First-level network adopts power communication private network. However, for the construction of power communication private network relies on power transmission line path, it actually hasn't covered some expressways, provincial highways. Power companies have built current communications network involving optical fiber cable, microwave, Power Line Carrier (PLC), satellite, which have fully covered headquarters, provincial and municipal companies. Therefore, the communication system of electric vehicle energy service network can further extend current enterprise network to solve the wide-area coverage problem.

Second-level network is the transmission network linking subordinate operation service centers and station level facilities. Second-level network mainly adopts provincial power communication private network. If the private network isn't available, the public communication network can be rented. Here, Access networks can apply WSN, Optical Fiber Composite Low-voltage Cable (OPLC), 3G/GPRS/4G/TD_LTE to implement information exchange between charging-swap devices, vehicle automatic identification equipment and monitoring equipment, and information interaction between inner equipment inside charging-swap stations, electric vehicles and battery packs, and information access of geographically distributed charging piles. On the whole, the communication network coverage is small for low-voltage side distribution network. We can adopt wired communications (optical fiber communications, etc.), IoT communication technologies and public wireless communications (GPRS/3G/4G) for different scenarios to aggregate a variety of perception information, for example, battery status information, identity information, electric vehicle status information, location information, the smart electric card identity information, charging-swap environment information. The aggregated information will be uploaded to the Second-level network layer through the IoT aggregation gateway. Meanwhile, charging-swap devices and various intelligent terminals also receive control instructions.

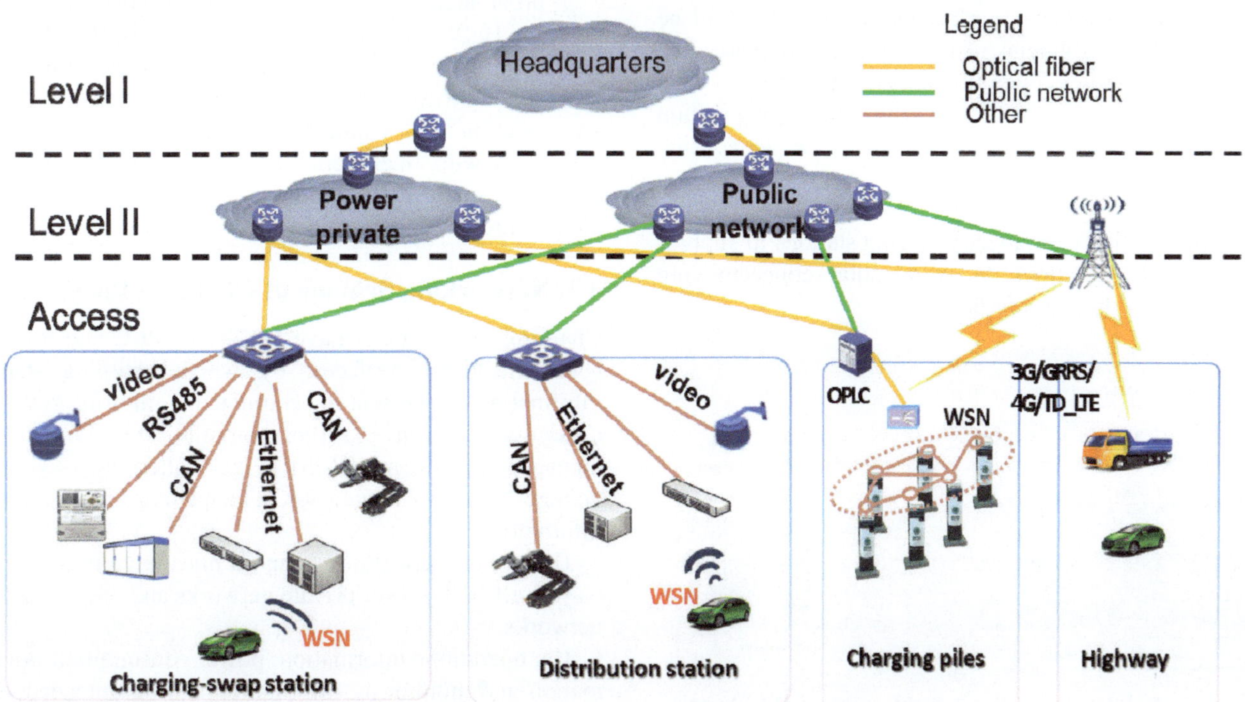

Figure 1. Three-layer communication network architecture.

4. Networking Schemes of Access Networks

Now the backbone telecommunication transmission networks and data communication network built by power grid companies can basically meet communication business demands of the First-level and Second-level network in the energy service network architecture. Here, we will mainly focus on designing networking schemes of access network layer and take low-voltage power communication network as main basic of access network. We will use self-built fiber network, electricity wireless broadband network, IoT technology together with renting public network to solve heterogeneous interconnection and collaborative communication problems for power wireless broadband network, mobile communication network, and WSN applied in different application scenarios. a particular IoT service can be applied in order to optimize application development and speed up application implementation [6]. Monitoring and automation applications will be central for IoT and rely on eventing and group communication [7]. These networking schemes will realize seamless network coverage for centralized charging stations, battery swap station, battery distribution stations, charging pile and electric vehicles, and network access within station sites.

4.1. Networking Scheme of Centralized Charging Stations

The communication network within a centralized charging station will connect with access network through the access gateway nodes. The operations, power utilization and monitoring information on the station LAN will be transmitted to provincial service center, and share data with electric vehicle operation management system. The networking structure inside a centralized charging station is shown in **Figure 2**. The following is networking description.

1) Switches will be core nodes interconnection of equipment inside centralized charging stations to support station-level operations. A workstation connects core switches via Ethernet interfaces.

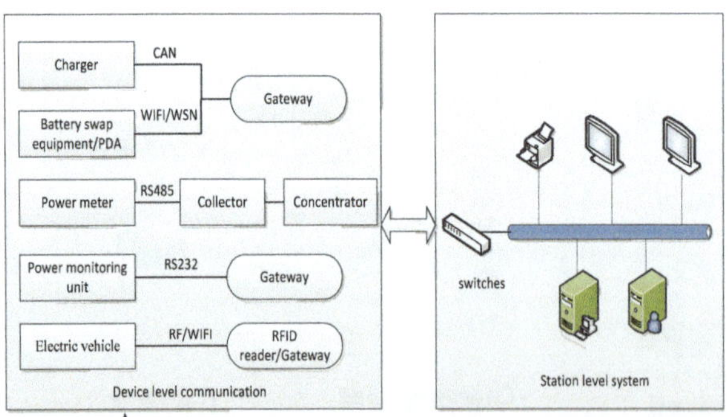

Figure 2. Networking scheme inside centralized charging station.

2) Charging devices access IoT communication gateway via industrial Ethernet or CAN bus, by which they upload battery information and charger information to the monitoring system, and also receive command instructions.

3) A power supply monitoring unit accesses IoT communications gateway via RS232 interface. A collector uses RS485 interface to communicate with electric energy meters and collect real-time consumption information.

4) Video terminals send surveillance data to station operation monitoring system via video cables.

5) Logistics vehicles can use micro-power wireless network to automatically communicate with vehicle identification readers inside a station.

In order to ensure station-level information network security, local centralized charging station business systems are logically isolated via Virtual LAN (VLAN) technology.

4.2. Networking Scheme of Battery Swap Stations

According to the practical communication requirements in a battery swap station, the swap devices usually use wireless communication technologies to link station-level monitoring system, for example, WIFI, ZigBee. The operations, power consumption and monitoring information of a battery swap station will be sent to regional operation management system via uplink network by renting special public communication network.

In urban areas, a battery swap station will connect the nearest 10kV switching substation, 35/110/220 kV substation with fiber optic cable resources. While in intercity highways or highway service areas, a battery swap station will connect the nearest 35/110/220 kV substation with fiber optic cable resources. The uplink communications network fiber optic cable can be laid along with power lines.

4.3. Networking Scheme of Charging Piles

Charging piles usually have CAN bus, PLC and RS485 communication interfaces, by which charging status information can be sent to centralized monitoring system. Charging piles can Real-time communicate with battery management system (BMS) to get battery type, single voltage, State-of-Charge (SOC), temperature and warning information.

The uplink networks of charge piles have two access ways: self-built power private networks and rented public networks.

The operations information, power consumption information and monitoring information can be uploaded via LAN deployed in10kV distribution room, switching cabinet, box-type substation. Also, they can use embedded wireless communication modules to connect public network.

4.4. Networking Scheme between Charging-swap Facilities and Operation Service Centers

The key communication requirements between charging-swap facilities and the operation service centers of energy network are wide communication range, rich data semantic content and various bandwidth applications, so we must flexibly use multiple communication technologies for different application scenarios. Power optical fiber, PLC, GPRS and other communication technologies can be mixedly used to communicate with operation service centers for charging-swap stations and distributed charging piles. For power fiber cables only rely on electric power communication systems resources, they can avoid technical conflict on frequency resources, routing coordination and electromagnetic compatibility with other users. PLC technology is unique communication way for power industry. The PLC scheme doesn't need to newly build communications infrastructure for it uses power line as communication media for data transmission and information exchange. Station-level management systems can real-time interact with electric vehicles energy service management system in operation service center, electric energy data acquire system and 95598 call center through power fiber-optic network.

4.5. Networking Scheme of Electric Vehicles

The main external interaction objects of electric vehicle are operation service centers, charging-swap stations and charging devices. For mobility of electric vehicles, their driving range, spatial location and speed are random. Obviously, different communication technologies will be applied inside electric vehicles or while they are moving in urban areas or on inter-city highways. The networking schemes of electric vehicles are described in **Figure 3**.

Interactive services of electric vehicles in different scenarios will adopt different communication modes. Careful monitoring and control of energy flows allows for minimum investment with respect to cost, weight and

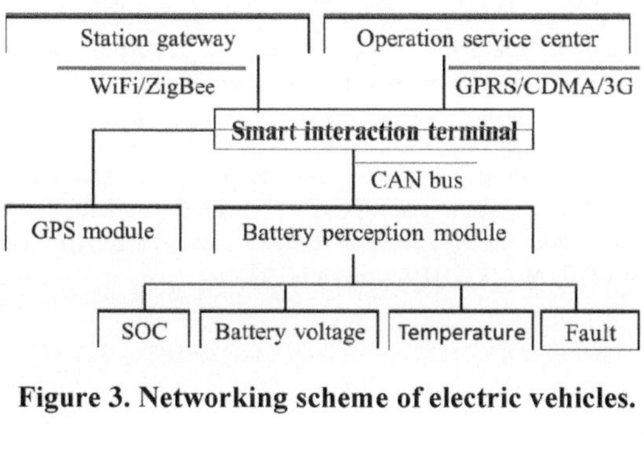

Figure 3. Networking scheme of electric vehicles.

volume [8]. Users can get real-time battery status data via vehicle CAN bus interface interacting with BMS while driving. The application of bus technology with CAN as the representative in vehicles not only reduces the har-ness of the car but also increases the reliability of the car [9]. IoT based smart interactive terminals can be embedded into electric vehicles to realize perception information transmission among different interior parts. Smart interactive terminals integrate battery energy sensor, temperature sensor, identification chip, communication unit, GPS module, navigation software to realize battery packs identification, power information collection, speech warning and map navigation, etc.

Electric vehicles can use WIFI, micro-power wireless network to interact with smart terminals via Iot gateways set in charging-swap stations. Furthermore, they generally interact with charging devices via CAN interface.

Users often adopt renting public wireless communication network to interact with service centers via GPRS/3G/4G outside charging-swap stations. In the future, the new generation power wireless broadband private network and White Space (WS) technology-based low-frequency wireless communication can be used to achieve region-scale coverage data transmission. In general, wireless mobile communication network can be used as the primary networking means, the roadside big WiFi AP networks and TD-LTE based new power wireless broadband communications serve effective supplement means.

4.6. Communication Protocols and Interfaces Security

The main communication interfaces of electric vehicles energy service Network are data communication interface of charging-swap devices, information interoperation interface of operation monitoring center and BMS interface of electric vehicles. In order to establish universal communication protocol, we must find out the common characteristics of electric vehicle charging demands and charging-swap device functions from too many manufacturers. However, the communication protocol for electric vehicles hasn't been already unified for the BMS interfaces from different manufacturers are special. To actually ensure real-time data upload of monitor electrical equipment operating condition, switch status and fault alarm in charging-swap facilities, the modified IEC 60870-5-104 protocol can be widely used.

In order to ensure wired communications security inside charging-swap stations, the monitoring systems of charging-swap and power supply management must only access the internal information network through communication front-end processor. Station-level security and protection subsystems access upwards the network to implement data transmission of video stream. Moreover, the devices inside a station must firstly be requested to traverse firewalls for ensuring network security if they will access to the internal information network. Under some special conditions, public wireless networks may be used in charging-swap stations if there are no wired power network links. In the above case, the communication front-end processor must be equipped with a radio signal transceiver and a SIM card. Meanwhile, The SIM card should have a static IP address. Thus, the equipment in a station will access the internal information network via public wireless communication network according to the APN (Access Point Name) special line of a station system

5. Conclusions

The reliable information collection, transmission, control and management of electric vehicles and charging-swap facilities are some important links to effectively supervise energy service procedure. In order to adapt mobility, diversity and universality requirements of electric vehicle energy services, we make deep demand analysis on the networking infrastructure and communications demands of diverse terminals, devices and monitoring systems distributed in wide area electric vehicle energy service network. According to energy service scenarios for electric vehicle and systematic requirements of intelligent operation monitoring, we propose multimode communication architecture for supporting wide area electric vehicle energy service network with the fusion of IoT technology. The architecture involves hybrid networking mode widely covering expressways, national highways, provincial highways, charging-swap facilities to implement seamless coverage and dynamic access of energy service information. We integrate Multimode communication technologies to design interactive networking schemes for EVs multi-business energy services. The networking schemes will partly solve some basic problems of perception data aggregation and information communication support for cross-regional energy services, and provide some technical means for friendly information interaction.

Electric vehicle energy service infrastructures will have rapid development in next decades in China. In future, it is urgently necessary to innovatively use IoT technology to build a demonstration platform of data acquisition and information management for electric vehicles energy service network. Moreover, the research on information interaction methods between electric vehicle and charging-swap networks should be further done. Wireless smart sensors, perception tags, GPS and other IntelliSense technologies need be integrated to fully collect data in physical world for reliably solving the problem of the electric vehicle performance assessment, fault diagnosis, safety pre-warning and error risks analysis.

6. Acknowledgements

This work is supported by National High-tech R&D Program of China (863 Program) (No. 2011AA05A116, 2012AA050804) and National Program on Key Basic Research Project of China (973 Program) (No. 2011CB302900).

REFERENCES

[1] C. Quinn, D. Zimmerle and T. H. Bradley, "The Effect of Communication Architecture on the Availability, Reliability, and Economics of Plug-in Hybrid Electric Vehicle-to-grid Ancillary Services," *Journal of Power Sources*, Vol. 195, No. 5, 2010, pp. 1500-1509.

[2] J. Moreno, M. E. Ortúzar and J. W. Dixon, "Energy-management System for a Hybrid Electric vehicle, Using Ultracapacitors and Neural Networks," *IEEE Transactions on Industrial Electronics*, Vol. 53, No. 2, 2006, pp. 614-629.

[3] N. Masuch, M. Lutzenberger, S. Ahrndt, A. Heßler and S. Albayrak. "A Context-aware Mobile Accessible Electric Vehicle Management System," *Proceedings of the Federated Conference on Computer Science and Information Systems*, Szczecin, 18-21 September 2011, pp. 305-312.

[4] M. Lukasiewycz, S. Chakraborty and P. Milbredt, "FlexRay Switch Scheduling-A Networking Concept for Electric Vehicles," *Design, Automation & Test in Europe Conference & Exhibition (DATE)*, Grenoble, 14-18 March 2011, pp. 1-6.

[5] T. Markel, M. Kuss and P. Denholm, "Communication and Control of Electric Drive Vehicles Supporting Renewables," *IEEE Vehicle Power and Propulsion Systems Conference*, Dearborn, 7-10 September 2009, pp. 27-34.

[6] M. Gigli and S. Koo, "Internet of Things: Services and Applications Categorization," *Advances in Internet of Things*, Vol. 1, No. 2, 2011, pp. 27-31.

[7] M. Kovatsch, "A User-Centered Application Layer for the Internet of Things", *ACM SenSys'11,* Seattle, 1-4 November, 2011.

[8] E. Meissner and G. Richter, "Battery Monitoring and Electrical Energy Management Precondition for future vehicle electric power systems", Journal of Power Sources, Vol. 116, No. 1, 2003, pp. 79-98.

[9] Q. Q. Zhang, Y. Wang and T. M. Yin, "Design of the Control System about Central Signals in Electric Vehicle," *Journal of Electromagnetic Analysis & Applications*, Vol. 2, No. 3, 2010, pp. 189-194.

2

Power Quality Consideration for Off-Grid Renewable Energy Systems

Mojgan Hojabri[1], Arash Toudeshki[2]
[1]Faculty of Electrical and Electronics Engineering, University Malaysia Pahang (UMP), Pekan, Malaysia
[2]Department of Electrical and Electronic Engineering, University Putra Malaysia (UPM), Serdang, Malaysia

ABSTRACT

Necessity of electricity access in remote area is the main reason for expanding decentralized energy system such as stand-alone power systems. The best electrical power supply must provide a constant magnitude and frequency voltage. Therefore, good power quality is an important factor for the reliable operation of electrical loads in a power system. However, the current drawn by most of electronic devices and non-linear loads are non-sinusoidal, which can result in a poor power quality, especially in off-grid power systems. Poor power quality is characterized by electrical disturbances such as transients, sags, swells, harmonics and even interruptions in the power supply. Off-grid power systems worldwide often struggle with system failures and equipment damage due to poor power quality. In this paper, MATLAB/Simulink is used to model and analyses power quality in an off-grid renewable energy system. The results show high voltage transient when the inductive loads were switched OFF. The voltage and current harmonics are also determined and compared for various types of loads.

Keywords: Off-Grid Renewable Energy System; Power Quality; Pulse Width Modulation (PWM) Inverters; Total Harmonic Distortion; Total Demand Distortion

1. Introduction

Energy demand increasing makes important problems such as grid instability or even outage. Therefore, more energy must be generated at the grid. However, energy generation by big plant is not economically. Moreover, implementation of distributed generation is rapidly increased. Because of increasing global warming, limitation and high cost of fossil fuel sources, governments tend to increase use and implementation of renewable energy sources. The main difference between renewable energy and fossil fuels systems is up-front cost versus lifelong energy cost. Currently, in most development countries, governments as well as utilities provide a variety of incentives, to help the renewable energy industry reach to a higher economic scale. **Figure 1** shows the annual growth rates of renewable energies in the world between 2006 and 2011 [1]. Solar/photovoltaic, wind, hydro, geothermal, tidal, wave and bio energy are examples of renewable energy sources which the solar/photovoltaic and wind are most popular among them.

However, environmental friendly is the principal advantage of renewable energies, high up-front cost and uncontrollability are the main disadvantages of it [2]. Renewable energy sources can be used as an off-grid or on-grid systems. An off-grid renewable energy system is typically a stand-alone power system located in a remote area to fulfill residential or commercial power needs. Because of the high installation price and technical problems for supplying utility power in remote area, this type of power system is useful and also isolated from the power grid system. The main objectives of this paper are to create a simulation of a typical off-grid system and study the power quality issues through the simulation. Moreover, it determines the issues and impacts of poor power quality issues on various types of linear and non-linear loads. The organization of this paper follows the structure of off-grid power system, power quality considerations, simulation with the concluded remarks.

2. Structure of Off-Grid Power System

A typical off-grid system is shown in **Figure 2**. As it is shown in this figure, an off-grid renewable energy

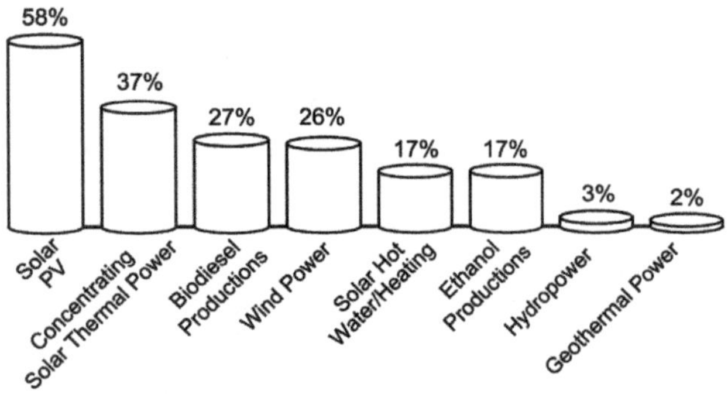

Figure 1. World average annual growth rates of renewable energy capacity between 2006-2011 [1].

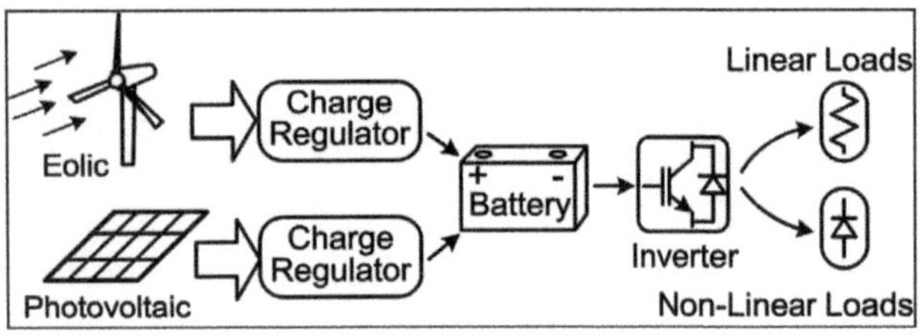

Figure 2. Block diagram of off-grid renewable energy system.

system is included a renewable energy source, charge controller, battery and inverter. The important of renewable energy sources are solar panel, wind turbine, hydro turbine, diesel or biofuel generator and geothermal source. Among these renewable energy sources, solar and wind are more common since their availability. Photo Voltaic (PV) cells have no moving parts. They require only sunlight to operate, and do so without depleting the materials they are made of. PV modules are encased in strong tempered glass and are tested to withstand wind, rain, snow, ice and hailstones. After installation, PV systems generate electricity for decades at no additional cost while producing no greenhouse gases. Off-grid systems with higher demand request might use another energy sources like wind turbine. Because much more energy is found in faster winds, it is wise to place wind turbines where the wind is strong, steady and smooth. Wind turbines placed just above the ground on a breezy knoll make lovely spinning pieces of artwork, but they don't produce much electricity. A good wind site has consistent, fast wind that pumps genuine energy into the turbine's generator for many hours of every day. Occasional gusty days or steady but light breezes just don't cut it.

Charge controller is another component of off-grid system which is needed to protect the batteries from over-charging voltage. Off-grid system also needs battery to provide power when needed. Price and performance, capacity, cycle life, self-discharge rate, safety, hazards, requirement maintenance, size and space requirement are the important features which must be considered to choose batteries for the system. The heart of the grid-direct system is a DC to AC inverter which adapts to the power grid voltage and frequency. Inverter technology has an important role to have safe and reliable grid interconnection operation of renewable energy systems. It is also necessary to generate a high quality power to the grid with reasonable cost [2-5]. Inverters are divided into three main types: grid-tied inverters, grid-tied inverters with battery back-up and stand-alone inverters. And also, inverter control techniques became interesting for power system researchers [6-8]. The important characteristics of good off-grid inverters are high efficiency, low standby losses, low harmonic distortion, easy maintenance and reliability. Therefore, some standards have been published to detect the power quality, grid-connected and unintentional islanding operation [9-12].

3. Power Quality Considerations

The term "power quality" refers to a wide variety of electromagnetic phenomena that characterizes voltage and current waveforms at a given time and at a given location on the power system [13]. Power quality is used to describe the electric power that drives an electrical load and the load's ability to function properly with that electric power. In the absence of high quality power, loads may malfunction, fail prematurely or not operate at all [14]. The power quality of a system is defined by a number of parameters, as long-duration, short-duration voltage variations, harmonic distortion and transients. Harmonics are current and voltage waveform components that represent multiples of the fundamental frequency. Harmonic distortion in an electrical power system is the alteration of the original shape or characteristics of the current or voltage waveforms due to the presence of harmonics. Harmonic distortion is the result of nonlinear loads and switching power electronic devices on the system. Some of the problems caused by harmonics include: very high neutral currents, flickering lights, random tripping of circuit breakers, malfunction of sensitive equipment, fire hazards, reduced power factor, overheated phase conductors, panels and transformers, reduced system capacity due to excessive heating, premature failure of transformers and UPSs. Total Harmonic Distortion (THD) is used to quantify the presence of harmonics in a power system. The THD for a voltage waveform is defined as the ratio of the sum of the voltage magnitudes of all harmonic components to the voltage magnitude of the fundamental frequency [15]:

$$\mathrm{THD} = \frac{\sqrt{V_1^2 + V_3^2 + \cdots + V_n^2}}{V_1} \tag{1}$$

The harmonic current distortion expressed as a function of the maximum demand load current using a

15- or 30-minute demand period is called Total Demand Distortion (TDD)[16]. The TDD is given by:

$$\text{TDD} = \sqrt{\frac{I_H}{I_L}} \quad (2)$$

where I_L is the maximum demand load current at the fundamental frequency, measured at the Point of Common Coupling (PCC) from a 15-minute or 30-minute billing demand load kW, and I_H is the square root of the sum of the squares of the harmonic currents, I_h [17]. I_H is given by:

$$I_H = \sqrt{\sum_{h=2}^{25} I_h^2} \quad (3)$$

Maintaining a good power quality is important for the reliable operation of loads. Different power quality standards are recommended to ensure every piece of equipment operates well with maximum efficiency, without causing any deterioration of the equipment itself. Voltage standards are also set by the National Electrical Code (NEC). According to NEC standards, a maximum voltage drop of 5% from the standard voltage is allowed at the equipment. This 5% voltage drop includes less than 3% voltage drop in a feeder and an addition of less than 3% voltage drop in individual branch circuits [18].

Table 1 specifies the harmonic current limits based on the size of the load with respect to the size of the power system to which the load is connected. The ratio Isc/IL is the ratio of the short-circuit current (Isc) available at the PCC to the maximum fundamental load current. It is recommended that the load current, IL, be calculated as the average current of the maximum demand for the preceding 12 months. As seen in **Table 1**, as the size of the user load decreases with respect to the size of the system, the percentage of harmonic current that the user is allowed to inject into the utility system increases. Most off-grid systems have a low short circuit current and would fit into the first, most restrictive category of TDD less than 5 percent.

Table 1. IEEE standard current distortion limits [6].

Isc/IL	<20	20 < 50	50 < 100	100 < 1000	>1000
H < 11	4	7	10	12	15
11 ≤ h < 17	2	3.5	4.5	5.5	7
17 ≤ h < 23	1.5	2.5	4	5	6
23 ≤ h < 35	6	1	1.5	2	2.5
35 ≤ h	0.3	0.5	0.7	1	1.4
Total Demand Distortion (TDD)	5	8	12	15	20

Isc: Maximum Short- Circuit Current
IL: Maximum Demand Load Current

Table 2 specifies the IEEE voltage distortion limit for utilities. At bus voltages of 69 kV and below, the THD should not exceed 5% and the individual harmonic distortion should not be more than 3% at the PCC. However, a THD well above the recommended maximum may not be a problem on a distribution circuit if the load demand is very low. Thus, the IEEE standard has defined TDD with its standard limits to take this situation into account. Different power conditioning devices provide essential protection against power quality problems. All of these devices provide isolation from the power quality disturbance. Power conditioning equipment can include one or more of the following: surge suppressor, noise filter harmonic filter, motor-generator set, and dual feeder with static transfer, isolation transformer, low-voltage line reactor, Uninterruptible Power Supply (UPS).

4. Simulation Results

A complete Simulink model for an off-grid power system with a pure sine wave inverter and a motor load is presented in **Figure 3**. The major parameters used in this Simulink model are shown in **Table 3**. A variable step discrete with a sampling time of 1 μs was chosen. A variable step solver shortens the simulation time of a Simulink model significantly by reducing the number of steps as necessary, by adjusting the step size for a given level of accuracy [19]. A switch was placed before the motor to analyze the motor performance at the instant of turning OFF and turning ON. A timer set a fixed timing for the switch to open and close. Scopes were placed to view the voltage and current waveforms at different nodes in the circuit.

Characteristics of a 560 W single-phase induction motor is shown in **Figures 4** and **5**. The three subplots represent the voltage waveform, current waveform and

Table 2. IEEE standard voltage distortion limits [6].

Bus Voltage	Individual Voltage Distortion (%)	Total Voltage Distortion (THD %)
V < 69 kV	3	5
69 kV< V < 161 kV	1.5	2.5
161 kV ≤ V	1	1.5

Table 3. Main parameters of Simulink model.

Device	Parameters
Battery	24 V
LC Filter	Cut-Off Frequency = 100 Hz, Switching Frequency = 2 kHz
Transformer	4 kVA, 60 Hz, 20:120
Induction Motor	560 W (0.75 hp), 120 V, 60 Hz

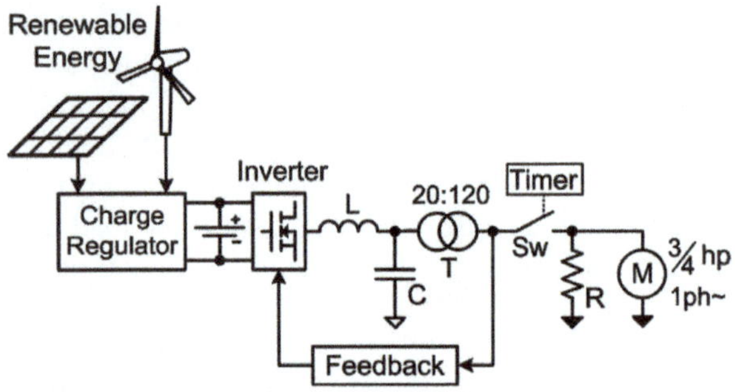

Figure 3. A Simulink model of an off-grid renewable energy system.

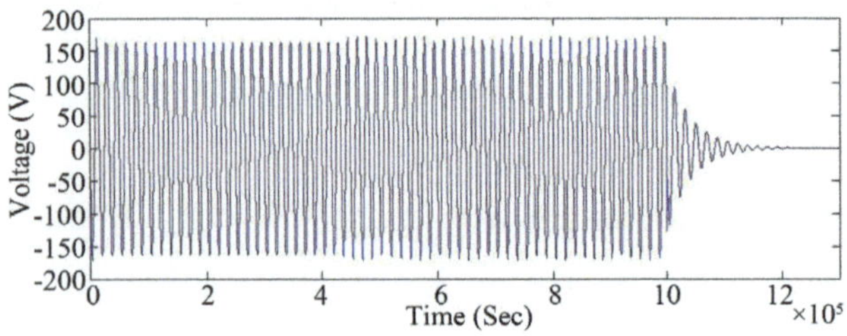

(a)

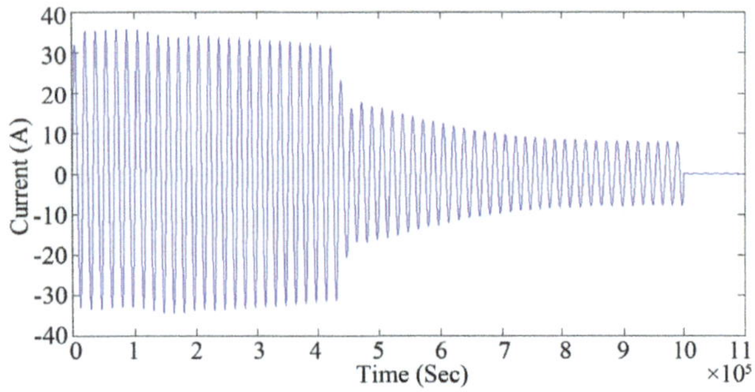

(b)

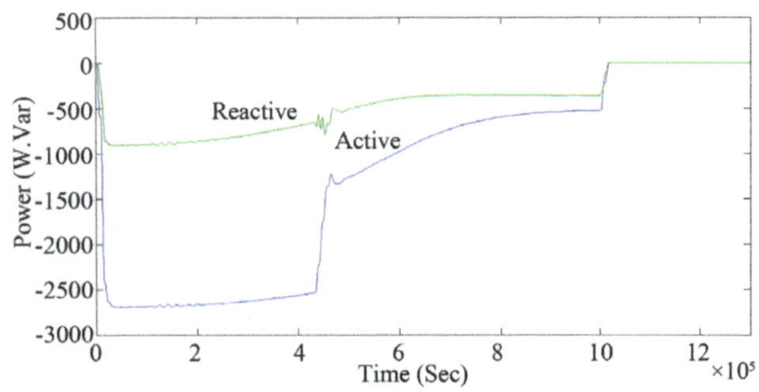

(c)

Figure 4. Output load characteristics of motor: (a) Voltage waveform; (b) Current waveform; and (c) Active and reactive power.

the power for the motor, respectively. When the switch was turned ON at T = 0.3 of a second, a current of almost 23.5 A was drawn by the motor while maintaining an output voltage of 117.2 V. A high amount of power (active power of 2632 W and reactive power of 801 VAR) was drawn by the motor at that time, compared to its

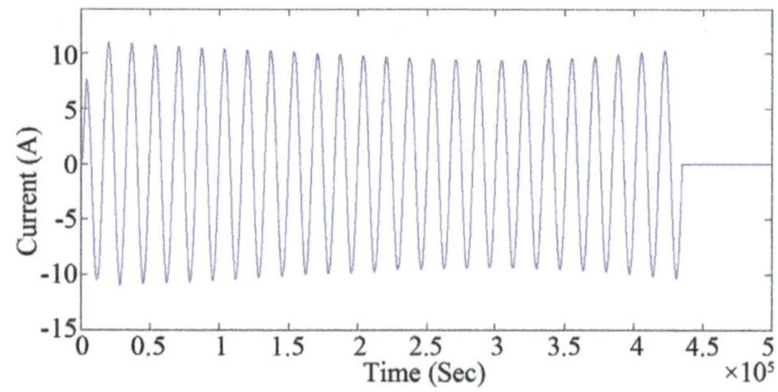

(a)

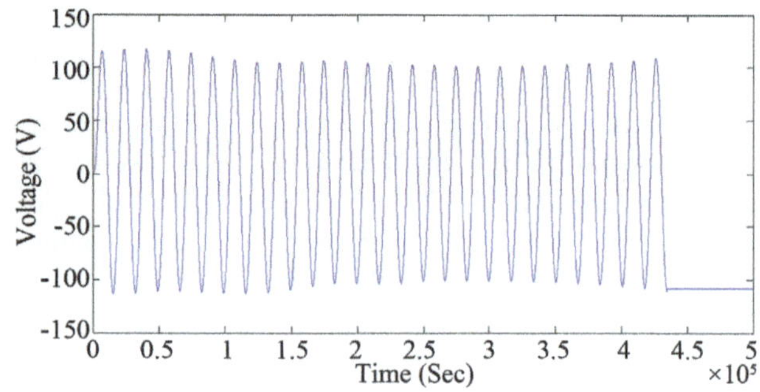

(b)

Figure 5. Motor characteristics: (a) Auxiliary winding curnt waveform; and (b) Capacitor voltage.

power rating of 560 W. The current drawn by the motor started decreasing after the auxiliary winding of the motor was disconnected (when the motor speed reached 75% of its synchronous speed of 1800 RPM), and the motor current settled to normal operating current of 3.25 A (**Figure 6**).

During the operating condition, more reactive power (375 VAR) was consumed by the motor than the active power (65 Watts). When the switch was turned OFF at (65/60)th of a second, the motor experienced a high voltage transient, the value even reaching 43 kV, due to the simulated instantaneous change in current.

The simulation results did not show any voltage sag caused by the induction motor during normal operation, but it indicated the high current at start up, almost six times the normal operating current, which was similar to the actual water pump drawing a start-up current seven times higher than its normal operating current. The simulation also predicted a much larger transient at turn-off due to the idealized instantaneous change in current. The gate pulses for each inverter switch were generated by the PWM generator as shown in **Figure 6**. Only a pair of switches (S1 and S4) or (S2 and S3) were turned ON at any instant of time. For example, at 0.035 sec, switches S1 and S4 were ON while S2 and S3 were OFF. The widths of the generated pulses were varied to ensure that their averaged output produced a 60 Hz sinusoidal waveform.

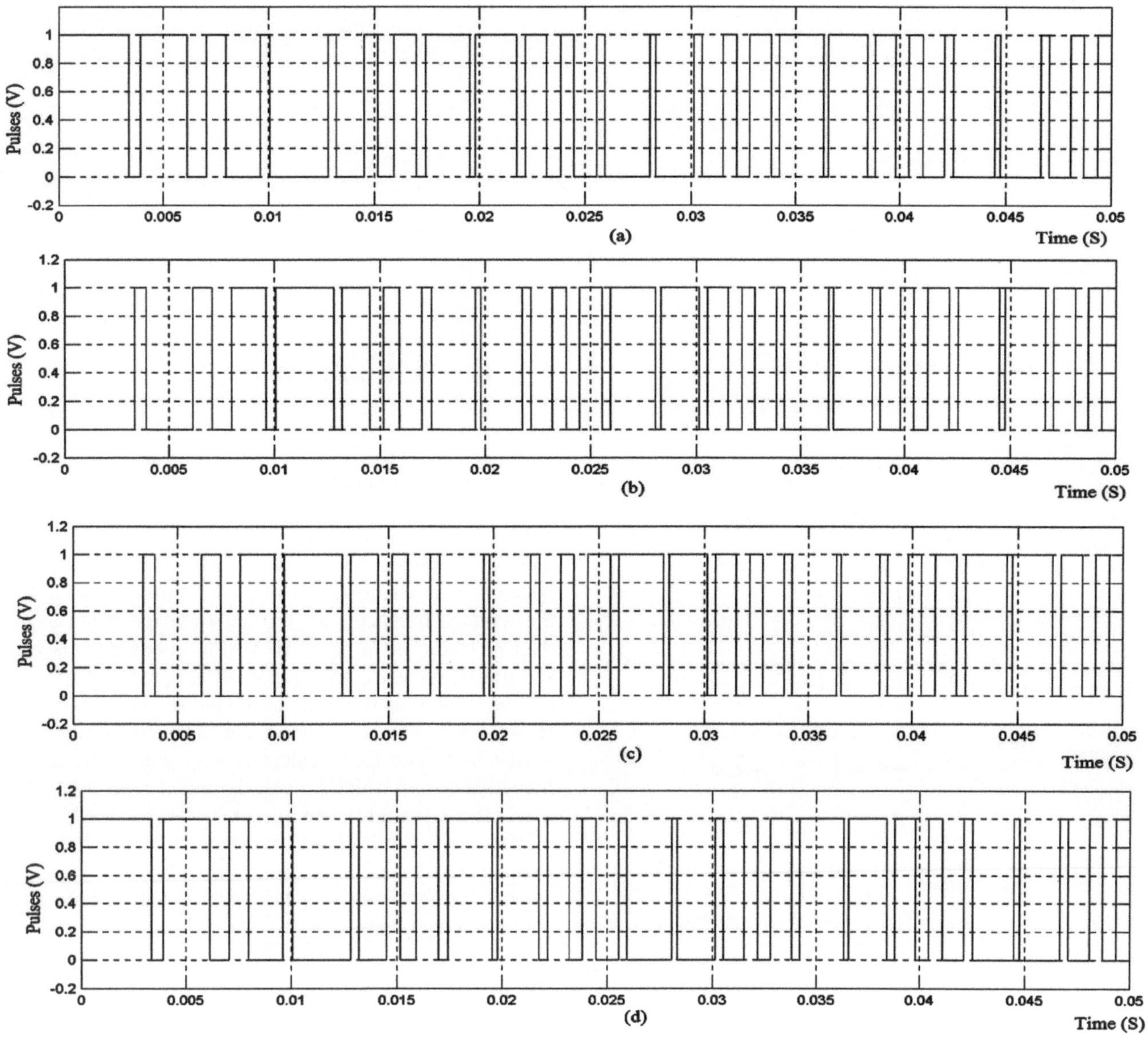

Figure 6. Gate pulses for inverter switches S1, S2, S3 and S4.

The higher frequency components in the PWM output of the pure sine wave inverter were filtered using a low pass LC filter with cut-off frequency 100 Hz, and then the resulting sine wave was passed through a step-up transformer. The primary and secondary voltage waveforms of transformer are shown in **Figure 7**. The peak value of the primary voltage was close to the battery voltage. The voltage transients that occurred at the instant of switching OFF the motor load travelled to the secondary side of the transformer. The secondary winding of the transformer experienced a voltage transient of nearly 1640 V on the positive half cycle and −2080 V on the negative half cycle at 1 sec. The transient voltage, however, did not travel to the transformer's primary side because of its nearly instantaneous duration, the absence of parasitic capacitance in the secondary winding in the model, and also because of the absence of a direct physical connection between the two windings [20].

Using the FFT analysis tool in Simulink, the THD of the output voltage and current waveforms were recorded. **Figure 8** shows the voltage harmonics for the 560 W motor supplied by the pure sine wave inverter, in which case the THD value was 8.34% for normal operation. As the low pass filter eliminated the low order harmonics well, some higher order harmonics were still seen, but with small magnitudes. The THD for the current waveform during the normal operating condition was 8.79% as shown in **Figure 9**. Both of these THD values were well above the IEEE standard of 5%. The generation of harmonic currents is a cause for equipment failure and is,

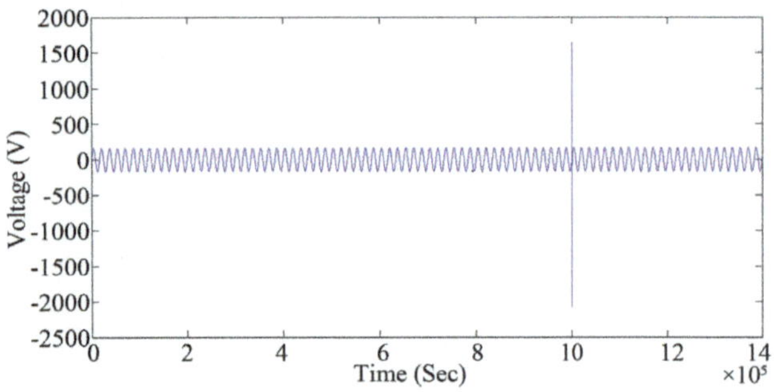

(a)

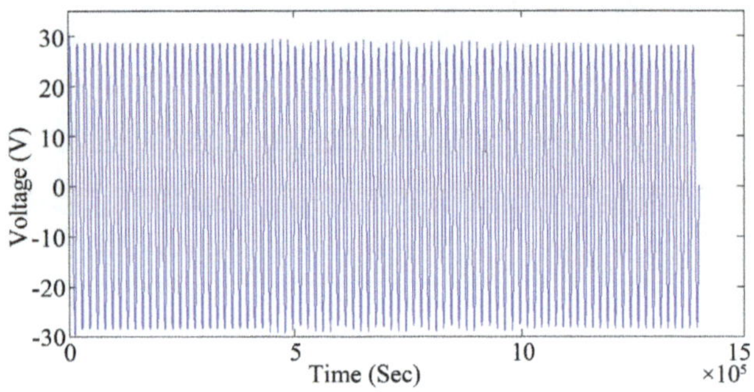

(b)

Figure 7. Secondary and primary voltage waveforms of a step-up transformer.

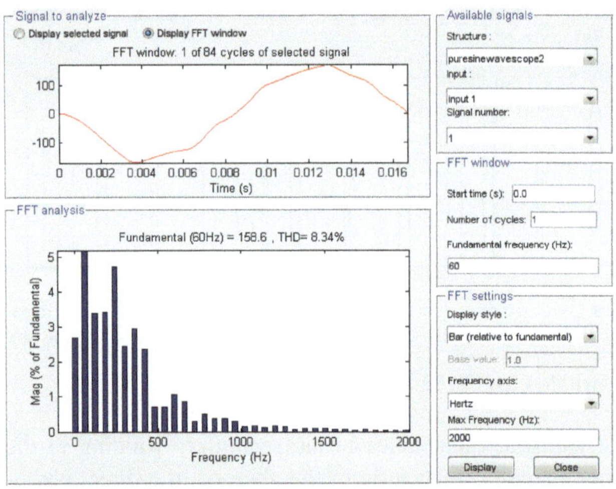

Figure 8. Voltage harmonics of an induction motor during normal operating conditions.

therefore, a serious issue for off-grid systems.

5. Power Quality Analysis for Resistive and Inductive Loads

The system power quality effects from other devices were observed by replacing the motor load with a parallel RL load of different power ratings. The loads were turned ON at 0.5 sec to allow time for the inverter transients to settle down, and the loads were turned OFF at 1.08 sec. When the load switch was turned OFF, large

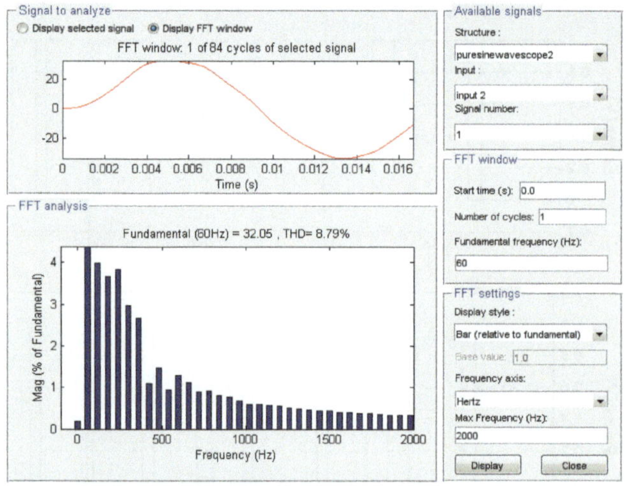

Figure 9. Current harmonics of an induction motor during normal operating conditions.

voltage transients were seen with the parallel RL load as they were with the motor load, due to the unrealistic, instantaneous change in the modeled current. For example, voltage transients of 350 kV occurred when an inductive load (500 W and 750 VAR) was used. This voltage transient caused a negative voltage transient of magnitude 1100 V at the secondary side of the transformer. However, there were no transients seen for a purely resistive load, due to its linear relationship with current. Inductive loads (which are the dominant reactive load in a residential setting) can therefore be assumed as the primary source of voltage transients. The simulated instantaneous change in current through the inductor is responsible for generating this high voltage transient according to the relation [21]:

$$V = L\frac{di}{dt} \tag{4}$$

where V is voltage induced across the inductor, L, due to changing current di/dt. High start-up currents are drawn by the inductive loads to energize their coils. However, none of the purely resistive loads or loads where active power dominated the reactive power drew high start-up currents, as expected. For purely resistive loads, the simulation results showed that the current and voltage harmonics decreased slightly with heavy loads. For inductive loads also, voltage harmonics decreased while the current harmonics increased with heavy loads because of the high fundamental current draw. **Table 4** summarizes the THD of the current and voltage waveforms for different simulated loads and the corresponding voltage transients that appeared in the system. This simulation shows the off-grid loads had significant current and voltage harmonics due to the inverter's voltage along with the power electronics used in those loads. So the load's voltage and current harmonic distortion increase by increasing the pure resistance load. For induc-

Table 4. Simulation results for current and voltage harmonics for different loads.

Load	Voltage THD%	Current THD %
No Load	7.45	-
150 W	7.68	7.68
500 W	8.17	8.17
700 W	8.56	8.56
100 W, 200 VAR	7.63	3.68
300 W, 500 VAR	7.87	3.91
500 W, 1000 VAR	8.18	3.93

tive loads, the total current harmonic distortion is in IEEE standard range. But the voltage THD is out of IEEE standard range (more than 5%).

6. Conclusion

In most cases, off-grid renewable energy systems are cheaper than extending the power grid to provide electricity for remote areas. Therefore, most countries are developing aims for electrification that includes renewable off-grid options and/or renewably powered mini-grids. However, good power quality is an important factor for the reliable operation of this system. In this paper, the MATLAB/Simulink software was used to model and analyze power quality for additional configurations of the typical off-grid system. The simulation results show the high voltage transient and high start-up current for inductive loads. Moreover, the results show the total harmonic distortion of current and voltage for pure resistance loads is significant. However, for inductive load, total demand distortion is in the acceptable range (less than 5%), but the total harmonic distortion of voltage is still more than IEEE standard limitation.

REFERENCES

[1] Global Status Report, "Renewable Energy Policy Network for the 21st Century," Renewables, Paris, 2012.

[2] M. Hojabri, A. Z. Ahmad, A. Toudeshki and M. Soheilirad, "An Overview on Current Control Techniques for Grid Connected Renewable Energy Systems," *2nd International Conference on Power and Energy Systems (ICPES)*, November 2012, pp. 119-126.

[3] J. H. R. Enslin and P. J. M. Heskes, "Harmonic Interaction between a Large Number of Distributed Power Inverters and the Distribution Network," *IEEE Transactions on Power Electronics*, Vol. 19, No. 6, 2004, pp. 1586-1593.

[4] A. Bhowmik, A. Maitra, S. M. Halpin and J. E Schatz, "Determination of Allowable Penetration Levels of Distributed Generation Re-Sources Based on Harmonic Limit Considerations," *IEEE Transactions on Power Delivery*, Vol. 18, No. 2, 2003, pp. 619-624.

[5] IEEE Standard 1547-2003, "IEEE Standard for Interconnecting Distributed Resources with Electric Power Systems," 2003, pp. 0_1-16.

[6] H. L. Jou, W. J. Chiang and J. C. Wu, "A Simplified Control Method for the Grid-Connected Inverter with the Function of Islanding Detection," *IEEE Transactions on Power Electronics*, Vol. 23, No. 6, 2008, pp. 2775-2783.

[7] M. Dai, M. N. Marwali, J. W. Jung and A. Keyhani, "A Three-Phase Four-Wire Inverter Control Technique for a Single Distributed Generation Unit In Island Mode," *IEEE Transactions on Power Electronics*, Vol. 23, No. 1, 2008, pp. 322-331.

[8] Q. Zhang, L. Qian, C. Zhang and D. Cartes, "Study on Grid Connected Inverter Used in High Power Wind Generation System," *41st IAS Annual Meeting, Industry Applications Conference*, Tampa, 8-12 October 2006, pp. 1053-1058.

[9] IEEE Standard 929-2000, "IEEE Recommended Practice for Utility Interface of Photovoltaic (PV) Systems," 2000.

[10] IEEE Standard Conformance Test Procedures for Equipment Inter-Connecting Distributed Resources with Electric Power Systems 1547. 1-2005, IEEE P1547.1.

[11] DIN VDE 0126-1-1, "Automatic Disconnection Device between a Generator and the Public Low-Voltage Grid," 1999.

[12] Grid Code, "High and Extra High Voltage," E.ON, Louisville, 2003.

[13] A. Moreno-Muñoz, "Power Quality: Mitigation Technologies in a Distributed Environment," Springer-Verlag London limited, London, 2007.

[14] M. K. Saini and R. Kapoor, "Classification of Power Quality Events—A Review," *International Journal of Electrical Power & Energy Systems*, Vol. 43, No. 1, 2012, pp. 11-19.

[15] G. R Slone, "The Audiophile's Project Sourcebook: 120 High-Performance Audio Electronics Projects," TAB Electronics, New York, 2001.

[16] IEEE Standard 519-1992, "IEEE Recommended Practices and Requirements for Harmonic Control in Electrical Power Systems," 1993.

[17] L.A. Schienbein and J.G DeSteese, "Distributed Energy Re-Sources, Power Quality and Reliability-Background," Pacific Northwest National Labratory, Richland, 2002.

[18] ANSI C84.1-2011, "American National Standard for Electric Power Systems and Equipment—Voltage Ratings (60 Hertz)," 2011.

[19] MathWorks, "Product Documentatio Choosing a Solver," MathWorks, 2012. http://www.mathworks.com/help/toolbox/simulink/ug/f11-69449.html

[20] ANSI C84.1-2011, "American National Standard for Electric Power Systems and Equipment Voltage Ratings (60 Hertz)," 2011.

[21] MathWorks Products, "Choose a Solver," Documentation Center, 2013. http://www.mathworks.com/help/gads/choosing-a-solver.html

3

High Voltage Stress Impact on P Type Crystalline Silicon PV Module

Han-Chang Liu*, **Chung-Teng Huang, Wen-Kuei Lee, Mei-Hsiu Lin**
Green Energy and Environment Research Laboratories, Industrial Technology Research Institute, Hsinchu, Taiwan

ABSTRACT

The effects of the high voltage stress and other environmental conditions on crystalline silicon photovoltaic module performance have not been included in the IEC 61215 or other qualification standards. In this work, we are to evaluate the potential induced degradation on p type crystalline silicon PV modules by three cases, one case is in room temperature, 100% relative humidity water bath, another is in room temperature, the front sheet coverage with aluminum foil and the other is in the 85°C, 85% relative humidity climate chamber. All the samples are applied with the −1000 V bias to active layers, respectively. Our current-voltage measurements and electroluminescence results showed in these modules power loss of 37.74%, 11.29% and 49.62%, respectively. These test results have shown that among high voltage effects the climate chamber is the harshest and fastest test. In this article we also showed that the ethylene vinyl acetate volume resistivity and soda-lime glass ingredients are important factors to PID failure. The high volume resistivity which is more than 10^{14} Ω·cm and Na less contents glass will mitigate the PID effect to ensure PID free.

Keywords: Potential Induced Degradation; High Voltage; Volume Resistivity

1. Introduction

In photovoltaic (PV) solar modules, reliability is the very important issue for solar power performance, as light induced degradation is a well-known phenomenon. It has long been included in the performance guarantees offered by producers in the industry or the calculations of project developers and system operators. Light induced degradation can cause an approximate 2% decrease in system performance in the first few hours of operation of any new PV installation. In 2005, a new form of performance degradation began to be noticed and now called potential induced degradation (PID) [1] which is high voltage stress effect in negative potential field relative to ground. With the more and more growing PV system and increasing system voltages the PID effects are more seriously and the leakage currents are the characterizations. Possible pathways for the leakage currents from the encapsulated cell to the frame are described by J. A. del Cueto [2]. The domain pathway to cause PID is via the front sheet as glass to the frame. Higher leakage currents can be caused by water entering the solar module causing the encapsulation material to become more conductive.

So far the potential degradation mechanism is not monitored by the typical PV tests listed in IEC 61215 [3]. Some researchers were trying to find out it. It is known that metal ions such as Na^+ formed from the oxides of the module glass can drift toward the cell if the cell is biased negatively [4]. Recently P. Hacke *et al.* found the increased Na concentrations in the surface and sub-surface area of PID affected samples were shown by secondary ion mass spectroscopy (SIMS) [5] and Na precipitates were found on the surface of such samples [6]. M. Schütze *et al.* show that PID can also be caused by other ions usually not present in photovoltaic modules indicating that the chemical nature of the ions is not relevant for PID [7]. The Dr. Liu *et al.* in their study directly verified that the PID caused by Na^+ from their saline water bath experiment [8]. Until now, there is no agreement with evidence that observed metal concentration increasing in the vicinity or inside the cell is responsible for shunting of PID affected modules. The effect of the high voltage and other environmental conditions on module performance has not been included in the IEC61215 or other qualification standards. In this work we discussed the PID effects in different environmental conditions. The case one is 100% relative humidity (RH) water bath, another is covered with conductive aluminum foil and the other is tested in the 85°C/85% RH chamber to provide

*Corresponding author.

the different test methods for PV manufacturer to ensure their solar cell modules to have PID free.

2. Experimental

Three commercial silicon PV modules with 6 × 10 mc-Si solar cells were applied to the PID test. Three methods are applied to the high voltage test, one is in the water baths with room temperature and 100% RH environmental factors for 168 hr, another one is in room temperature and full area coverage with aluminum foil of the modules' front surface for 168 hr [9], the third method is using 85˚C and 85% RH damp heat chamber for 48 hr [10], then application of −1000 V between the cells (via the junction box) and the aluminum frame, respectively. In addition, a 4 × 3 mini module which without soda-lime glass was made and applied −1000 V between the ells (via the junction box) and the aluminum frame to identify the PID formation issue.

For characterization of these modules prior and after the PID test a Berger PSS30 flash tester [11] and a high resolution electroluminescence (EL) camera were used. The volume resistivity test was according to ASTM-D257 [12] "Standard Test Methods for DC Resistance or Conductance of Insulating Materials" using 500 ± 5 V as the applied direct voltage and charged with 60 sec to measure the volume resistivity variety.

3. Results and Discussion

Three commercial silicon PV modules were produced in order to compare different options for PID test conditions. The degree of PID was measured in terms of the standard test condition (STC) module power as shown in **Table 1**.

In **Table 1** all the samples after PID testing are shown the power loss more than 5% during 168 hr or 48 hr test. From our results the PID test under water bath had about three times power loss than that was covered with aluminum foil that means water acceleration the PID rate. In another method by chamber test we find the same result as test under water bath. The chamber PID test reducing 70% time is more effective than water bath.

Table 1. PID test data under different conditions.

Module NO	Test time	Voc	Isc	Vmp	Imp	Pmax	ΔPmax
	[hr]	[V]	[A]	[V]	[A]	[W]	[%]
Water bath	0	36.670	8.290	29.160	7.700	224.430	−37.737
	168	34.184	8.356	21.774	6.410	139.737	
Al foil	0	36.690	8.420	28.860	7.770	224.310	−11.289
	168	36.562	8.467	27.372	7.270	198.988	
Chamber	0	37.410	8.530	30.071	8.024	241.284	−49.623
	48	33.600	8.203	23.381	5.199	121.551	

Furthermore we used electroluminescence technology to check these cells performance. **Figure 1** was EL image after PID test in water bath condition.

We can see that serious dark areas more than two strings after PID test, relevant the Voc decreased about 2.5 V and power loss 38% that means weakening of cell's depletion zone, resulting in a reduction of Voc and Pmax. In the **Figure 2**, there was EL image after PID test coverage with aluminum foil.

From the **Figure 2** there are random dark areas distribution in the EL images, and relevant Voc decreased slightly about 0.1 V and loss power 11%. Compared to the two testing results humidity played an important role on acceleration PID testing. Humidity can enhance the leakage currents flow from module-cells through module insulation and packaging materials, to the module frames, to earth-ground via module supports, moreover moisture entrancing the module can induce encapsulant degradation and reduce insulation of encapsulant to form PID

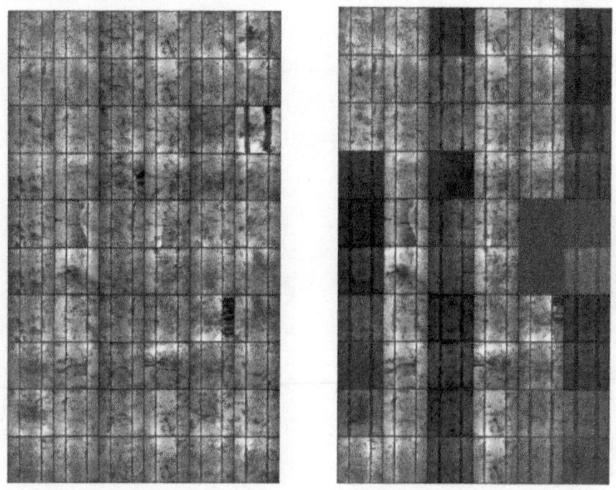

(a) (b)

Figure 1. EL images of a module (a) pre and (b) post PID test in water bath.

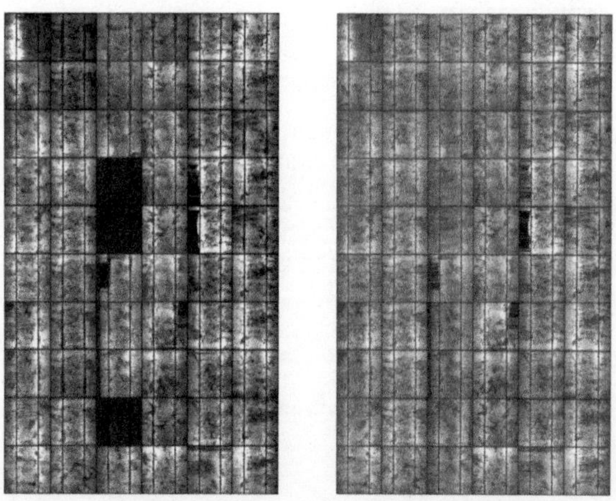

(a) (b)

Figure 2. EL images of a module (a) pre and (b) post PID test with front sheet coverage with aluminum foil.

prone, the encapsulant especially meaning ethylene vinyl acetate (EVA). **Figure 3** was EL image after PID test in 85˚C/85% RH climate chamber which also had the seriously dark areas after 48 hr PID test. In the chamber except moisture we added heat to accelerate the PID test. It can more quickly and effectively check the photovoltaic solar cell module whether it have PID prone.

Furthermore we studied the volume resistivity relative to PID effect. Two different EVA films A and B were aged in 85˚C and 85% RH climate chamber monitored one week as showed in **Figure 4**.

The film A showed that volume resistivity from 10^{14} order decreased to 10^{13} order, opposite to A the film B appearance good performance during aged period. From the results we can deduce that increasing EVA film conductivity caused the module to PID prone. The **Figure 4** also showed that the excellent encapsulant can restrain moisture entrance to module and mitigate PID occurrence. Because no prediction of PID effect without testing is possible so far, hence test can protect manufacturer from massive customer complains.

In order to study the PID mechanism we designed a 4 × 3 mini module without soda-lime glass as front sheet for the PID testing. The testing was performed in the climate chamber with 85˚C and 85% RH conditions applying −1000 V until 48 hr. The test result was showed in **Table 2**.

There is only 1% power loss after PID test which shows good performance than that with soda-lime glass specimens testing previously. EL image without glass module had no obvious difference before and after PID test as showed in **Figure 5** that means the module was PID free.

From the result we can speculate some compounds in the soda-lime glass which also plays important role for PID effect. The compounds especially sodium ions can migrate easily from glass to cell if strong negative electric field exists between cell and module aluminum frame. Above all result, we can suppose the PID process as follows: moisture permeate into module which causes EVA degradation to become more conductive and helpful Na migration to cell surface or P/N junction, finally causes the cell shunting to loss output power.

4. Conclusions

The lifetime of PV modules are reduced by various degradation factors such as: harsh environments, high system voltages and material failures over long periods of

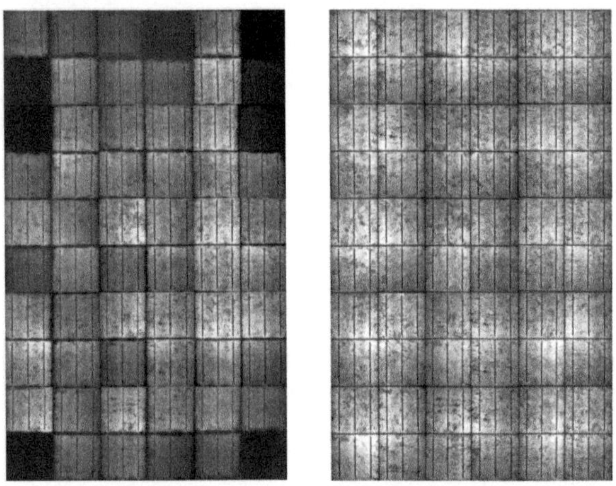

(a) (b)

Figure 3. EL images of a module (a) pre and (b) post PID test in 85˚C/85% RH climate chamber.

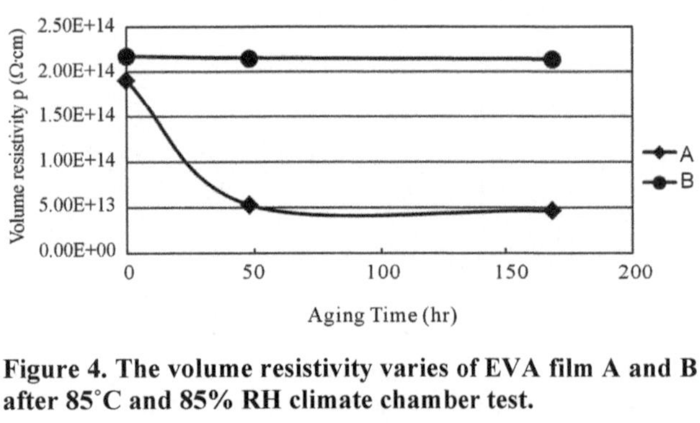

Figure 4. The volume resistivity varies of EVA film A and B after 85˚C and 85% RH climate chamber test.

Table 2. PID test data without Soda-lime glass module.

Module NO.	Test Time [hr]	Voc [V]	Isc [A]	Vmp [V]	Imp [A]	Pmax [W]	ΔPmax %
Without Soda-lime glass	0	7.485	8.552	5.799	7.888	45.742	−0.955
	48	7.480	8.449	5.767	7.856	45.305	

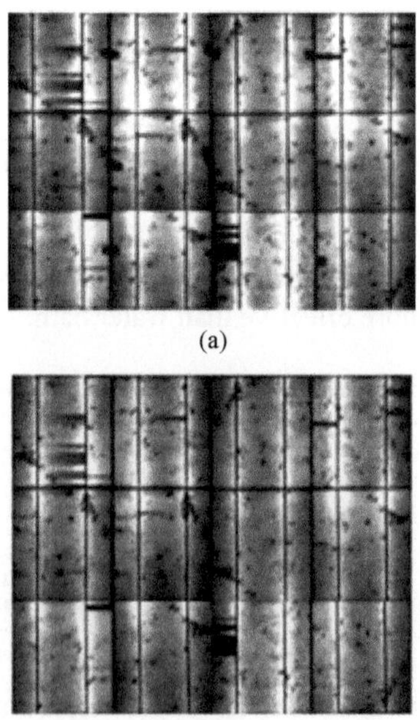

(a)

(b)

Figure 5. EL images of a module (a) pre and (b) post PID test without Soda-lime glass module.

operation. In this article we investigated the high system voltage effect or PID effect. Our PID test results showed that PID in chamber is the harshest and fastest test method than in water bath and coverage with aluminum foil. Our results also verified that the volume resistivity and soda-lime glass ingredients are important factors to PID failure.

Nowadays, PID effects are focused by every PV manufacturer to assure their products PID free. In the future how to reduce module PID effect to increase reliability and performance will lead PV modules to more competitiveness and wide acceptance of PV technologies.

5. Acknowledgements

The financial support provided by Bureau of Energy (contract No.: 102-D0305) is gratefully acknowledged.

REFERENCES

[1] R. Swanson, *et al.*, "The Surface Polarization Effect in High-Efficiency Silicon Solar Cells," *Proceedings of the 15th International Photovoltaic Science & Engineering Conference*, Shanghai, 11-13 October 2005, pp. 410-413.

[2] J. A. del Cueto, "Degradation of Photovoltaic Modules under High Voltage Stress in the Field," *Proceedings of SPIE*, Vol. 7773, 2010.

[3] IEC 61215, "Crystalline Silicon Terrestrial Photovoltaic Modules—Design Qualification and Type Approval," International Electrotechnical Commission, Geneva, 2005.

[4] D. E. Carlson, *et al.*, "Corrosion Effects in Thin-Film Photovoltaic Modules," *Progress in Photovoltaics: Research and Applications*, Vol. 11, No. 6, 2003, pp. 377-386.

[5] P. Hacke, *et al.*, "Characterization of Multicrystalline Silicon Modules with System Bias Voltage Applied in Damp Heat," *Proceedings of the 25th European Photovoltaic Solar Energy Conference and Exhibition*, Valencia, 6-10 September 2010, pp. 3760-3765.

[6] P. Hacke, *et al.*, "System Voltage Potential-Induced Degradation Mechanisms in PV Modules and Methods for Test," *Proceedings of the 37th IEEE Photovoltaic Specialists Conference*, Seattle, 19-24 June 2011, pp. 000814-000820.

[7] M. Schütze, *et al.*, "Investigations of Potential Induced Degradation of Silicon Photovoltaic Modules," *Proceedings of the 26th European Photovoltaic Solar Energy Conference and Exhibition*, Hamburg, 5-8 September 2011, pp. 3097-3102

[8] H. C. Liu, C. T. Huang and W. K. Lee, "Study of Potential Induced Degradation Mechanism in Commercial PV Module," *38th IEEE PVSC*, Austin, 2012, pp. 002442-002444.

[9] H. Nagel, A. Metz and K. Wangemann, "Crystalline Si Solar Cells And Modules Featuring Excellent Stability Against Potential-Induced Degradation," *Proceedings of the 26th European Photovoltaic Solar Energy Conference and Exhibition*, Hamburg, 5-8 September 2011, pp. 3107-3112.

[10] S. Koch, *et al.*, "Potential Induced Degradation Effects and Tests for Crystalline Silicon Cells," *Photovoltaic Module Reliability Workshop*, Golden, 28 February-1 March 2012.

[11] IEC 2007, "Photovoltaic Devices-Solar Simulator Performance Requirements," IEC 60904-9 ed. 2, 2007.

[12] ASTM-D257, "Standard Test Methods for DC Resistance or Conductance of Insulating Materials," 2012.

4

Numerical Study on the Hydrogen Fueled SI Engine Combustion Optimization through a Combined Operation of DI and PFI Strategies*

Medhat Elkelawy, Hagar Alm-Eldin Bastawissi
Department of Mechanical Power Engineering, Tanta University, Tanta, Egypt

ABSTRACT

As the practicability of a hydrogen-fueled economy emerges, intermediate technologies would be necessary for the transition between hydrocarbon fueled internal combustion engines and hydrogen powered fuel cells. In the present study, the hydrogen engine efficiency and the load control are the two main parameters that will be improved by using the combined operation of in-cylinder direct fuel injection (DI) and port fuel injection (PFI) strategies to obtain maximum engine power outputs with acceptable efficiency equivalent to gasoline engines. Wide open throttle (WOT) operation has been used to take advantage of the associated increase in engine efficiency, in which the loads have been regulated with mixture richness (qualitative control) instead of volumetric efficiency (quantitative control). The capabilities of a 3D-CFD code have been developed and employed to simulate the whole engine physicochemical process which includes the hydrogen injection through the intake manifold (PFI) and/or the hydrogen DI in the engine compression stroke. Conditions with simulated PFI, PFI + DI and DI have been analyzed to study the effects of mixture preparation behaviors on the hydrogen ignition and its flame propagation inside the engine combustion chamber. Numerically, the CFD code has been intensively validated against experimental engine data which provided remarkable agreement in terms of in-cylinder pressure history evaluation.

Keywords: Hydrogen Fuel; SI Engine; Port Fuel Injection; Direct Injection; Wide Open Throttle; Kiva-3vr2

1. Introduction

Hydrogen engine research topics introduce the engine obstacles such as suffering from abnormal combustion, which has been considered quite challenging, and the related measures to avoid its drawbacks are important for engine design, mixture formation, and load control [1]. For such engines, the hydrogen introducing methods into the engine cylinder determine the engine performance and durability. In this regard, two methods have been used and tested previously, the external and internal mixture formation [2]. The injection of water into both suction and compression strokes [3], the late external hydrogen supply and direct injection are all suggested methods to delay or even prevent backfiring by either adding a cooling effect or avoiding a combustible mixture during the intake phase. However, external mixture formation by means of port fuel injection (PFI) has been demonstrated to result in higher engine efficiencies, extended lean operation, lower cyclic variability and lower NOx production compared to direct fuel injection in compression stroke [4,5]. Thus, it was attributed to enhancement of the mixture homogeneity due to longer mixing time in the case of PFI fuel supply. However, the mixture preparation is decreased for DI method where the generated turbulence behind the intake valves does not contribute to the mixing process. Hydrogen DI strategy is used in high loads conditions coupled with the delaying of injection timing to reduce the significancy of the NOx emissions [6,7].

Basically, when we consider the engine assembly and retrofitting costs, the external mixture formation provides a greater degree of freedom concerning storage methods. Also, DI injectors designs still require more development to obtain high flow rates and reach the robustness needed for the harsh combustion chamber environment conditions [8]. Consequently, both external and internal mix-

*Optimizing the Hydrogen Engine Performance Techniques.

ture formations have advantages and disadvantages. DI is better for full load performance; PFI is suggested at part and moderate load conditions [5,9]. Contemporary reviews of mixture formation techniques for hydrogen engines can be found in references [10,11] for more details. The optimized operation of the hydrogen-fueled engines can be achieved with a dual injection (PFI + DI) system and throttle valve control [5,12].

In the present study, the engine efficiency and the load control are the two main parameters that will be improved by using the hybrid control strategy of PFI + DI, and lean combustion near the engine cylinder wall to obtain maximum engine power outputs with an acceptable efficiency equivalent to gasoline. New simulation capabilities have been implemented into a 3D-CFD Kiva-3vr2 code to simulate the whole engine physico-chemical process which includes the hydrogen jet interaction with the in-cylinder flow field and its effects on the mixture distribution and homogeneity behind the spark ignition location and the in-cylinder wall.

1.1. Experimental and Numerical Analysis

The pressure of the in-cylinder data was measured by a piezoelectric transducer and amplified with the charge amplifier and finally processed by a data acquisition system. The dynamic top-dead-center (TDC) was established through the tested engine motoring. The crank angle signal was obtained from an angle-encoder device which mounted on the crank shaft. The signal of the cylinder pressure was acquired for every one crank degree, and the acquisition process covered 20 complete successive cycles; the average value of these 20 cycles was outputted as the pressure data used for the calculation of combustion parameters.

Wall function for the velocity and temperature within the region in proximity to the walls has been activated in our multidimensional model. Wall function is an analytic solution to simplify turbulence equations which used to calculate wall shear stresses and heat transfer losses. It is used instead of numerical solution of the complete turbulence equations near the wall because of the difficulties of generating a sufficient engine mesh resolution. The heat transfer derived model is based on the assumptions of incompressible flows with the perturbation theory [13]. However, improved wall function has been used to solve the governing equations which include continuity, momentum, and energy near the cylinder wall [14]. It is employed to the logarithmic law-of-wall to define the velocity and temperature profiles in the near-wall regions.

1.2. Model Governing Equations and Numerical Methods

The KIVA-3vr2 code for computational fluid dynamics has been modified and employed to investigate—numerically—the hydrogen engine for both the underlying engine physics, which include the hydrogen PFI or/and DI spray dynamics, chemical energy released by fuel combustion, heat transfer flux across the engine boundary, and the effects of the in-cylinder flow field on the mixture preparation prior combustion events. However, the code integrates the important principle of physical models which include all the in-cylinder flow field governing equations, gas or liquid spray, combustion behaviors, turbulence equations, and moving pistons or valve boundaries to simulate the whole internal combustion engine characteristics. The Arbitrary Lagrangian description simulation of the hydrogen gas injection, by using an original KIVA spray sub-model, proved to be a comparable difficulty to the liquid fuel injection in both diesel and gasoline engines. The primary challenge is the large change of length scale from the flow of gas in the nozzle orifice to the fuel penetration in either intake manifolds or in-cylinder direct injection. The experimental work was carried out by using fuel injection system capable to supply the hydrogen fuel into the intake manifold and the engine cylinder directly from the fuel bottles after using the proper pressure regulator and valves to adjust the fuel line pressure and flow rate.

The original liquid fuel injection model is replaced by developing a new sub-model program. The new sub-model program simulates the hydrogen injection process as a Eularian phase description through a new boundary condition which represents the nozzle configuration and its location. It is adjusted to set the inlet velocity of the gas fuel injection boundary as a value known by either experimental tests or the gas dynamic calculation of the injector nozzle. The modification depends on the use of regions where each region has the same or different inlet condition. So, in order to get multiple inlets, it is necessary to define a new region which is used as the injector nozzle configuration. This modification enables the use of the suggested model to simulate the gas fuel supply with the hybrid suggested techniques to study the interaction of the intake manifold and the in-cylinder flow on the homogeneity of fuel/air mixture prior to the combustion event. Further details about KIVA-3vr2 code can be found in the original technical report of Los Alamos National Laboratory [13,14]. **Table 1** lists the main specifications of our engine.

During our simulation, the engine speed was kept constant at 1500 rpm and the engine load was controlled qualitatively with the mixture richness in which the throttling valve was removed (WOT). More information about the location of the ignition point, the location of the nozzle outlet of the injector used the figure of the combustion chamber and the strength of the swirl, squish and tumble in the combustion chamber can be found in

Table 1. Engine specifications.

Bore × Stroke	76.0 mm × 88.0 mm
Connecting-Rod length	131.0 mm
Displacement Volume	415 cm^3
Compression Ratio	14
Number of Valves	2 for Intake & 2 for Exhaust
Intake Valve Open/Close	12° BTDC/48° ABDC
Intake Valve Diameter	29.0 mm
Exhaust Valve Open/Close	45° BBDC/10° ATDC
Exhaust Valve Diameter	27.0 mm
Hydrogen Nozzle Diameter	5.0 mm
Piston Shape	Spherical Piston Dish
Combustion Chamber Shape	Pent-Roof

the engine reference manual [15].

2. Results and Discussion

At the base case study, the three examined strategies of hydrogen fuel supply systems are simulated at fixed engine speed and air-fuel equivalence ratio of 1500 rpm and 0.52, respectively. The results of the in-cylinder mixture formation in the cases of port fuel injection (PFI), in-cylinder direct injection (DI), and the hybrid technique (PFI + DI) types are represented as a function of hydrogen concentration and air-fuel equivalence ratio distribution in **Figure 1**. The designed hybrid technique is adjusted in our case to supply 30% of the total amount of hydrogen which corresponds to the air-fuel equivalence ratio of 0.52 with a wide-open-throttle by using a direct injection. The hydrogen injector is held open a period of 65 crank angle degree and the end of injection [EOI] timing is adjusted at 40 crank angle degrees BTDC in the engine compression stroke. The port fuel injection strategy is adjusted to open the hydrogen injector after closing the exhaust valve through suction stroke with 10 crank angle degrees and the injection duration is adjusted to 45 crank angle degrees.

As shown in **Figure 1**, the flow structure through the intake manifold and behind the intake valves inside the cylinder enhances further mixing properties of the hydrogen air mixture. The manifold hydrogen injection timing and duration have been studied in order to optimize the hydrogen air mixture homogeneity and the engine efficiency. The earlier timing and shorter duration of hydrogen injection will produce insufficient mixture homogeneity inside the engine cylinder. Also, the engine volumetric efficiency and the possibility of engine backfire will be reduced. In addition, the late timing and prolonging the injection duration will cause a high-quality of mixture preparation while the engine volumetric efficiency and back fir probability will be increased. That problem is related to the buoyancy and diffusion of hydrogen in a confined space such as the engine cylinder where the hydrogen clouds can be accumulated together as shown in **Figure 1** in the case of PFI and PFI + DI strategy.

The demonstration shown in **Figure 1** presents the simulation results from kiva-3vr2 code for hydrogen clouds and plumes being released into the engine cylinder which is filled with air. The portion of the 3-D plot of the cylinder volume has been removed to make the clouds and injected plumes visible. At the beginning of the simulation, the hydrogen clouds in the case of PFI flew into the cylinder to react with the flow field which was created behind the intake valves. Those clouds collected together to form a single cloud which rose quickly to the upper region of engine cylinder and then spread out and began to fill the cylinder volume with hydrogen from the top to the bottom. The final cloud consisted of a rich hydrogen region under the intake valves position (**Figure 1** PFI Case) on the upper left side of the chamber. As a result, the equivalence ratio at the spark plug location is almost less than 0.4 and the hydrogen concentration were visibly decreased near the spark plug location and the cloud itself deflected toward the cylinder wall.

In the case of hydrogen DI, the hydrogen fuel is released into the engine cylinder as a plume (or jet). That plume starts to expand vertically along the cylinder volume until it reaches the piston bowl to get depressed vertically and expand in the horizontal direction towards the cylinder wall. By this method, the air-hydrogen equivalence ratio distribution is concentrated at the central region of the cylinder volume to form a higher value of an equivalence ratio than a whole cylinder average value which is adjusted herein to be 0.52. So, the hybrid technique will take the benefits of other types (PFI, and DI) through injecting a certain percentage of the hydrogen fuel by DI method while the remaining will be supplied by using PFI method. However, by using the hybrid technique, the hydrogen fuel concentration at the spark plug location is quite enough to form a sufficient hydrogen flame. The flame of hydrogen fuel is characterized as a higher burning velocity and shorter quenching gap (or quenching distance), which is the main cause of reducing the thermal efficiency of hydrogen-fueled engines in comparison with conventionally fueled engines. Thus, the quenching gap of hydrogen is approximately three times less than that of other fuels, such as gasoline. Therefore, hydrogen flames will travel closer to the cylinder wall before they are extinguished and this makes it more difficult to quench than gasoline flames. This smaller quenching distance can also increase the tendency for backfire since the flame from a hydrogen-air mixture can get past a nearly closed intake valve more readily than the flame from a hydrocarbon-air mixture.

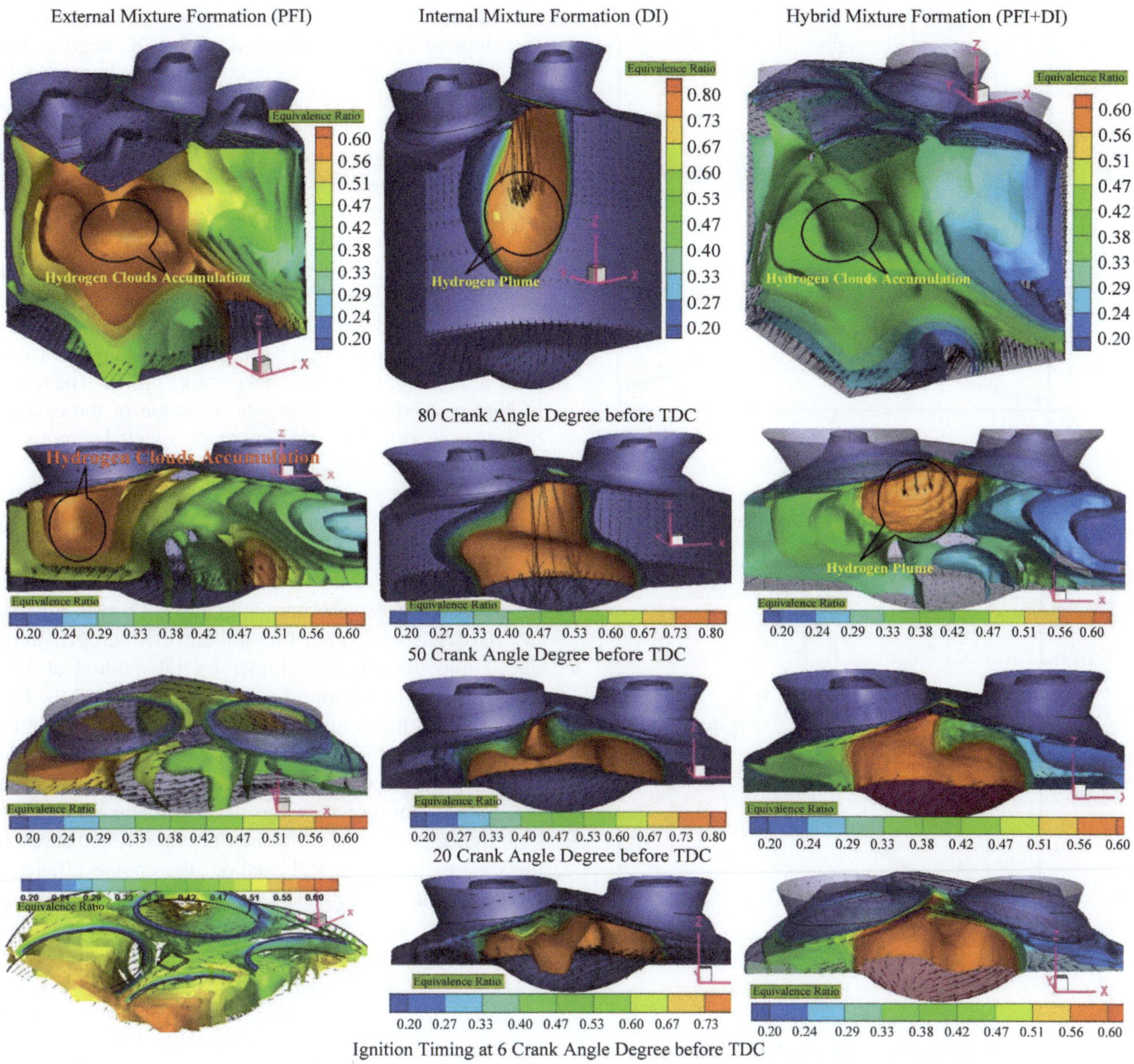

Figure 1. Mixture formations as a function of an air-fuel equivalence ratio of the three tested cases at different crank angle positions, PFI + 30% DI ratio, total equivalence ratio 0.52, EOI at 40 crank angle degrees BTD, DI duration is 65 crank angle degrees, and engine speed of 1500 rpm.

So, our proposed techniques of hybrid fuel supply will reduce the hydrogen concentration close to the cylinder wall (as shown in **Figure 1**) and enhance a further distribution of the fuel inside the combustion chamber prior to the start of ignition events.

Figure 2 shows the relationship between the in-cylinder pressure and temperature with the engine crank rotation angle at three different operating methodologies which are introduced in **Figure 1**. The in-cylinder average pressure curves show a pronounced difference in both the in-cylinder pressure trace during the engine compression stroke and the crank angle corresponding to a maximum cylinder pressure where hydrogen DI method causes a significant rise of the in-cylinder pressure during the compression stroke. This results, is reflecting the effect of hydrogen direct injection on increasing the cylinder charge mass and its pressure which will increase the pumping loss of the cycle when we compare it with the PFI case. However, the engine pumping losses were calculated by consider the required piston work only, which is done against the direct injection mass flow and its pressure rise effect into the engine during compression stroke. The tested engines are designed to minimize the restriction of air flowing into the engine during suction stroke so that the air can be drawn into the cylinder as close as possible to atmospheric pressure.

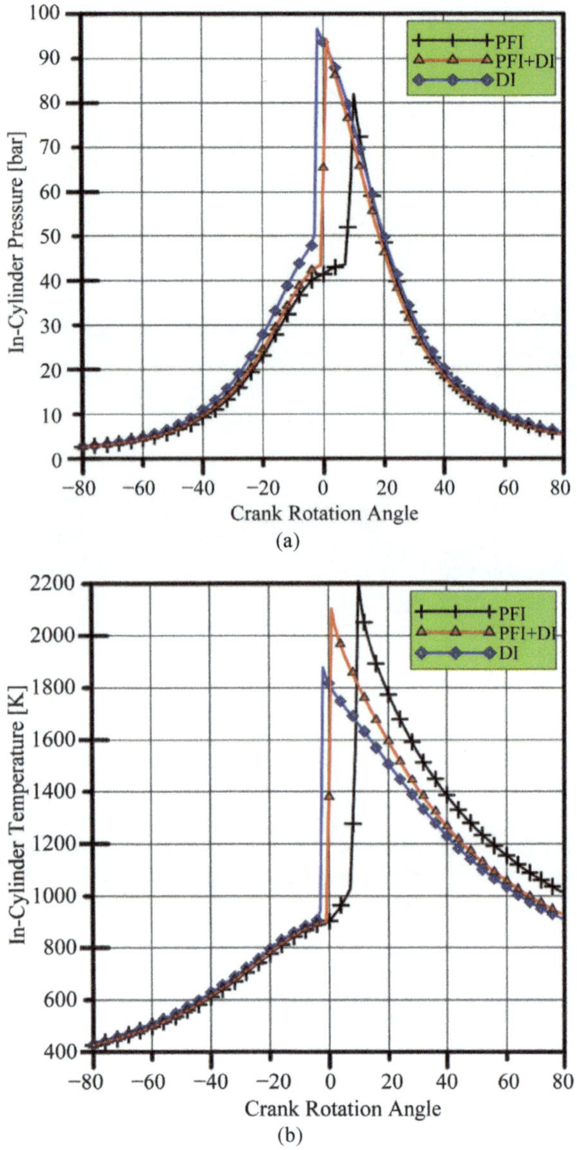

Figure 2. In-Cylinder pressure (Figure 2(a)) and temperature (Figure 2(b)) versus crank rotation angle at different cases, start of ignition of −6 BTDC, PFI + 30% DI ratio, total equivalence ratio of 0.52, EOI at 40 crank angle degrees BTD, DI duration is 65 crank angle degrees, WOT, and 1500 rpm.

Thus, wide open throttle (WOT) operation has been used to take advantage of the associated increase in the engine efficiency in which the loads have been regulated with mixture richness (qualitative control) instead of volumetric efficiency (quantitative control). That is why the operations, with PFI in the case of part load, are suggested previously in the case of throttle manifold due to its effect on the engine pumping losses and then the engine efficiency. However, the result of **Figure 2** indicates that the maximum cylinder pressure decreases and the crank angle corresponding to the maximum cylinder pressure moves further from TDC if the operation changes from DI towards a full PFI strategy. The maximum cylinder temperature increases where the crank angle corresponding to its value is located after TDC degree. This effect is contributed to the effect of mixture homogeneity with the charged air where the lower cylinder temperature means lower fuel-to-air utilization. Those results are presented in **Figure 1** in the case of direct injection methodology where the hydrogen plume is concentrated at the central region of cylinder volume while the surrounding boundary near the cylinder wall is mostly free from the hydrogen fuel. However, the suggested technique of the hydrogen supplied (PFI + DI) produces a result almost identical to the results of both of DI method in the in-cylinder pressure trace and PFI method in terms of in-cylinder temperature profile. The results of **Figure 3** which represent the variation of the cylinder heat energy with crank rotation angle versus hydrogen injection strategies at the same condition **Figure 2**.

A further study on the effect of ignition timing on the engine indicated thermal efficiency was undertaken in our three tested cases of DI, PFI, and DI + PFI at a fixed load or equivalence ratio equal to 0.52, 1500 rpm engine speed. The proposed technique of PFI + DI was adjusted in this case to maintain the amount of hydrogen introduced into the engine cylinder by DI method at 15% from the total fuel introduced (85% PFI + 15% DI). **Figure 4** shows the results of varied ignition timing effect on the calculated indicated thermal efficiency at different hydrogen supplied strategies. The results indicated that the ignition timing has a significant influence on the engine indicated thermal efficiency for all of tested cases. In the case of hydrogen PFI method, the engine efficiency increases with retarding the ignition timing angle from −11 to −6 BTDC to be more than 40% while more retarding will reduce the engine efficiency to reach its minimum value when the ignition timing is close to the engine TDC.

However, the engine efficiency trend is completely different in the cases of hydrogen DI and PFI + DI than that in the case of PFI in which the engine efficiency is enhanced by applying our proposed hybrid method to be more than 41% if we add 15% of the hydrogen fuel by using DI method and adjusting the ignition timing.

To clarify the difference between the intake port injection and the suggested hybrid method at the moderate load condition, the engine simulation at varying percentage of direct injection ratio with wide open throttling (WOT) has been recorded at the condition of a fixed amount of the total hydrogen inside the cylinder. The ratio of the direct injection is defined as the fraction of in-cylinder direct injection to the amount of the total supplied hydrogen. The direct injection ratio of 0% means that all of the hydrogen is supplied through the intake port fuel injection and 100% direct injection ratio means that all of the hydrogen is supplied through the direct injection.

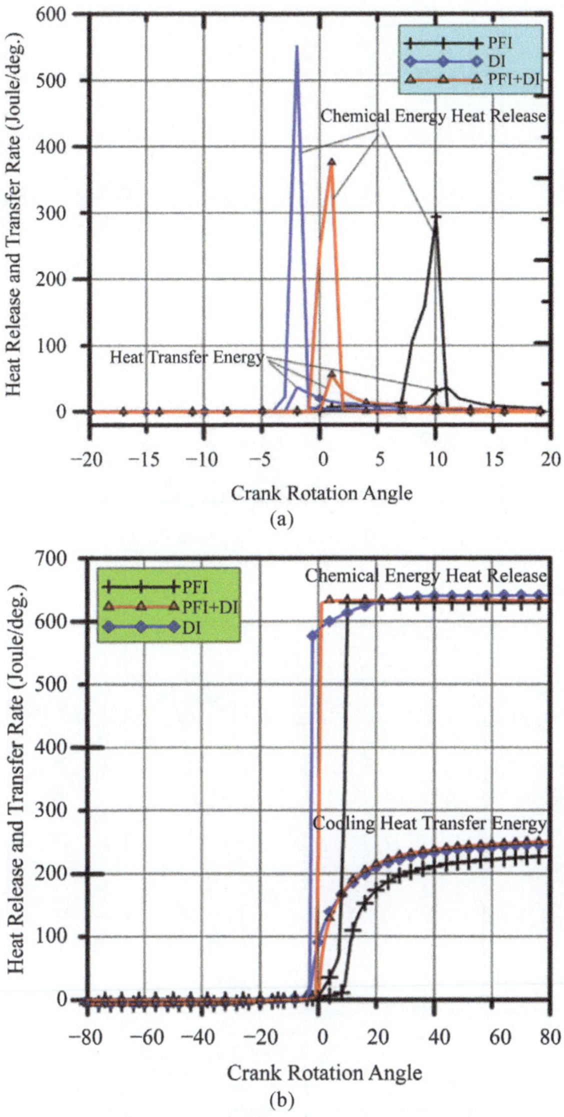

Figure 3. (a) and (b) Variation of the cylinder heat energy with crank rotation angle versus hydrogen injection strategies, start of ignition of −6 BTDC, PFI + 30% DI ratio, total equivalence ratio 0.52, EOI at 40 crank angle degrees BTD, DI duration is 65 crank angle degrees, WOT, and engine speed of 1500 rpm.

Figure 5 shows the indicated thermal efficiency variations of the direct injection ratio at different ignition timing. The results represented in that figure were conducted to determine the optimized hydrogen percentage that will produce the maximum engine efficiency. As we can see, the obtained results indicated that both the direct injection ratio and ignition timing have a significant effect on the engine performance in which the indicated thermal efficiency increases with increasing the percentage of the direct injection ratio and decreases with advancing the ignition timing. In general, both of the indicated thermal efficiency and the engine output were increased by retarding the injection timing. This can be explained by the formation of a stratified mixture as a

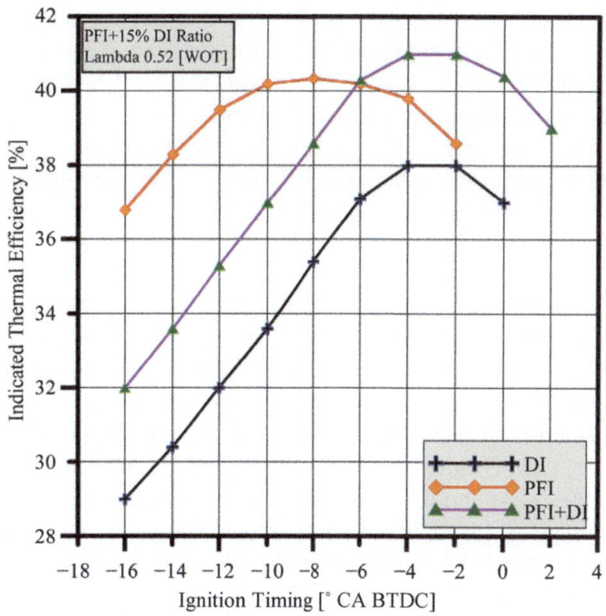

Figure 4. Indicated thermal efficiency variation in ignition timing at different hydrogen injection strategies at total equivalence ratio 0.52, engine speed 1500 rpm, EOI at 40 crank angle degrees BTD,DI duration is 65 crank angle degrees, WOT, and PFI + 15% DI ratio.

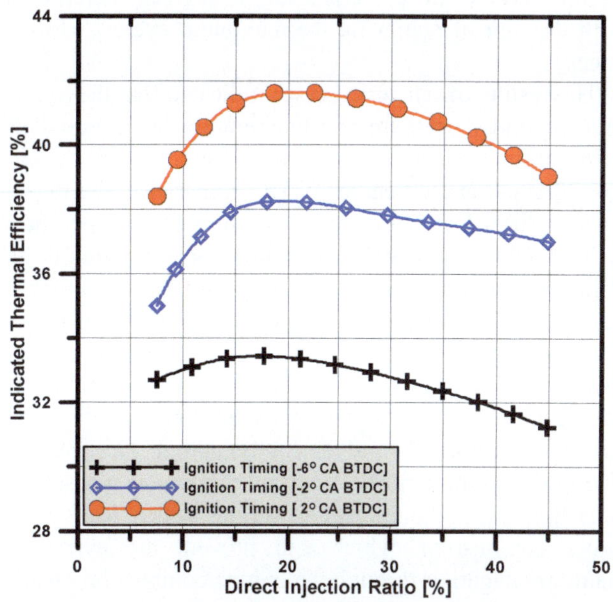

Figure 5. Indicated thermal efficiency variation in direct injection ratio at different ignition timing, 1500 rpm, WOT, EOI at 40 crank angle degrees BTD, DI duration is 65 crank angle degrees, and total equivalence ratio 0.6.

result of retarding the end of injection timing which leads to a faster and turbulent combustion which reduces the heat transfer from the burning gas to the combustion chamber walls and increases the apparent heat release of the hydrogen combustion. However, the maximum indicated thermal efficiency was observed at a direct injection ratio between 15% to 20%, and this may be explained due to a transition from a homogenous mixture, which is represented by PFI method, to a stratified mixture in our proposed hybrid technique.

The time intervals between the end of fuel injection and the ignition timing are very sensitive to direct-injection of hydrogen gas engine combustion. The turbulence created inside the combustion chamber by the fuel jet may maintain high and relatively strong mixture stratification near the spark plug location. However, when decreasing the time intervals between the end of injection and the ignition timing or even overlapping the injection timing with the ignition timing, this may give fast or slow burning rate, which will affect the engine performance. Furthermore, the effect of varying the time periods between the end of injection and the ignition timing has been investigated in our proposed technique at equivalence ratio 0.52, engine speed 1500 rpm, WOT, 15% direct injection ratio. **Figure 6** shows the calculated results of the in-cylinder pressure and temperature variation crank rotation angle at different end of injection (EOI) timing in the case of PFI + 15% DI ratio at a fixed ignition timing of −6 crank angle degrees BTDC. The end of injection timing varied from −60 to 0 crank angle degrees BTDC in 20° crank angle degrees increments with the aim of achieving the maximum cycle performances.

The results of our tested cases indicated that the maximum cylinder pressure and temperature being equal or held constant when the end of injection varied from −40 to −20 crank angle degrees BTDC. Otherwise, the maximum value of the in-cylinder pressure and temperature will be reduced and the crank angles corresponding to its value will be retarded after TDC. Also, the combustion duration will increase, as seen in the case of end of injection 0 crank angle degrees BTDC, to produce lower engine efficiency. However, retarding the end of injection timing increases the engine indicated thermal efficiency (as seen in **Figure 7**) and the engine output. This may be due to the formation of a stratified mixture as a result of retarding the end of injection timing which leads to a faster combustion, which will prevent the hydrogen flame propagation through the engine combustion chamber, and reduce the heat transfer from the burning gas to the combustion chamber walls and increase the apparent heat release fraction in hydrogen combustion.

To clarify the effect of the end of hydrogen injection and ignition timing on the engine indicated thermal efficiency at the condition of maximum engine efficiency, which occurs in our case at air-fuel equivalence ratio 0.41, the results of the engine simulation data are summarized in **Figure 7**.

When the end of injection varied from −40 crank angle degrees to −20 crank angle degrees BTDC, the maximum

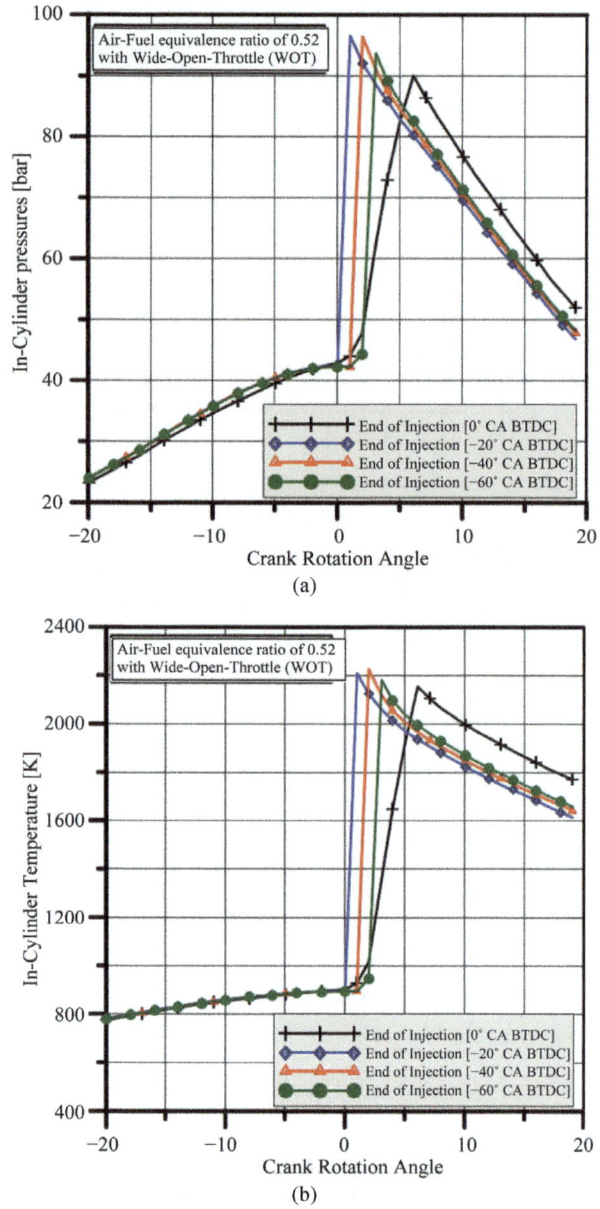

Figure 6. In-cylinder pressure (Figure 6(a)) and temperature (Figure 6(b)) variation crank rotation angle at different end of injection timing in the case of PFI + 15% DI ratio, ignition timing of −6 crank angle degrees BTDC, 1500 rpm, WOT, DI duration is 65 crank angle degrees, and total equivalence ratio 0.52.

ITE were observed to be greater than 41%. However, the results of varying the end of injection timing in the range of −60 to −20 crank angle degrees BTDC show that the spark timing range of −8 to −4 crank angle degrees gets the maximum ITE, otherwise any values outside that ranges will produce lower engine ITE. So, the values of the end of injection as well as the spark ignition timing of −30 and −6 CA degree BTDC, respectively, were selected to study the basic engine performance, which use the common methods for feeds the hydrogen into the

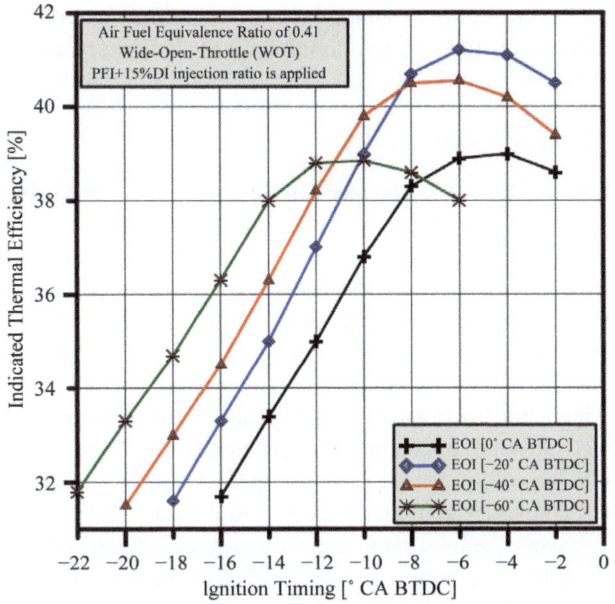

Figure 7. Indicated thermal efficiency variation in ignition timing at a different end of injection timing (EOI) in the case of PFI + 15% DI ratio, 1500 rpm, WOT, DI duration is 65 crank angle degrees, and total equivalence ratio 0.41.

engine and our proposed technique.

3. Engine Performance

The basic engine performances for our three tested hydrogen supply cases are compared to various air-fuel equivalence ratios. **Figure 8** shows the variation of Indicated Mean Effective Pressure (IMEP) in terms of the air-fuel equivalence ratio at different hydrogen supply strategies. As the mixture becomes leaner, the fuel-air mixing of the intake port injection is improved and the indicated thermal efficiency increases (as seen in **Figure 9**). On the contrary, the intake air flow rate is reduced due to the hydrogen injection in the intake port. Thus, the maximum IMEP is as small as the IMEP of the in-cylinder direct injection at the equivalence ratio of 0.46, as the mixture becomes richer. In the region above the equivalence ratio of 0.7, the combustion duration was prolonged and the backfire occurred in the intake port in our experimental activity and the engine could not be stabilized with the intake port injection. Nevertheless, by using the hybrid method of the hydrogen supply, the operating regions have been extended in both rich and lean sides.

Figure 8 shows a pronounced increase in the engine indicated mean effective pressure in the case of using the hybrid method (PFI + DI) more than the other two techniques in all engine equivalence ratios. **Figure 9** shows the results of indicated thermal efficiency in terms of the air-fuel equivalence ratio. The results indicated that the best thermal efficiency for the tested cases will occur at

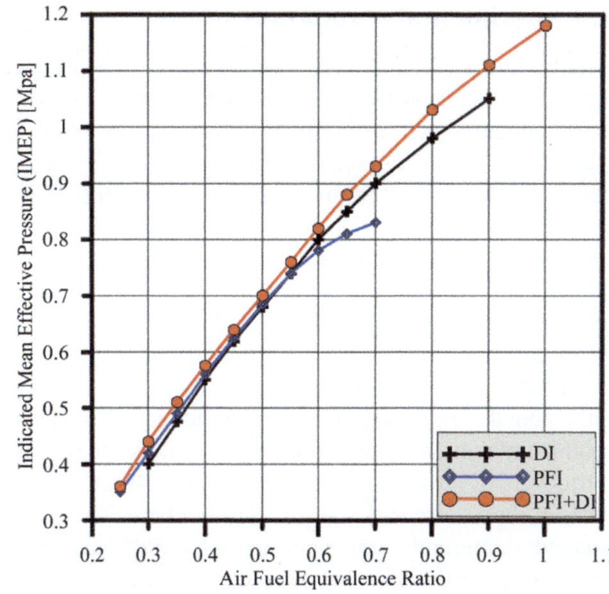

Figure 8. Indicated mean effective pressure as a function of air-fuel equivalence ratio in the cases of port fuel injection (PFI), the in-cylinder direct injection (DI), and PFI + 20% DI ratio at 1500 rpm, ignition timing of −6 crank angle degrees BTDC, DI duration is 65 crank angle degrees, EOI of −30 crank angle degrees BTDC, WOT.

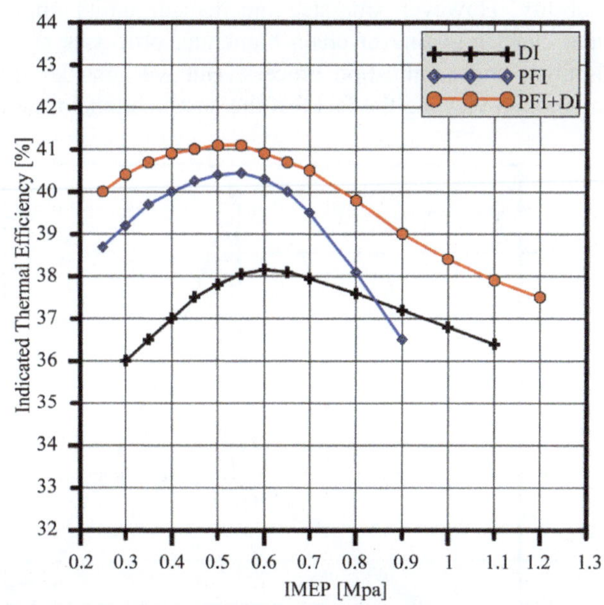

Figure 9. Indicated thermal efficiency as a function of indicated mean effective pressure in the cases of port fuel injection (PFI), the in-cylinder direct injection (DI), and PFI + 20% DI ratio at 1500 rpm, DI duration is 65 crank angle degrees, EOI of −30 crank angle degrees BTDC, WOT.

the air-fuel equivalence ratio in the range 0.37 to 0.5. The thermal efficiency of the hybrid method is higher than that of the conventional methods of port fuel injection and the in-cylinder injection at all the engine load conditions.

4. Experimental Validation

Finally, the CFD code has been intensively validated against experimental engine data which provided a remarkable agreement in terms of the in-cylinder pressure history evaluation as seen in **Figure 10**. Three tested cases of the proposed hydrogen supply (PFI + DI) technique have been selected to evaluate the uncertainty of the achieved data from our CFD code. These cases are introduced in the legend to **Figure 10** which includes the total air-fuel equivalence ratio of 0.33, 0.63, and 0.85 to represent low, moderate, and high load condition, respectively. The selected percentages of direct injection ratio and the ignition timing are optimized in order to produce maximum cycle efficiency and maximum brake torque conditioning. As seen in **Figure 10**, the comparison shows a remarkable coincidence between accomplished numerical and experimental data. There is a pronounced difference between the numerical and experimental data for the in-cylinder pressure peak values at moderate and high load conditions. This disagreement or inconsistency is caused by the kinetic model of an early one-step global reaction mechanism. This model has been activated instead of that based on the elementary reactions which offer the best accuracy and reliability. However, one-step mechanism omits important chain initiating or chain branching processes at the ignition and combustion process. But we consider this method in viewing the fact that the computational time is the main limiting factor when comparing it with the calculation time of the coupled CFD/elementary chemical kinetics.

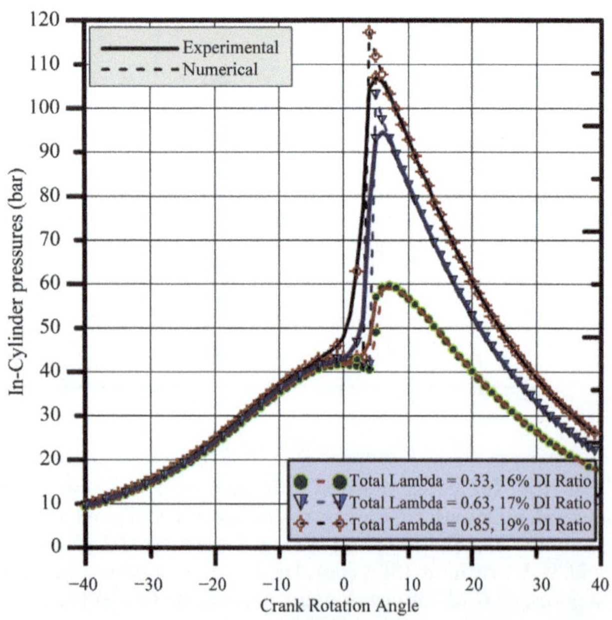

Figure 10. Experimental and numerical results of the in-cylinder pressure data variation in crank rotation angle at different load condition of the hybrid method of PFI + DI at engine speed 1500 rpm, ignition timing of −6 crank angle degrees BTDC, DI duration is 65 crank angle degrees, EOI of −30 crank angle degrees BTDC, WOT.

5. Conclusions

The objective of the present research is to perform a numerical investigation to study the effect of using the hybrid control strategy of PFI + DI, and lean combustion to obtain maximum engine power outputs with an acceptable efficiency which is equivalent to gasoline engines. Wide open throttle operation has been used to take advantage of the associated increase in engine efficiency, in which the loads have been regulated with mixture richness instead of volumetric efficiency. The main conclusions of the study are as follows.

- The hydrogen fuel supply simulation results in the case of intake port as well as the direct injection inside the engine cylinder show that the hydrogen has unique characteristics in a confined space which will affect the flow and mixing process with the charged air.
- Indicated thermal efficiency will decrease with retarding the spark ignition timing in the case of intake port injection (PFI) while it will increase in the case of DI or the hybrid PFI + DI method.
- A stratified charge by direct fuel injection into a lean mixture can be achieved if the direct injection ratio is more than 15% and less than 20% of the total supplied hydrogen.
- To achieve a remarkable consistency between the simulation and experimental results is the CFD solution with accurate detailed chemistry. The chemistry module should address the problem of CFD's computational stiffness by providing efficient and accurate solution algorithms that assure robust coupling of the chemistry and the flow behaviors.
- The proposed hybrid method can satisfy the low and high load conditions without abnormal combustion at the wide-open-throttle condition. The indicated thermal efficiency and engine operation stability of the hybrid method of PFI + DI are superior to those of the port fuel injection and the in-cylinder direct injection at the low and high load regions, respectively, if the end of injection (EOI) timing in the range of −40 to −20 crank angle degrees BTDC and the spark ignition timing range of −8 to −4 crank angle degrees.

REFERENCES

[1] S. Verhelst, R. Sierens and S. Verstraeten, "A Critical Review of Experimental Research on Hydrogen Fueled SI Engines," SAE Technical Paper 2006-01-0430, Society of Automotive Engineers (SAE), USA, 2006.

[2] M. K. Mahesh, Neelu, C. Prakash and G. Viswanathan, "Review of Fuel Induction Technologies for Automotive Hydrogen Propulsion," SAE Technical Paper 2005-26-350, Society of Automotive Engineers (SAE), USA, 2005.

[3] A. M. Nande, T. Wallner and J. Naber, "Influence of Water Injection on Performance and Emissions of a Direct-Injection Hydrogen Research Engine," SAE Technical Paper 2008-01-2377, Society of Automotive Engineers (SAE), USA, 2008.

[4] J. R. Smith, S. Aceves and P. Van Blarigan, "Series Hybrid Vehicles and Optimized Hydrogen Engine Design," SAE Technical Paper 951955, Society of Automotive Engineers (SAE), USA, 1995.

[5] H. S. Yi, K. Min and E. S. Kim, "The Optimised Mixture Formation for Hydrogen Fuelled Engines," *International Journal of Hydrogen Energy*, Vol. 25, No. 7, 2000, pp. 685-690.

[6] D. Messner, A. Wimmer, U. Gerke and F. Gerbig, "Application and Validation of the 3D CFD Method for a Hydrogen Fueled IC Engine with Internal Mixture Formation," SAE Technical Paper 2006-01-0448, Society of Automotive Engineers (SAE), USA, 2006.

[7] A. Wimmer, T. Wallner, J. Ringler and F. Gerbig, "H2-Direct Injection—A Highly Promising Combustion Concept," SAE Technical Paper 2005-01-0108, Society of Automotive Engineers (SAE), USA, 2005.

[8] H. Rottengruber, M. Berckmüller, G. Elsässer, N. Brehm, *et al.*, "Direct-Injection Hydrogen SI-Engine—Operation Strategy and Power Density Potentials," SAE Technical Paper 2004-01-2927, Society of Automotive Engineers (SAE), USA, 2004.

[9] H. Rottengruber, *et al.*, "A High-Efficient Combustion Concept for Direct Injection Hydrogen Internal Combustion Engine," *15th World Hydrogen Energy Conference*, Yokohama, 27 June-2 July 2004.

[10] W. Peschka and W. Escher, "Germany's Contribution to the Demonstrated Technical Feasibility of the Liquid-Hydrogen Fueled Passenger Automobile," SAE Technical Paper 931812, Society of Automotive Engineers (SAE), USA, 1993.

[11] S. Verhelst and T. Wallner, "Hydrogen-Fueled Internal Combustion Engines," *Progress in Energy and Combustion Science*, Vol. 35, No. 6, 2009, pp. 490-527.

[12] S. Toshio, "Improving Thermal Efficiency by Reducing Cooling Losses in Hydrogen Combustion Engines," *International Journal of Hydrogen Energy*, Vol. 32, No. 17, 2007, pp. 4285-4293.

[13] A. A. Amsden, P. J. O'Rourke and T. D. Butler, "KIVA-II: A Computer Program for Chemically Reactive Flows with Sprays," Los Alamos National Laboratory Report 1989.

[14] A. A. Amsden, "KIVA-3V: A Block-Structured KIVA Program for Engines with Vertical or Canted Valves," Los Alamos National Laboratory Report, 1997.

[15] Orbital Engine Company (Australia) Pty Ltd, "Orbital Project: UTP001. OCP UTP Single Cylinder Research Engine," Orbital Engine Company, Australia, 2004.

Energy and Cost Analysis of Cement Production Using the Wet and Dry Processes in Nigeria

Olayinka S. Ohunakin, Oluwafemi R. Leramo, Olatunde A. Abidakun, Moradeyo K. Odunfa, Oluwafemi B. Bafuwa
Mechanical Engineering Department, Covenant University, Ota, Nigeria.

ABSTRACT

The study evaluates the energy consumption of both wet and dry processes cement manufacturing plant in Nigeria. Energy consumption data collected for the period 2003 to 2011 were used to estimate the energy consumption of the crushing, milling, agitation, burning, grinding and bagging operations. The total energy evaluation was based on the three primary energy sources which include electrical, combustion and human. The total estimated energy intensities were 6545 MJ/ton and 4197 MJ/ton for wet and dry processes respectively. The percentage consumption of energy in each operation is 93.68 and 90.34% (burning), 2.11and 4.33% (milling), 0.43 and 0.67% (crushing), 1.39 and 0% (agitation), 2.12 and 3.90% (grinding), and 0.27 and 0.75% (bagging) of the total energy inputs for the wet and dry processes respectively. Furthermore, the average total energy cost of production showed that wet process is approximately 40% more cost intensive in cement production than the dry process while at the same time it is cost effective to run production on energy through gas powered plant than the national grid.

Keywords: Wet Process; Dry Process; Cement; Crushing; Milling; Nigeria

1. Introduction

Cement and/or clinker (cement primary input) is a commodity being produced in over 150 countries of the world [1]. It is an essential input into the production of concrete needed for building purposes and other construction related activities. According to Madlool *et al.* [2], world demand for cement was predicted to increase from 2283 million tonnes in 2005 to about 2836 million tonnes in the year 2010 [2]. The growth witnessed in recent days is largely driven by rising production in emerging economies and developing countries, especially in Asia. In 2006, almost 70% of the world production was in Asia (47.4% in China, 6.2% in India, 2.7% in Japan and 13.2% in other Asian countries) and about 13.4% in Europe [3].

In Nigeria, cement production grew rapidly from 2 million tonnes in 2002 to 17 million in 2011 [4]. This has led to the Nigeria cement industry accounting for 63.6% of the West African region's cement output in 2011. Daily production is in excess of sales having recorded a zero importation from January 2012 to date and in the process of formalizing the exportation of cement to Economic Community of West African States (ECOWAS) and other neighboring countries. With the new Ibeshe cement factory by Dangote Group (commissioned in February, 2012), the country's production capacity is expected to hit 39.4 million metric tonnes per annum thereby recognizing Nigeria as a cement producing country.

The cement sub-sector is one of the most energy consuming industries and it consumes approximately 12% - 15% of total industrial energy use [2,5]; since the industry sector plays a significant role in global energy consumption, its demand can be said to be majorly determined by population and socio-economic activities of a country. Large volumes of CO_2 are however being emitted during cement production and it is believed that this sector represents 5% - 7% of the total CO_2 anthropogenic emissions [6,7]. Since the associated energy used in the item production is extensively based on fossil fuels, environmental issues are further heightened and are of great importance. Therefore, a detailed review on the energy use and savings is necessary to identify energy wastage so that necessary measures could be implemented to reduce energy consumption in this sub-sector [2].

The escalating production of cement in the Nigeria

thus calls for a proportionate rise in energy need and cost, and environmental issues relating with the CO_2 emission. According to Fadare *et al.* [8], energy cost constitutes a major component of the overall production cost in manufacturing industries; it accounts for about 60.75% of the direct manufacturing cost of cement [9], hence energy utilization efficiency is a major determinant of the profitability of manufacturing system.

In Nigeria, approximately 40 to 50 per cent of cement manufacturing cost is energy related; each tonne of cement requires 60 - 130 kg of fuel oil or its equivalent and about 105 kWh of electricity, depending on the cement variety and process type employed [10]. Cement production spreads across five geo-political zones due to the vast deposit of raw materials (**Table 1**). Kilns are majorly being fired by the use of heavy fuel oil (LPFO), coal and natural gas. However, the dearth of natural gas supply in the northern part of the country has restricted its use in kiln firing to plants located in the southern region.

In Adeloye [11], the unit cost of fuel component for cement production is as low as $6 per tonne in China as opposed to $30 per tonne in Nigeria; this has contributed largely to the high and persistent rise in unit cost of cement production. There is thus the need for the adoption of energy efficiency in cement production in Nigeria. Recently, there has been an increasing interest in using energy analysis techniques for energy-utilization assessments in order to attain energy saving, and hence financial savings [5]. In this study, in-depth energy evaluation is carried out on a large scale cement production firm, whose mode of operation is based on both the wet and dry processes by evaluating specific energy cosumption including electric, combustion, human as well as total energy of its various units of operation with a view to optimizing the plants' energy consumption. Various energy savings measures peculiar to the industry were also presented.

2. Methodology

The plant adopted for the study has a wet production capacity of 1 million tonnes per year and a dry process output of 1.2 million tonnes per year. Six operation units are identified for the wet process while five units are identified for the dry process. The operation units considered for the purpose of this work include crushing, milling, agitation, burning, grinding and bagging (**Figures 1** and **2**). For each of these operation units, energy input was accounted for by noting and quantifying the type of energy that was used. The primary energy sources being utilized in the plant are electrical, combustion and manual energy; combustion energy is consumed only during the burning operation in cement processing. An inventory of the electrical motors with their respective power ratings, power ratings of the other machines and heaters, personnel involved, time required for production and material flow in each of the units operation along-

Table 1. Locations, capacities and status of cement companies in Nigeria.

Company name	Location (state)	Region
Sokoto cement	Sokoto	North-West
Ashaka cement	Gombe	North-East
Bauchi-Gwana cement	Bauchi	North East
Benue cement company	Benue	North-Central
Obajana cement plc	Kogi	North-Central
Unicem cement	Cross River	South-South
Wapco cement plc	Ogun	South West
Purechem cement	Ogun	South West
Wapco cement plc	Ogun	South West
Dangote cement plc	Ogun	South West
Ibeshe cement company	Ogun	South West
Ava cement	Edo	South-South

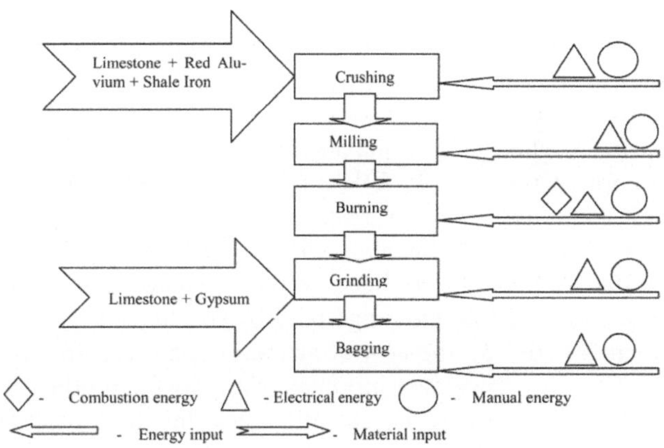

Figure 1. Material and energy flow diagram for dry process manufacturing of cement.

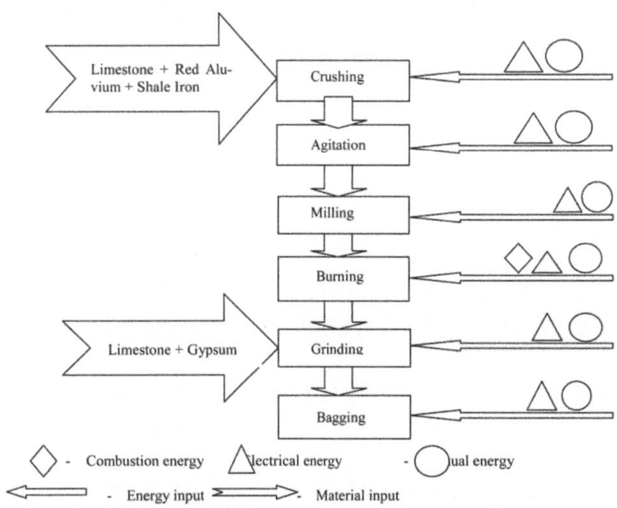

Figure 2. Material and energy flow diagram for wet process manufacturing of cement.

side with the units capacities were all collated from the respective departments in the production plant. The production processes were monitored and data for an eight year period ranging from 2003 to 2010 were collected.

2.1. Estimation of Energy Input

Energy inputs which comprise electrical, combustion and manual energy for each unit of operations were cal- culated yearly for the eight year period. According to Fadare *et al.* [8], the electrical energy usage by the equipment in kWh, was obtained as the product of the rated power of each motor and the number of hours of operation expressed in Equation (1); a motor efficiency of 80% was assumed to compute the electrical inputs mathematically as:

$$E_p = nPt \quad (1)$$

where E_p is the electrical energy input, n is the electric motor efficiency, t is time taken in hours and P is power rating of each electric motor.

In Odigboh [12], at maximum continuous energy consumption rate of 0.30 kW and conversion efficiency of 25%, the physical power output of a normal human labourer in tropical climate is approximately 0.075 kW and sustained for an 8 - 10 hours workday. Hence, employing the current minimum wage paid by the federal government (**Table 2**), the cost of manual energy per unit operation was calculated in Equation (2) as the product of the manual energy consumption and the unit cost of manual energy [12].

$$E_m = 27Nt \quad (2)$$

where E_m is the manual energy in MJ, 27 is the average power of a normal human labour in MJ, N is the number of persons involved in an operation while t in hours, is the useful time spent to accomplish a given task.

Combustion energy was estimated based on the volume of natural gas consumed in the burning operation and converted to appropriate energy units for analysis.

2.2. Estimation of Energy Intensity (EI_i)

The energy consumed per unit product (energy intensity) for each of the unit operation (EI_i) and the average energy intensity (EI_{tt}) for cement production by either the dry or wet process is expressed in (3) and (4) as given in [8]:

$$EI_i = \frac{Et_i(MJ)}{\text{Total weight of product output per kg}} \quad (3)$$

$$EI_{tt} = \frac{E_{tt}(MJ)}{\text{Total weight of product}(kg)} \quad (4)$$

where Et_i and E_{tt} are the sums of energy inputs per unit operation and sum of energy inputs for all operations respectively.

3. Results and Discussion

Production of cement by the dry and wet processes follows the energy and mass flow diagrams shown in **Figures 1** and **2** respectively. The electrical, manual and combustion energy consumption together with the material mass flow are allotted to each unit operations with the dry cement operation having five unit operations (crushing, milling, burning, grinding and bagging) and the wet cement process involves six units which includes crushing, agitation, milling, burning, grinding and bagging as indicated in the figures. Comprehensive description of cement manufacturing is given in [2]. The unit operations were carried out in continuous process and the energy inputs into each of the operations were accounted for by noting and quantifying the type of energy that was used. The energy consumption data that were obtained provided useful information on the source of energy requirement for each unit of operation.

Tables 3 and **4** show the computed total amount of energy requirement needed for cement production using the wet and dry processes respectively for year 2003; similar step is employed for the remaining study period (2004-2010) as summarized in **Tables 5** and **6**. The respective average total energy intensities were computed as 6545 MJ/ton and 4197 MJ/ton for the wet and dry processes from 2003 to 2010.It can further be deduced within the study period, that the overall combustion energy intensity ranged from approximately 90 to 92% for the wet process and 84 to 87% for the dry process whereas the proportion of electrical energy is between 7.28 and 9.46%, 11.73 and 14.8% of the total energy required by the wet and dry production processes respectively (**Table 5**).

The lowest energy intensities came from manual energy which occupied about 0.42 to 0.83% and 0.45 to 2.21% for the wet and dry processes respectively with the study period.

As shown in **Table 6**, among all the operations undertaken in the wet process of cement manufacturing, burning operation has the highest consumption of the total energy required for manufacturing, ranging from 91.60%

Table 2. Manual energy cost per kWh.

Years	Naira (₦)
2003	5500
2004	5500
2005	5500
2006	7500
2007	7500
2008	7500
2009	7500
2010	7500

Table 3. Time and energy requirement for the wet process year 2003.

			TIME AND ENERGY REQUIREMENT				
			Year 2003				
S/N(i)	Process	Time (h)	Electrical Energy	Manual Energy	Combustion Energy	Total Energy	Percentage energy
			$E_p = nPt(MJ)$	$E_m = 27Nt(MJ)$	$E_c(MJ)$	$Et_i(MJ)$	(Eti/Ett) × 100
1	Crushing	6853	17609030	5777404	0	23393288	0.43%
2	Milling	16018	1.09E + 08	6740305	0	1.16E + 08	2.11%
3	Agitation	8640	71457583	4814504	0	76280727	1.39%
4	Burning	13595	82991537	3851603	5.05E + 09	5.13E + 09	93.68%
5	Grinding	12213	1.09E + 08	7703206	0	1.16E + 08	2.12%
6	Bagging	8422.759	9079599	5777404	0	14865426	0.27%
	Total Etti		3.99E + 08	34664426	5.05E + 09		
	Ett(%)		7.28%	0.63%	92.09%		

Table 4. Time and energy requirement for the dry process year 2003.

			TIME AND ENERGY REQUIREMENT				
			Year 2003				
S/N(i)	Process	Time (h)	Electrical Energy	Manual Energy	Combustion Energy	Total Energy	Percentage energy
			$E_p = nPt(MJ)$	$E_m = 27Nt(MJ)$	$E_c(MJ)$	$Et_i(MJ)$	(Eti/Ett) × 100
1	Crushing	1124.39	4308677	6126693	0	10435370	0.67%
2	Milling	2731.06	63385954	3676016	0	67061969	4.33%
3	Agitation	0	0	0	0	0	0.00%
4	Burning	3422.88	63681998	4901354	1.33E + 09	1.4E + 09	90.34%
5	Grinding	6195.13	53769764	6739362	0	60509126	3.90%
6	Bagging	3441.739	3710139	7964701	0	11674840	0.75%
	Total Ett		1.89E + 08	29408125	1.33E + 09		
	Ett(%)		14.18%	2.21%	85.92%		

Table 5. Primary energy consumption pattern for wet and dry processes for year 2003-2010.

Energy Input			2003	2004	2005	2006	2007	2008	2009	2010
Wet	Electrical	Total X 108 (MJ)	3.99	3.81	3.81	4.34	4.0	3.67	3.1	3.24
		Percentage (%)	7.28	7.82	8.77	8.37	8.40	8.56	9.46	8.60
	Manual	Total (MJ)	34664426	38460177	35988149	27690052	20216244	27580086	25273040	24487280
		Percentage (%)	0.63	0.79	0.83	0.53	0.42	0.64	0.77	0.65
	Combustion	Total X 109 (MJ)	5.05	4.45	3.93	4.72	4.34	3.89	2.94	3.42
		Percentage (%)	92.09	91.39	90.40	91.09	91.17	90.79	89.76	90.75
Dry	Electrical	Total X 108 (MJ)	1.89	2.72	3.44	4.15	3.6	3.43	3.23	3.85
		Percentage (%)	14.18	11.73	12.55	13.08	14.14	13.83	13.83	12.17
	Manual	Total (MJ)	29408125	28467870	27211923	27370359	20718599	18229617	17008677	14282744
		Percentage (%)	2.21	1.23	0.99	0.86	0.81	0.74	0.73	0.45
	Combustion	Total X 109 (MJ)	1.33	2.02	2.37	2.73	2.16	2.12	2.0	2.76
		Percentage (%)	85.92	87.04	84.46	86.05	85.04	85.44	85.45	87.38

of the total energy in 2009 to 93.68% in 2003. Crushing operation ranged from about 0.19 to 0.48% while milling, agitation, grinding, and bagging operations ranged from 1.94 to 2.34%, 1.39 to 2.30%, 2.12 to 3.25% and 0.27 to 0.45% of the total energy respectively within the period. Similarly with the dry process (**Table 6**), burning consumed 89.02 to 90.34% of the energy representing the highest share, whereas milling and grinding consumed 3.95 to 4.35% and 3.70 to 5.78% respectively whilst 0.54 to 0.70% and 0.42 to 0.75% of the energy used for cement production were used for crushing and bagging respectively.

In addition, the wet cement processing is also found to consume approximately 5995.59 MJ/ton of overall energy intensity and employ about 35% of combustion energy per tonnage of cement more than the dry operation

Table 6. Energy consumption pattern of different operations for the wet and dry processes in the cement plant for the year 2003 to 2010.

	Process	Energy	2003	2004	2005	2006	2007	2008	2009	2010
Wet	Crushing	Total (MJ)	23393288	23255197	24980322	25582237	20039612	3126011	3086831	7341077
		Percentage (%)	0.43	0.48	0.57	0.49	0.42	0.07	0.09	0.19
	Milling	Total (MJ)	1.16 × 108	1 × 108	97639578	1.22 × 108	1.06 × 108	1.04 × 108	75114668	73269419
		Percentage (%)	2.11	2.06	2.25	2.36	2.23	2.43	2.30	1.94
	Agitation	Total (MJ)	76280727	76325960	76013045	75665798	75004293	76231060	75409746	75248501
		Percentage (%)	1.39	1.57	1.75	1.46	1.58	1.78	2.30	2.00
	Burning	Total x 109 (MJ)	5.13	4.54	4.01	4.81	4.42	3.97	3.00	3.49
		Percentage (%)	93.68	93.16	92.28	92.88	92.83	92.66	91.60	92.70
	Grinding	Total x 108 (MJ)	1.16	1.17	1.21	1.3	1.26	1.15	1.06	1.05
		Percentage (%)	2.12	2.41	2.79	2.51	2.64	2.69	3.25	2.78
	Bagging	Total (MJ)	14865426	15969908	15886454	15521178	14486826	15586156	14742614	14525860
		Percentage (%)	0.27	0.33	0.37	0.30	0.30	0.36	0.45	0.39
Dry	Crushing	Total (MJ)	10435370	14619994	17498493	22252421	16265416	14476562	12934127	17224976
		Percentage (%)	0.67	0.63	0.64	0.70	0.64	0.58	0.55	0.54
	Milling	Total (MJ)	67061969	94422875	1.16 × 108	1.38 × 108	1.08 × 108	1.07 × 108	94189047	1.25 × 108
		Percentage (%)	4.33	4.07	4.24	4.35	4.24	4.33	4.03	3.95
	Agitation	Total (MJ)	0	0	0	0	0	0	0	0
		Percentage (%)	0	0	0	0	0	0	0	0
	Burning	Total x 109 (MJ)	1.4	2.11	2.49	2.86	2.27	2.22	2.08	2.88
		Percentage	90.34	91.03	90.73	90.31	89.12	89.42	89.02	91.10
	Grinding	Total (MJ)	60509126	85866789	1.06 × 108	1.31 × 108	1.37 × 108	1.26 × 108	1.35 × 108	1.26 × 108
		Percentage (%)	3.90	3.70	3.87	4.12	5.39	5.09	5.78	3.98
	Bagging	Total (MJ)	11674840	13223341	14458648	16411750	15459365	14339406	14556926	13222404
		Percentage (%)	0.75	0.57	0.53	0.52	0.61	0.58	0.62	0.42

whereas the dry cement process consumes 3609.75 MJ/ton. The high energy consumption of the wet process over the dry can be attributed to the mix preparation method adopted prior to burning of clinker in the kiln (water being added to the raw materials to form raw thick slurry), whereas the dry process is only based on the preparation of fine powdered raw meal by grinding raw material followed by drying. The required evaporation of wet slurry before calcinations makes the wet process more energy intensive and expensive than the dry process. This is further reflected in **Figure 3**.

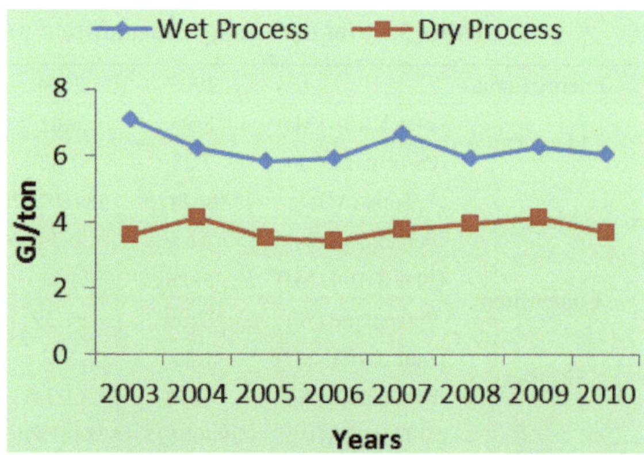

Figure 3. Combustion energy intensities.

3.1. Electrical Energy Intensities Per Unit Operation

Figures 4 and **5** depict the electrical energy intensities per unit operation for the wet and dry processes respectively. It can be observed that grinding operation consumes the highest electrical energy input of 146 MJ per tonnage of product in the wet manufacturing process of cement production, followed by burning, milling, agitation, crushing and bagging with 118, 90, 22, 19 and 15 MJ/ton in that order. However, in the dry process, burning has the highest electrical intensities among all the operations with an approximate value of 170 MJ/ton whereas grinding, milling, crushing and bagging consume 141, 115, 18 and 15 MJ/ton of electrical energy input respectively. Dry process is not subjected to agitation operation.

Furthermore, going by **Figures 4** and **5**, electrical energy intensity consumed in the burning operation of dry process is about 31% higher than that in the wet process of cement. The dry process cement kiln consumes more electrical energy per ton because of the multiple induced draft fans used in the control of air movement through the cyclones and the length of the kiln; the wet process

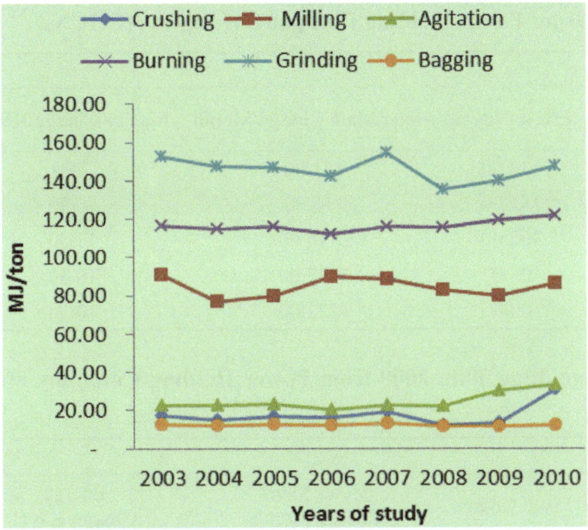

Figure 4. Electrical energy intensities for different process in wet process plant.

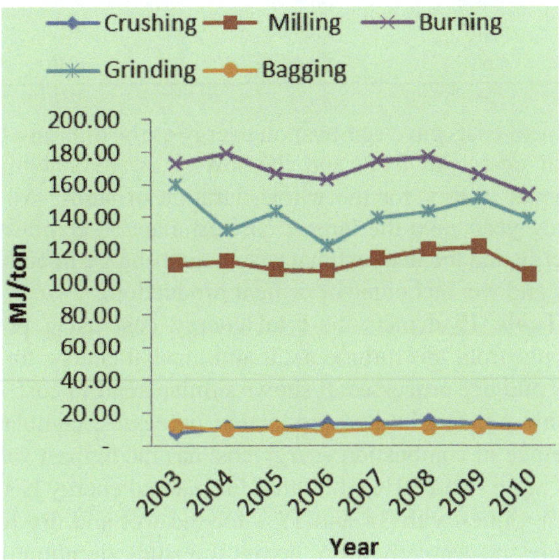

Figure 5. Electrical energy intensities for different process in dry process plant.

cement kiln only makes use of one induce draft fan along the length of the kiln with no cyclones attached.

3.2. Manual Energy Intensities

Manual energy is required for both wet and dry manufacturing process of cement. **Figure 6** shows manual energy intensities for the different operating units in cement production; over the study period, crushing operation requires the highest manual energy intensities for the wet process and bagging for the dry process, followed by the grinding operation and the lowest being the crushing for the dry process and agitation for the wet process. It was also observed during the course of the research work that kiln operation has the highest level of skilled labour, because of the level of system and process sophistication, while bagging operation has the highest level of unskilled labour.

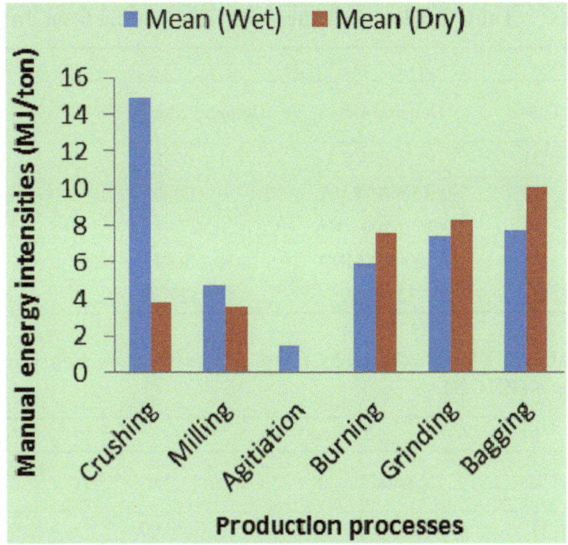

Figure 6. Manual energy intensities for the wet and dry processes within the study period.

3.3. Cost Analysis

The national grid and gas turbine power plant are the two primary sources of electrical energy utility in the cement plant under study.The use of natural gas fired power plant by manufacturers during production is very necessary to ensure smooth production operations and protection of sensitive manufacturing equipment due to the persistent power outages (energy crisis) being witnessed in all sector of the country. A situation whereby 0% power outage involving the use of national grid for the total production time, and situation of 100% power outage involving only the use of electrical power from the gas turbine power plant are considered in calculating the cost of energy inputs of the plant.

The cost of electrical energy based on the national grid was calculated as the product of the energy consumption per unit operation and unit cost of energy (**Tables 7** and **8**) while the electrical energy cost based on the gas powered plant was computed as the product of the gas consumption of the power plant in standard cubic feet (scf) and the unit cost of gas (Naira per standard cubic feet) as listed in **Table 9**. The current minimum wage per month by the Federal Government (**Table 2**) was used to compute the unit cost of manual energy and the appropriate cost/kWh generated in accordance with Equation (2).

The energy costs of the different unit operations for the production of cement by wet and dry process, with the use of national grid and gas turbine power plant as separate sources of electrical energy for year 2003 are given in **Tables 10-13** while **Tables 14-17** depict the summary of cost values for the duration of study.

Table 7. Electrical energy tariff schedule from July 1st. 2009 from Power Holding Company of Nigeria (PHCN).

Industrial Class						
Class	Demand Level	Demand Charge/KVA	Minimum Charge /Month	Fixed Charge	Meter Main Charge/Month	Energy Charge/KWH
D1	>5 < 15 kVA	00.00	136.00	136.00	151.00	7.90
D2	>15 < 45 kVA	00.00	181.00	181.00	755.00	10.30
D3	>45 < 500 kVA	278.88	7550.00	362.00	2416.00	10.30
D4	>500 < 2 MVA	303.13	47,188.00	0.00	3322.00	10.30
D5	>2 MVA	327.38	2,265,011.00	0.00	3322.00	10.30

Table 8. Electrical energy tariff schedule from February 1st. 2002 to June 30th. 2009 from Power Holding Company of Nigeria (PHCN).

Industrial Class						
Class	Demand Level	Demand Charge/KVA	Minimum Charge /Month	Fixed Charge	Meter Main Charge/Month	Energy Charge/KWH
D1	>5 < 15 kVA	200.00	90.00	90.00	100.00	6.50
D2	>15 < 45 kVA	200.00	120.00	120.00	500.00	8.50
D3	LV: >45 < 500 kVA	230.00	5000.00	240.00	1600.00	8.50
D4	HV: >500 < 20 MVA	250.00	31,250.00	0.00	2200.00	8.50
D5	>20 MVA	270.00	1,500,000.00	0.00	2200.00	8.50

Table 9. Price of natural gas in Naira per standard cubic feet.

Years	₦/standard cubic feet (scf)
2003	0.2364
2004	0.4955
2005	0.4959
2006	0.4431
2007	0.7300
2008	0.7340
2009	0.9534
2010	0.6383

3.4. Total Energy Cost Using Combustion Energy, Manual Energy and Electrical Energy from National Grid and Gas Turbine Plant for Wet and Dry Processes

It can be observed as computed in (**Table 14**) that the average total energy cost of cement production by the wet process using power from national grid and manual energy during the study period has the highest in 2009 and the least in 2003. Among the constituents of the estimated total cost of energy, combustion cost records the highest average followed by electrical cost and the least being manual energy as 63, 25 and 13% respectively. It can also be seen from the table that the highest in combustion energy cost is obtained in 2009 representing 73.31% for the year while manual energy cost representing 14.22% in 2003 gave the lowest in the study period. In the same vein, for the cost of producing cement using the dry process, the largest portion is obtained in 2010 and the lowest in 2003. However, cumulative average of the constituent costs gave combustion energy as the highest of the total cost with 51% and the lowest as 15% with the manual energy for the whole duration of study. All the study year gave the largest cost estimate as combustion energy and the least estimate as manual energy in both the dry and wet technique of cement production.

Table 15 depicts the total energy cost using power supply from gas turbine plant and manual energy for the wet and dry processes. It shows similar trend in cost with **Table 14**; for the two production processes, cumulative average of combustion energy cost has the highest values having 67 and 59% while that for manual energy has the least values with 14 and 17% for the wet and dry techniques respectively; also, across the study duration, cost of combustion energy is the highest in 2009 and electrical energy is the least in 2003 for the two processes. Furthermore, the average total energy cost of production showed that wet process is approximately 40% more cost intensive in cement production than the dry process. However, comparing the two tables, it is cheaper to run production on energy based on gas powered plant than the national grid.

Table 16 consists of energy cost for each production operations for the wet and dry processes using power supply from national grid. It can be observed that burning operation consumes the largest energy as reflected in the cost with 70 and 64% for the wet and dry processes in that order, followed by grinding, milling, agitation (wet), bagging whereas the least is in crushing operation with 2.5 and 3.3% for the wet and dry techniques respectively for the study period; the dry process is not employ agitation operation. On the yearly basis within the study period,

Table 10. Total energy cost using combustion energy, manual energy and electrical energy from national grid supply for wet process.

Year 2003									
S/N	Process	Energy from National Grid		Combustion Energy		Manual Energy		Total Energy	
		Electrical Energy $E_p = nPt(MJ)$	Energy Cost N	Energy Consumption SCF	Energy Cost N	Energy Consumption Em = 27 Nt(KWh)	Energy Cost N	Energy Cost N	Percentage of energy cost
1	Crushing	4,891,397.28	41,576,876.88			1,604,834.54	55,784,048.77	97,360,925.65	4.14
2	Milling	30,272,738.56	257,318,277.76			1,872,306.97	65,081,390.23	322,399,667.99	13,70
3	Agitation	19,849,328.64	168,719,293.44			1,337,362.12	46,486,707.31	215,206,000.75	9.14
4	Burning	23,053,204.64	195,952,239.44	4,556,863,878.96	1,077,242,620.99	1,069,889.70	37,189,365.85	1,310,384,226.27	55.68
5	Grinding	30,182,719.68	256,553,117.28			2,139,779.39	74,378,731.69	330,931,848.97	14.06
6	Bagging	2,522,110.84	21,437,942.15			1,604,834.54	55,784048.77	77,221,990.92	3.28
	Total		941,557,746.95		1,077,242,620.99		334,704,292.63	2,353,504,660.56	
	Percentage Total		40.01%		45.77%		14.22%		100

Table 11. Total energy cost using combustion energy, manual energy and electrical energy from national grid supply for dry process.

Year 2003									
S/N	Process	Energy from National Grid		Combustion Energy		Manual Energy		Total Energy	
		Electrical Energy $E_p = nPt(MJ)$	Energy Cost N	Energy Consumption SCF	Energy Cost N	Energy Consumption Em = 27 Nt(KWh)	Energy Cost N	Energy Cost N	Percentage of energy cost %
1	Crushing	1,196,854.69	10,173,264.84			1,701,859.09	59,156,622.00	69,329,886.84	6.84
2	Milling	17,607,209.37	149,661,279.61			1,021,115.45	35,493,973.20	185,155,252.81	18.26
3	Agitation								0.00
4	Burning	17,689,443.84	150,360,272.64	1,202,122,588.97	284,181,780.03	1,361,487.27	47,325,297.60	481,867,350.27	47.52
5	Grinding	14,936,045.42	126,956,386.08			1,872,045.00	65,072,284.20	192,028,670.28	18.94
6	Bagging	1,030,594.29	8,760,051,49			2,212,416.82	76,903,608.60	85,663,660.09	8.45
	Total		445,911,254.65		284,181,780.03		283,951,785.60	1,104,044,820.29	
	Percentage Total		43.9%		28.02%		28.00%		100

Table 12. Total energy cost using combustion energy, manual energy and electrical energy from gas turbine power plant supply for wet process.

Year 2003									
S/N	Process	Energy from National Grid		Combustion Energy		Manual Energy		Total Energy	
		Electrical Energy $E_p = nPt(MJ)$	Energy Cost N	Energy Consumption SCF	Energy Cost N	Energy Consumption Em = 27 Nt (KWh)	Energy Cost N	Energy Cost N	Percentage of energy cost %
1	Crushing	45,046,356..35	10,648,958.64			1,604,834.54	55,784,048.77	66,433,007.41	4.02
2	Milling	278,790,801.62	65,806,145.50			1,872,306.97	65,081,390.23	130,987,535.74	7.92
3	Agitation	182,798,468.41	43,213,557.93			1,337,362.12	46,486,707.31	89,700,265.24	5.43
4	Burning	212,303,931.10	50,188,649.31	4,556,863,878.96	1,077,242,620.99	1,069,889.70	37,189,365.85	1,164,620,636.15	70.45
5	Grinding	277,961,790.54	65,710,167.28			2,139,779.39	74,378,731.69	140,088,898.98	8.4
6	Bagging	23,226,881.24	5,490,834.72			1,604,834.54	55,784,048.77	61,274,883.50	3.71
	Total		241,158,313.40		1,077,242,620.99		334,704,292.63	1,653,105,227.01	
	Percentage Total		14.59%		65.16%		20.25%		100

Table 13. Total energy cost using combustion energy, manual energy and electrical energy from gas turbine power plant supply for dry process.

Year 2003										
S/N	Process	Energy from National Grid		Combustion Energy		Manual Energy		Total Energy	
		Electrical Energy $E_p = nPt(MJ)$	Energy Cost N	Energy Consumption SCF	Energy Cost N	Energy Consumption Em = 27 Nt(KWh)	Energy Cost N	Energy Cost N	Percentage of energy cost %
1	Crushing	11,147,448.88	2,635,256.92			1,701,859.09	59,156,622.00	61,791,878.92	9.04
2	Milling	163,992,729.09	38,767,881.16			1,021,115.45	35,493,973.20	74,261,854.36	10.86
3	Agitation								0.00
4	Burning	164,758,657.16	38,948,946.55	1,202,122,588.97	284,181,780.03	1,361,487.27	47,325,297.60	370,456,024.19	54.19
5	Grinding	139,113,632.35	32,886,462.69			1,872,045.00	65,072,284.20	97,958,746.89	14.33
6	Bagging	9,598,907.31	2,269,181.69			2,212,416.82	76,903,608.60	79,172,790.29	11.58
	Total		115,507,729.00		284,181,780.03		283,951,785.60	683,641,294.64	
	Percentage Total		16.90%		41.57%		41.54%		100.00

Table 14. Total energy cost using combustion energy, manual energy and electrical energy from National grid supply for wet and dry processes.

		Energy Input	2003	2004	2005	2006	2007	2008	2009	2010
Wet	Electrical	Total X 108 (N)	941,557,756.95	899,044,769.39	900,356,387.01	1,024,596,014.60	944,855,808.32	865,898,402.97	731,067,479.99	927,042,555.51
		Percentage (%)	40.01	27.56	29.94	29.72	21.25	20.25	16.01	26.75
	Manual	Total (N)	334,704,292.63	371,354,375.70	347.485,576.61	534,110,331.87	389,948,885.02	531,989,215.44	487,488,850.80	567,451,891.67
		Percentage (%)	14.22	11.39	11.55	15.49	8.77	12.44	10.68	16.37
	Combustion	Total X 109 (N)	1,077,242,620.99	1,991,217,397.96	1,759,565,452.97	1,888,732,302.40	3,111,655,564.78	2,877,123,473.11	3,346,853,110.42	1,970,958,208.83
		Percentage (%)	45.77	61.05	58.51	54.79	69.98	67.30	73.31	56.87
		Total energy cost (N)	2,353,504,660.56	3,261,616,543.06	3,007,407,416.59	3,447,438,648.87	4,446,460,258.12	4,275,011,091.52	4,565,409,441.21	3,465,452,656.01
Dry	Electrical	Total X 108 (N)	445,911,254.65	642,657,551.69	812,580,451.96	978,683,285.38	849,561,050.34	809,637,699.10	762,476,425.67	1,100,770,853.21
		Percentage (%)	43.97	35.29	39.76	37.68	30.22	31.56	27.15	36.40
	Manual	Total (N)	283,951,785.60	274,873,104.47	262,746,230.04	527,943,813.60	399,638,744.40	351,629,056.80	328,078,480.80	330,978,776.04
		Percentage (%)	28.00	15.09	12.86	20.33	14.21	13.71	11.68	10.95
	Combustion	Total X 109 (N)	284,181,780.03	903,467,378.11	968,565,978.51	1,090,693,044.66	1,562,479,185.18	1,403,808,051.83	1,718,015,421.18	1,592,050,821.00
		Percentage (%)	28.02	49.61	47.39	41.66	55.57	54.73	61.17	52.65
		Total energy cost (N)	1,014,044,820.29	1,820,998,034.27	2,043,892,660.51	2,597,320,143.64	2,811,678,979.91	2,565,074,807.73	2,808,570,327.65	3,023,800,450.26

Table 15. Total energy cost using combustion energy, manual energy and electrical energy from gas turbine power plant supply for wet and dry processes.

Process	Energy Input		2003	2004	2005	2006	2007	2008	2009	2010
Wet	Electrical	Total X 108 (N)	241,158,313.40	482,650,524.11	483,744,859.09	497,473,042.06	755,794,167.99	865,898,402.97	889,043,662.52	622,868,966.55
		Percentage (%)	14.59	16.96	18.67	17.03	17.75	20.25	18.82	17.97
	Manual	Total (N)	334,704,292.63	371,354,375.70	347,485,576.61	534,110,331.87	389,948,885.02	531,989,215.44	487,488,850.80	567,451,891.67
		Percentage (%)	20.25	13.05	13.41	18.29	9.16	12.44	10.32	16.37
	Combustion	Total X 109 (N)	1,077,242,620.99	1,991,217,397.96	1,759,565,452.97	1,888,732,302.40	3,111,655,564.78	2,877,123,473.11	3,346,853,110.42	1,970,958,208.83
		Percentage (%)	65.16	69.98	67.92	64.68	73.09	67.30	70.86	56.87
		Total energy cost (N)	1,653,105,227.01	2,845,222,297.77	2,590,795,888.67	2,920,315,676.34	4,257,398,617.78	4,275,011,091.52	4,723,385,623.75	3,465,452,656.01
Dry	Electrical	Total X 108 (N)	115,507,729.00	348,930,082.50	242,302,380.76	475,180,992.56	679,567,486.95	651,181,300.93	796,556,722.86	635,358,151.41
		Percentage (%)	16.90	22.85	16.44	22.69	25.72	27.06	28.02	24.83
	Manual	Total (N)	283,951,785.60	274,873,104.47	262,746,230.04	527,943,813.60	399,638,744.40	351,627,056.80	328,078,480.80	330,978,776.04
		Percentage (%)	41.54	18.00	17.83	25.21	15.13	14.61	11.54	12.94
	Combustion	Total X 109 (N)	284,181,780.03	903,467,378.11	968,565,978.51	1,090,693,044.66	1,562,479,185.18	1,403,808,051.83	1,718,015,421.18	1,592,050,821.00
		Percentage (%)	41.57	59.16	65.73	52.09	59.15	58.33	60.44	62.23
		Total energy cost (N)	683,641,294.64	1,527,270,565.08	1,473,614,589.31	2,093,817,850.83	2,641,685,416.53	2,406,618,409.57	2,842,650,624.84	2,558,387,748.46

burning operation had the highest cost in 2009 representing 77.96% while crushing represents 0.74% for 2010 for the wet process. However, using power supply from turbine power plant based on natural gas for the two processes (**Table 17**), burning operation consumes the largest energy as reflected in the total average cost for the study period with 72 and 69% respectively for the wet and dry processes while the least energy is consumed in bagging (wet) and crushing operation (dry) with 3.3% and 3.2% respectively. Furthermore, on the yearly basis within the study period, burning operation had the highest cost in 2009 with 75.96% while crushing represents 1.16% of the cost computed for 2010 for the wet process; dry process has the highest portion of cost in burning as 72.39% in 2010 while the least cost is given by the crushing operation as 1.69% in 2010.

3.5. Energy Cost Per Tonnage of Cement Produced by the Wet and Dry Process

The average energy cost per tonnage of cement by wet and dry processes for the period of study (2003-2010) is represented in **Figures** 7 and **8**.

Approximate values of ₦5551.72and ₦3174.13 per ton for electrical energy supplied from national grid were computed for wet and dry processes in that order while ₦5,382.46 and ₦2,960 per ton for the electrical energy supplied from gas turbine power plant were calculated for wet and dry processes respectively. The comparative analysis of average cost per tonnage gave reduction in ₦/ton of approximately ₦169.26/ton and ₦214.13/ton for the wet and dry process based on electrical energy from national grid and gas turbine power plant respectively; a resulting change in source price of electrical energy by 55.85% more on the dry process than wet process was obtained.

3.6. Energy Efficiency Options Available in the Industry

Various aspects of the cement manufacturing processes has been found to consume and/or waste energy. Opportunities for either reducing this energy consumption or recycling them where appropriate are being continuously sort out for. The following are considered necessary in optimizing energy consumption in a cement manufacturing process: 1) plant retrofits (process modification and design modification) as included in **Table 18**) energy efficient best practices in cement manufacturing. Burning operation takes place in the kiln and has been discovered from the study as the most energy intensive operation in this research work.This is followed by grinding, milling, agitation (wet), bagging and crushing respectively; these operations can be targeted for energy consumption reduction with the application of economically viable energy efficient tech-

Table 16. Total energy cost using combustion energy, manual energy and electrical energy from National grid supply for wet and dry processes.

Process		Energy	2003	2004	2005	2006	2007	2008	2009	2010
Wet	Crushing	Total (N)	97,360,925.65	115,278,680.40	115,471,574.21	165,831,370.21	101,349,870.54	52,272,163.30	33,696,770.66	45,977,246.38
		Percentage (%)	4.14	3.53	3.84	4.81	2.28	1.22	0.74	1.33
	Milling	Total (N)	322,399,667.99	286,167,929.72	277,059,281.97	359,750,545.67	305,203,317.56	326,556,618.25	256,579,336.91	303,282,985.54
		Percentage (%)	13.70	8.77	9.21	10.44	6.86	7.64	5.62	8.75
	Agitation	Total (N)	215,206,000.75	215,726,176.44	212,704,809.47	249,891,076.40	237,131,378.53	260,794,349.96	244,952,117.84	292,296,285.78
		Percentage (%)	9.14	6.61	7.07	7.25	5.33	6.10	5.37	8.43
	Burning	Total x 109 (N)	1,310,384,226.27	2,223,148,574.46	1,979,083,016.89	2,164,668,995.82	3,345,873,592.52	3,138,209,855.72	3,559,220,301.42	2,263,751,459.88
		Percentage (%)	55.68	68.16	65.81	62.79	75.25	73.41	77.96	65.32
	Grinding	Total x 108 (N)	330,931,848.97	333,871,508.18	339,057,660.21	385,166,568.92	350,651,601.69	352,638,326.10	330,518,286.88	393,716,528.29
		Percentage (%)	14.06	10.24	11.27	11.17	7.89	8.25	7.24	11.36
	Bagging	Total (N)	77,221,990.92	87,423,673.85	84,031,073.84	122,130,091.86	106,250,497.28	144,539,778.19	140,442,627.50	166,428,150.12
		Percentage (%)	3.28	2.68	2.79	3.54	2.39	3.38	3.08	4.80
Dry	Crushing	Total (N)	69,329,886.84	77,962,379.85	81,184,712.44	134,182,615.84	79,762,925.31	52,789,026.60	47,900,866.56	63,171,020.24
		Percentage (%)	6.84	4.28	3.97	5.17	2.84	2.06	1.71	2.09
	Milling	Total (N)	185,155,252.81	249,008,669.49	300,058,555.30	386,546,383.12	304,358,557.99	299,855,768.21	265,795,694.92	398,859,001.61
		Percentage (%)	18.26	13.67	14.68	14.88	10.82	11.69	9.46	13.19
	Agitation	Total (N)	-	-	-	-	-	-	-	-
		Percentage (%)	-	-	-	-	-	-	-	-
	Burning	Total x 109 (N)	481,867,350.27	1,156,669,956.81	1,278,883,175.82	1,490,827,832.35	1,873,870,527.06	1,698,970,760.51	1,972,715,946.21	1,990,648,431.57
		Percentage	47.52	63.52	62.57	57.40	66.65	66.23	70.24	65.83
	Grinding	Total (N)	192,028,670.28	249,659,415.13	294,490,005.06	414,344,809.45	409,653,667.29	378,807,603.16	393,743,371.35	436,072,906.82
		Percentage (%)	18.94	13.71	14.41	15.95	14.57	14.77	14.02	14.42
	Bagging	Total (N)	85,663,660.09	87,697,612.98	89,276,211.89	171,418,502.88	144,033,302.27	134,651,649.26	128,414,448.61	135,049,906.02
		Percentage (%)	8.45	4.82	4.37	6.60	5.12	5.25	4.57	4.47

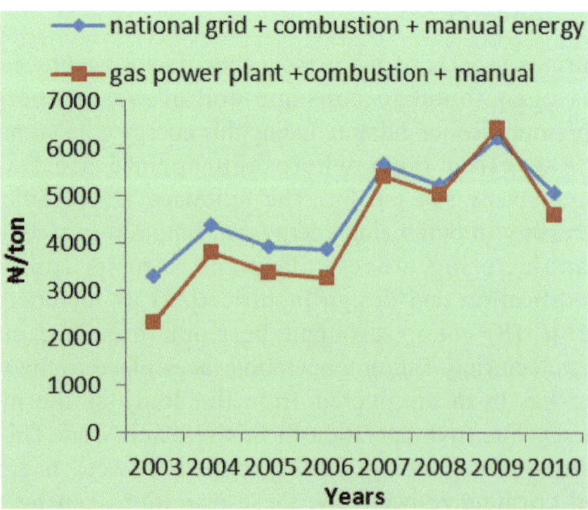

Figure 7. Energy cost per tonnage of wet process cement in Naira (₦).

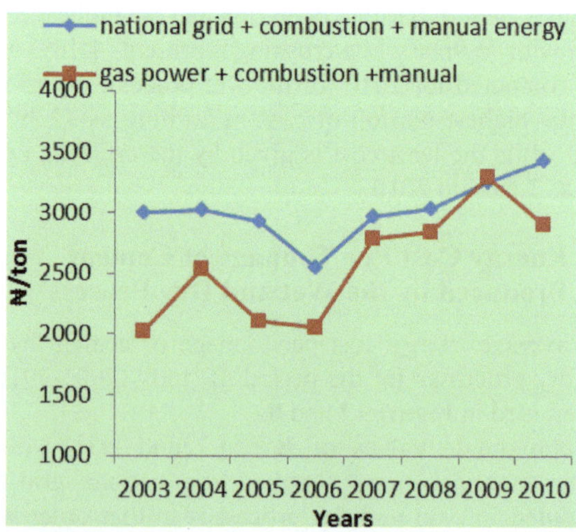

Figure 8. Energy cost per tonnage of dry process cement in Naira (₦).

Table 17. Total energy cost using combustion energy, manual energy and electrical energy from gas turbine power plant supply for wet and dry processes.

Process		Energy	2003	2004	2005	2006	2007	2008	2009	2010
Wet	Crushing	Total (N)	66,4433,007.41	98,898,398.62	96,640,692.30	142,321,608.88	93,390,262.54	52,272,163.30	34,475,745.29	40,239,904.02
		Percentage (%)	4.02	3.48	3.73	4.87	2.19	1.22	0.73	1.16
	Milling	Total (N)	130,987,535.74	184,108,578.12	177,352,910.33	216,430,384.95	256,453,499.31	326,556,618.25	292,515,843.93	238,828,516.19
		Percentage (%)	7.92	6.47	6.85	7.41	6.02	7.64	6.19	6.89
	Agitation	Total (N)	89,700,265.24	137,583,524.09	134,635,276.51	163,090,216.14	203,371,363.41	260,794,349.96	281,410,626.30	225,214,459.40
		Percentage (%)	5.43	4.84	5.20	5.58	4.78	6.10	5.96	6.50
	Burning	Total x 109 (N)	1,164,620,636.15	2,133,146,281.48	1,893,790,514.51	2,056,116,396.29	3,309,958,732.92	3,138,209,855.72	3,587,770,534.55	2,198,843,656.85
		Percentage (%)	70.45	74.97	73.10	70.41	77.75	73.41	75.96	63.45
	Grinding	Total x 108 (N)	140,088,898.98	214,072,422.05	214,734,076.10	232,946,687.29	292,807,774.52	352,638,326.10	382,432,242.21	299,589,747.91
		Percentage (%)	8.47	7.52	8.29	7.98	6.88	8.25	8.10	8.65
	Bagging	Total (N)	61,274,883.50	77,413,092.40	73,642,418.92	109,410,382.78	101,416,985.09	144,539,778.19	144,780,631.21	158,562,782.91
		Percentage (%)	3.71	2.72	2.84	3.75	2.38	3.38	3.07	4.59
Dry	Crushing	Total (N)	61,791,878.92	68,612,231.74	58,301,835.34	113,010,737.07	73,232,657.17	46,607,388.52	49,157,618.70	43,161,172.44
		Percentage (%)	9.04	4.49	3.96	5.40	2.77	1.94	1.73	1.69
	Milling	Total (N)	74,261,854.36	150,968,231.74	115,009,408.46	223,575,916.15	254,773,647.19	251,544,794.58	275,465,261.49	250,316,704.00
		Percentage (%)	10.86	9.88	7.80	10.68	9.64	10.45	9.69	9.78
	Agitation	Total (N)	-	-	-	-	-	-	-	-
		Percentage (%)	-	-	-	-	-	-	-	-
	Burning	Total x 109 (N)	370,456,024.19	1,061,969,516.29	1,095,183,682.07	1,332,831,783.49	1,826,650,368.00	1,655,036,483.36	1,981,152,705.37	1,851,893,344.61
		Percentage	54.19	69.53	74.32	63.66	69.15	68.77	69.69	72,39
	Grinding	Total (N)	97,958,746.89	163,937,526.62	127,374,501.41	263,396,422.30	347,298,012.00	322,652,802.03	407,510,642.38	288,171,811.55
		Percentage (%)	14.33	10.73	8.64	12.58	13.15	13.41	14.34	11.26
	Bagging	Total (N)	79,172,790.29	81,782,761.59	77,745,162.04	161,002,991.82	139,730,732.17	130,776,941.06	129,364,396.91	124,844,715.86
		Percentage (%)	11.58	5.35	5.28	7.69	5.29	5.43	4.55	4.88

nologies.

A lot of energy has been observed to be lost in the form of waste heat through the kiln exhaust gas and the air discharged from the clinker cooler stack where the temperature could be as high as 315 and 215°C respectively [2]. Re-direction of these high temperature gases through a wasteheat recovery steam generator for electricity production has been considered as a means of making the process more efficient. Energy consumption can also be reduced through the addition of pre-calciner to lower the specific fuel consumption and reduce thermal NOx emission. Improvement in the efficiency of electric motors, replacement of all old/obsolete devices together with the use of adjustable speed drives (ASDs) for various electric motors can result into large energy and cost savings. Various other energy efficient best practices adopted to save energy and hence GHG emissions reduction in a cement industry were comprehensively discussed in

Table 18. Process and design modifications.

Raw material preparation	Replace traditional mills with roller mills and high pressure pressing milling
	Improve ball mills
	Apply efficient separators
	Insulate the hot air piping entering mills in order to dry materials
	Utilize pre-crusher
	Continuous control of raw material level in mills
	Adjust raw material size
Clinker production	Combustion system improvement
	Heat loss reduction by insulation
	Improve cooler operation and replace satellite coolers by grates ones
	Adopt mechanical transportation system (elevator instead of airlift)
	Add multistage pre-heaters and pre-calcinator
	Re-adapt wet type to dry type or semi-dry type
	Use low pressure cyclones
	Install eefficient fans
	Recycle waste heat
	Ensure air-tight system (good air draft control)
	Utilize variable speed drives for proper motor utilization
Finished grinding	Improve ball mills
	Replace traditional mills with roller mills and high continuous control of cement level in mills
	Control the size of input clinker and softness of cement according to standards
	Apply grinding aid materials to increase grinding capacity and reduce energy intensity
	Immediate grinding of clinker from kiln is encouraged to reduced the crushing strength, thereby reducing the electrical energy per ton required for grinding
Preventive maintenance	Correct plant maintenance schedule should be encouraged in order to operate plant in optimal capacity

[2,13,14].

4. Conclusions

An investigation of the energy consumption for the production of cement by wet and dry processes in Nigeria was carried out. From the study, the following observations were made:

- Combustion, electric and human energy constituted the major portion of the energy input in the production of cement by the wet and dry processes.
- The total estimated energy intensities were 6545 and 4197 MJ/ton for wet and dry process respectively. Furthermore, energy utilization is less intensive for the dry process than the wet cement manufacturing technique.
- The most energy intensive operation of all the operations in cement production was the burning operation consuming about 93% and 90% of the total energy input for the wet and dry processes respectively. Significant amount of heat is wasted in burning. Therefore, improvements can be made in this section to reduce heat loss or recycle heat.
- It has been observed that dry process is more efficient compared to wet process. It has been proved to reduce the energy consumption by 40% than wet process because extra energy is needed in wet techniques to remove moisture contained in wet slurry.
- Wet process plants which are energy inefficient and still available in the country should be replaced or upgraded to reduce its overall energy consumption.

REFERENCES

[1] T. Engin and V. Ari, "Energy Auditing and Recovery for Dry Type Cement Rotary Kiln Systems—Acase Study," *Energy Conversion and Management*, Vol. 46, No. 4, 2005, pp. 551-562.

[2] N. A. Madlool, R. Saidura, M. S. Hossaina and N. A. Rahim, "A Critical Review on Energy Use and Savings in the Cement Industries," *Renewable and Sustainable Energy Reviews*, Vol. 15, No. 4, 2011, pp. 2042-2060.

[3] US Department of Energy, "Energy and Emission Reduction Opportunities for the Cement Industry," 2012. http://www1.eere.energy.gov/industry/imf/pdfs/eeroci_dec03a.pdf

[4] http://tribune.com.ng/index.php/news/48440-nigeria-has-not-imported-cement-in-9-months-cman

[5] A. Avami and S. Sattar,"Energy Conservation Opportunities: Cement Industries in Iran," *International Journal of Energy*, Vol. 3, 2007, pp. 101-110.

[6] M. Schneider, M. Romer, M. Tschudin and H. Bolio, "Sustainable Cement Production—Present and Future," *Cement and Concrete Research*, Vol. 41, No. 7, 2011, pp. 642-650.

[7] M. G. Rasul, W. Widianto and B. Mohanty, "Asessment of Thermal Performance and Energy Conservation Opportunities of a Cement Industry in Indonesia," *Applied Thermal Engineering*, Vol. 25, No. 17-18, 2005, pp. 2950-2965.

[8] D. A. Fadare, O. A. Bamiro and A. O. Oni, "Energy and Cost Analysis of Organic Fertilizer Production in Nigeria," *Energy*, Vol. 35, 2010, pp. 332-340.

[9] G. O. Unachukwu, "Energy Efficiency Measures Investigation in Cement Company: BCC Case Study," *Nigerian Journal of Renewable Energy*, Vol. 10, No. 1-2, 2003, pp. 85-92.

[10] http://www.proshareng.com/news/download.php?Item=2619.pdf

[11] http://www.nigerianbestforum.com/blog/?p=59984

[12] E. U. Odigboh, "Machines for Crop Production," In: B. A. Stout and B. Cheze, Eds., *CIGR Handbook of Agricultural Engineering Plant Production*, American Society of Agricultural Engineers, 1997, pp. 1-21.

[13] E. Worrell and C. Galitsky, "Energy Efficiency Improvement and Cost Saving Opportunities for Cement Making," Ernest Orlando Lawrence Berkeley National Laboratory, 2008.

[14] E. Worrell, N. Martin and L. Price, "Potentials for Energy Efficiency Improvement in the US Cement Industry," *Energy*, Vol. 25, 2000, pp. 1189-1214.

6

Optimum Distribution of Two Different Phase Change Materials between Various Components of Roof Air-Conditioned Room, Suitable to Reduce Annual Energy Consumption

Bouchra Abouelkhayrat*, Hamid Hamza, Jawad Lahjomri, Abdelaziz Oubarra
Laboratory of Mechanics, Faculty of Sciences Ain Chock, Hassan II University, Casablanca, Morocco

ABSTRACT

Obviously, the outside annual climate change caused either by a major solar input during the hottest period or by a temperature drop during the coldest period leads to discomfort inside buildings. This effect can be reduced by storing heat transmitted in phase change materials (PCM) as latent heat, in order to ensure a good situation of thermal comfort during all months of the year. In this work, thermal behavior of two roofing systems is studied. One roof is constituted only by usual materials in building. In the other, two phase change materials (PCM) are introduced according to three configurations. Study is interested to assess incorporation effect of two PCMs within the roof and to evaluate the optimum locations to reduce the energy consumption of air-conditioned room. Mono-dimensional numerical model validated analytically and experimentally, is used to carry out a parametric analyzes to determine characteristics of the layers in which the roofs are formed regardless of external climate effect. Numerical calculations are performed for three configurations of roof. Results show that insertion of phase change materials in roof provides best energy consumption saving regardless annual climate change. Generally, the three configurations lead to different results, depending on the combination of PCMs. This difference becomes less important when selection of PCMs takes account the thermal comfort level and temperatures of hottest and coldest periods.

Keywords: Phase Change Material (PCM); Thermal Comfort; Multilayer Roof; Energy Consumption

1. Introduction

Recently, modern lifestyle leads to a huge problem of energy consumption worldwide. Especially, the excessive climatic conditions produce discomfort situation, which requires frequently using air conditioning systems, subsequently results in increasing of domestic energy consumption. Hence, this involves an increasing interest in thermal correction of building designs to create comfort situations. In order to save this energy exhaustion, among best known solutions proposed are: Use insulated walls/roofs in building [1], use of reflective color outdoor surfaces [2] and use of energy storage devices in the building structure [3]. As it is predicted in the literature, the mismatch that is often found between the supply and demand of energy can be corrected by these systems which provide a valuable solution. This is concerned against waste energy by storing and to restore it after, which are as important as developing new sources of energy. Hence the thermal energy storage systems such as phase change materials (PCMs) provide the potential to attain energy savings, which in turn reduce the environment problems due to conventional energy use.

Currently, the research about use of the phase change material (PCM) was increasingly under consideration especially for latent heat accumulator [4,5], solar heating system [6], building energy conservation and waste heat recovery [3,7,8]. The phase change material use, aims to reduce waste of energy by storing and restoring it after,

*Corresponding author.

when it is needed to consumption. Due to its important thermal energy storage as latent heat of fusion, that is advantageous over sensible heat because of its high storage density and the isothermal characteristics of charging and discharging process. Moreover, latent heat storage has the capacity to store heat of fusion at a constant or near temperature which corresponds to the phase change material (PCM). In fact, there are large numbers of PCMs [9,10] that melt and solidify at a wide range of temperatures, making them attractive in a number of building applications [10].

Generally building thermal performance evaluation of roof envelope with or without PCM has focused to optimize a suitable roof for air-conditioned rooms. Scientist research about this problem is carried out in the periodic external climatic conditions, depending on the building material choice, and also the internal level of comfort required [11-15]. The procedure applied by many authors is numerical study of one dimensional heat transfer through a wall/roof under convection boundary conditions, for air-conditioned rooms [5,7,16,17]. For roof with PCM, the works carried out a parametric study using enthalpy formulation taking into account heat latent of phase change material. Among technical procedure to operate the PCM in building there are two principal ways. Whatever, composites such as the PCM mixture with other construction compound called shape-stabilized PCM (SSPCM) [18] or well PCM coated in a thin metal container [19].

The main reason of using PCM in building [8,20,21] as energy storage devices in multilayer wall/roofs is to decrease the external climate effect on the inner surface of roof including PCM. Thus the heat energy will be absorbed both latent and sensible heat. But, this is not checked during all months of the year because the melting temperature of one PCM is not suitable for annual variations outside temperature. Hence, providing constant temperature in air-conditioned room without spending a large energy amount throughout the year requires to investigate the effect of two different PCMs in roof/walls building. This trend is expected narrowing of the heat transferred to local for all seasons. This technique was investigated by Pasupathy and Velraj [15]. The authors studied, firstly one layer of PCM kept in a residential building roof, after that, two PCMs with different melting points are incorporated in the roof. This solution leads to roof with two PCMs thicker than one without PCM; however, the effect of PCMs location is not studied.

Hence, this work studies the improvement of classical roof taken as reference, constituted by usual building materials. Two different phase change materials are introduced as several combinations of PCMs between referential roof layers in different positions. Thus, energy consumption to maintain the constant indoor air temperature is used as evaluation parameter. This parameter has also been used as criterion for insulation efficiency studied by G. Barrios *et al.* [1,22], and the effect of PCM ceiling panels investigated by N.A. Yahaya *et al.* [23]. Then, the energy consumption will assess for roofs kept in same total thickness with and without PCMs. The numerical simulation of roof thermal behavior over all year allows the achievement of annual energy consumption amount of air-conditioning room. Further, in order to determine optimal position of PCMs, this work is performed for three different configurations composites. The results obtained are taken for selecting the suitable combination of PCMs and also the optimal configuration. This is allows determining the criteria for selecting the PCMs likely to ensure the energy economy specified before.

2. Illustration of the Problem Studied

2.1. Statement of Roofs

The roof geometry used as reference is shown schematically in **Figure 1**. It is multilayer structure of thickness L = 30 cm, consisting of Plaster $e_P = 5$ cm, Cement Rendering $e_{CR} = 5$ cm and Concrete of thickness $e_C = 20$ cm.

The roof is studied under external climatic conditions (convection, solar radiation) while its internal surface is in contact with a local temperature conditioned. Latitude and longitude coordinates used to identify roof site designed to be in Casablanca city are respectively: Latitude 33°32'N and Longitude 7°41'W.

The improvement of reference roof undergoes this conditions during all year, requires to incorporate two phase change materials (PCM_1 and PCM_2) of thicknesses $e_{m1} = e_{m2} = 3$ cm. The integration of these materials is achieved without changing the basic structure. For this, the thickness of the concrete is maintained constant $e_C =$ 20 cm, but the layers of Plaster and Cement Rendering are reduced from 5 cm to 2 cm. When PCMs are inserted, the thickness of Plaster and Cement Rendering becomes respectively $e_P = 2$ cm and $e_{CR} = 2$ cm.

Hence three configurations schematized in **Figure 2** are considered according to location of phase change materials integrated in reference roof. In configuration 1 the both PCMs are separated by concrete, while configurations 2 and 3 are respectively for the cases where the PCMs are above and below concrete. Each PCMs pair, leads to two ways of insertion for each configuration by inverting the PCMs which result six combinations. These different combinations are considered and studied in order to ensure minimum energy consumption provided by the air conditioner, knowing that this investigation occurs during 12 months. Thus, it is expected that the effect of the external environment will be less noticeable on the roof due to the presence of PCM.

Requirements determined by studying of this problem

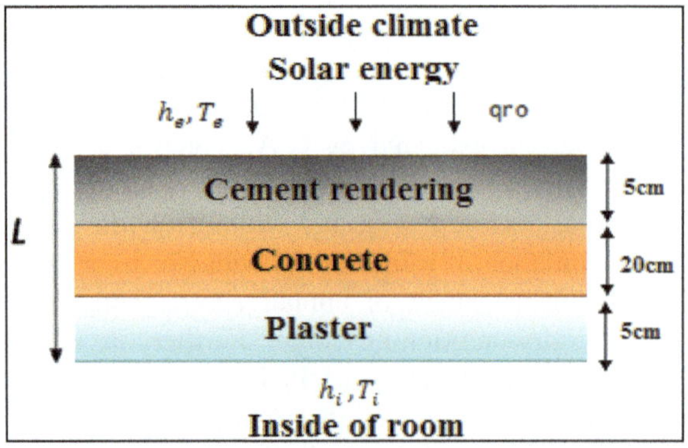

Figure 1. Scheme of reference roof.

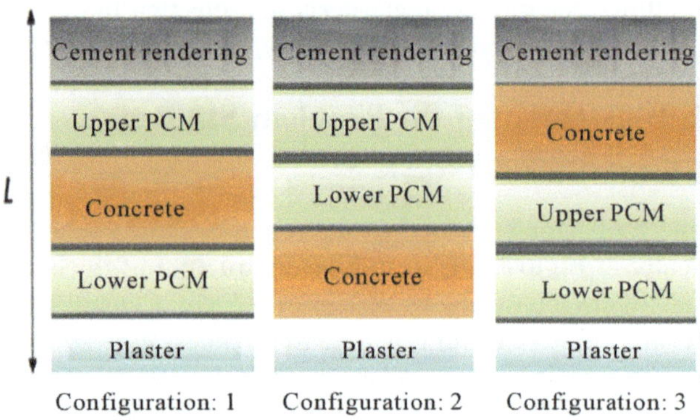

Figure 2. Positions of PCMs integration in referential roof.

lead to development of the following assumptions to define the numerical model:

- In your paper title, if the words "that uses" can accurately replace the word "using", capitalize the "u"; if not, keep using lower-cased.
- The heat conduction in the composite wall is one dimensional and the end effects are neglected.
- The PCM is homogeneous and isotropic.
- The interfacial resistances are negligible.
- All physical properties of both liquid and solid material are constant.
- The volume expansion is neglected during melting of the PCM.
- The effects of natural convection in the liquid phase of the PCM are negligible. This hypothesis is admitted for thin PCM thickness [5,24]. In this study, the aspect ratio is ~0.01.

2.2. Mathematical Model

Solving the equation of problem for different roof materials and PCM is given by Equation (1). Hence the dimensional equation of energy in different parts of system is written as:

$$\frac{\partial T}{\partial t}+\beta\frac{\partial f}{\partial t}=\alpha\frac{\partial^2 T}{\partial x^2} \qquad (1)$$

α is the thermal diffusivity of i-th material layer, T the temperature, t the time, and x the position which varies inside the layer.

The term $\beta=\lambda_f/C$ is present only for phase change material PCM during melting at $T=T_m$, characterized by melting ratio $f\in]0,1[$, such λ_f is the latent heat of PCM and C is the specific heat.

The melting thermal diffusivity of PCM α during melting is calculated using the following equations: $\alpha=\alpha_s(1-f)+\alpha_l f$.

Such as the indices s, and l indicate respectively solid and liquid phase.

Continuity of temperature and flux are applied in all unions between *i-th* and (i+1)-th layers according Equations (2) and (3) respectively.

$$(T_i)_{i/i+1}=(T_{i+1})_{i/i+1} \qquad (2)$$

$$K_i\left.\frac{\partial T_i}{\partial x}\right|_{i/i+1}=K_{i+1}\left.\frac{\partial T_{i+1}}{\partial x}\right|_{i/i+1} \qquad (3)$$

where: T_i, K_i and T_{i+1}, K_{i+1} are respectively temperatures and thermal conductivities of the *i-th* and $(i+1)$-*th* layers respectively. $i/i+1$ denotes that the calculation is evaluated in the interface between the *i-th* and $(i+1)$-*th* layers.

The boundary conditions on both sides of the roof are respectively:

- Indoor boundary condition :

$$K_1\left.\frac{\partial T}{\partial x}\right|_{x=0}=h_{int}\left(T_{int}-T_{x=0}\right) \qquad (4)$$

where K_1 is the thermal conductivity of the indoor layer and h_{int} is the indoor heat transfer coefficient, T_{int} is the indoor air temperature and $T_{x=0}$ is the temperature on the indoor surface.

- Outdoor boundary condition:

$$K_N\left.\frac{\partial T}{\partial x}\right|_{x=L}=h_{out}\left(T_{out}-T_{x=L}\right)+\varepsilon I_S-R_{ES} \qquad (5)$$

where K_N is the thermal conductivity of the outdoor layer and h_{out} is the outdoor convection coefficient, $T_{x=L}$ is the temperature on the outdoor surface, T_{out} is the outdoor air temperature. ε is the solar absorptivity of the outside surface, I_S is the solar radiation and R_{ES} is the radiation exchange with the sky. These last parameters affecting outdoor boundary condition are calculated for horizontal roofs in Casablanca by using the ASHRAE model [25,26].

The previous system of Equation (1) associated to interfaces (Equations (2) and (3)) and boundary conditions (Equations (4) and (5)), has been solved by implicit finite difference scheme. This stable and convergent scheme is able to occur the evolution of temperature profile at each point of the roof.

2.3. Validation of Numerical Solution

The Before presenting the numerical results for the phase change system, the numerical code must be validated by comparison with analytical solution available in the literature.

First, the phase-change model was checked against the classical Newman problem. The PCM is initially at melting temperature $T_i = T_m$ in semi-infinite long rectangular with a uniform cross section. At $t > 0$, the surface is kept at a temperature $T_e > T_m$ and liquefaction of PCM takes place. Position of solid-liquid interface versus time is calculated by numerical code and compared to its analytical values given by the function [17]: $y(\tau) = 2\gamma\sqrt{\tau}$ where γ is an unknown to be determined by solving the following Equation (6) [17]:

$$\gamma \exp(\gamma^2) erf(\gamma) = \frac{C_l}{\lambda\sqrt{\pi}}(T_e - T_m) \qquad (6)$$

Result of comparison between numerical and analytical solutions is reported in **Figure 3**. It can be seen that the present model agrees well with the results of this study.

The second validation reported in **Figure 4** is a comparison of the numerical code with the numerical and experimental results obtained by the study carried out by Amir *et al.* [3] using the paraffin n-octadecane such as phase change material of melting temperature T_m = 28°C. The operation of storing electrical energy during about four hours is followed by a release of energy during sixteen hours. The end of the process of storage and the beginning of energy releasing appear in the **Figure 4** by a peak representing the maximum value of the PCM temperature. The numerical code is suitable to the same operating conditions carried on a building configuration used in thermal analysis electric heating floor panels with daily heat storage [3]. The results of the comparison between the experimental and numerical temperatures profiles within the PCM show good agreement between the results of this study and the current numerical model. The results depicted in **Figure 4** prove that their reliability and accuracy are assured; furthermore this numerical model can give the results to a multilayer roof/wall.

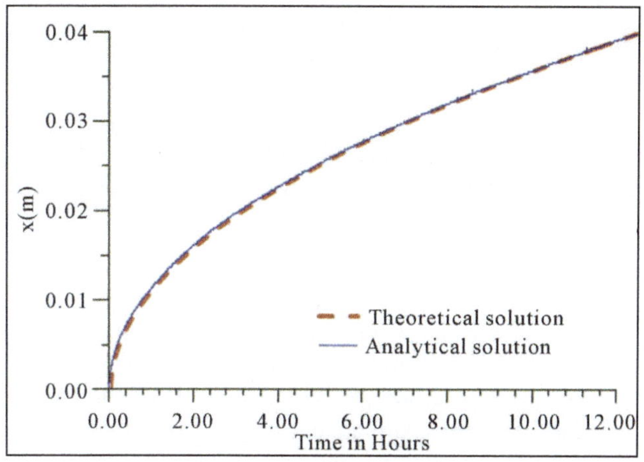

Figure 3. Evolving of melting front position with time.

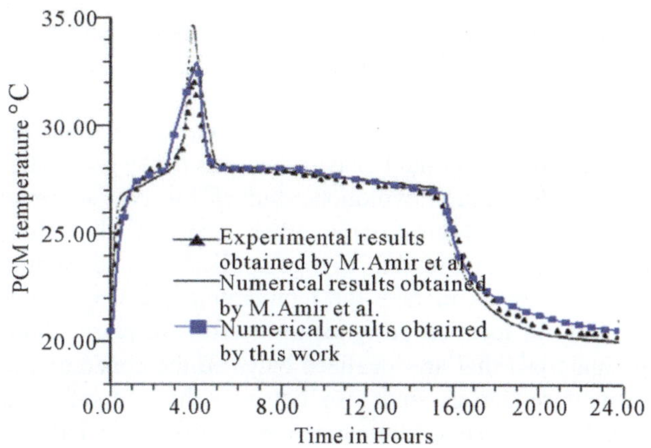

Figure 4. Temperature profile during a cycle of storage-release.

3. Parametric Study and Discussion

Simulation results of the developed numerical code are relating to all months of year. The various configurations shown schematically above (**Figure 2**) as well as reference roof (**Figure 1**) have been treated for similar thermal conditions imposed.

The internal temperature is expected to be maintained at a constant temperature $T_{int} = 20^\circ\text{C}$ whereas the external temperature follows a sinusoidal temporal evolution which profile is given for extreme period as follows:

In summer (hottest period):

$$T_{ext} = 27 + 13 \times \sin\omega(t - \tau/4) \qquad (7)$$

In winter (coldest period):

$$T_{ext} = 13.5 + 8.5 \times \sin\omega(t - \tau/4) \qquad (8)$$

With: τ, ω are respectively daily period and pulsation.

In addition to the effect of the variation of the outdoor temperature, the terrace is subjected to the action of solar radiation, approached by the Mediterranean ASHRAE model [25,26].

The effect of these outside conditions at inside roof surface appeared as a heat flow fluctuations. The fluctuations transfer leads to important energy consumption of air conditioner able to ensure an inside constant temperature. This energy consumption is calculated as an average daily at inside roof surface by Equation (9) and also its annual accumulation defined by Equation (10).

$$\Phi_{\text{Daily}} = \int_{\text{Daytime}} h_{int}\left(T(x=0) - T_{int}\right)\mathrm{d}t \qquad (9)$$

$x = 0$ corresponds to the inner section of the roof

$$\Phi_{\text{Annual}} = \sum_{i=1}^{m=12} \left(\Phi_{\text{Daily}}\right)_i \quad (10)$$

Such, that i means the month that varies from January until December.

3.1. Study of Reference Roof

The referential roof is constituted by three different building materials of physical properties summarized in **Table 1**. This system of roof without PCM undergoes outdoor periodic climate conditions which affects interior comfort. Therefore the annual energy consumption, defined by Equation (10), needed to keep the indoor air temperature constant 20˚C, reached 3.686 kWh/m^2. Hence it is interesting to check whether the inclusion of PCMs layers at the interfaces of the roof schematized above in **Figure 1** can reduce this energy consumption. In order to determine phase changes materials most suitable to reduce the consumption of energy, it requires carrying out a thermal behavior analysis of the referential roof. Temperature evolution studied for two extreme periods; one is hottest, while other is coldest, allows determination of temperature range applicable in all months of year. Levels to take the temperature are chosen according for location designed to insertion of phase change materials that are localized between the concrete and cement rendering, and also between the concrete and plaster. The results summarized in the following **Table 2** show that the choice of the most suitable PCM for the warm period is that which at a melting temperature between 24˚C and 34˚C, while the coldest period requires a choice between 17˚C and 23˚C. Hence, it is provided that choice of two PCMs having melting temperature between 17˚C and 34˚C, can achieve most suitable result during all the year. Finally, PCMs added are then a combination of PCMs of melting temperature in range of hottest period with other corresponds coldest period.

Although the literature [9,10] presents a variety of phase change materials, an unavailability of certain physical properties necessary for the numerical study limits the choice of operating PCMs. The phase change materials selected are given in **Table 3**. There are two kinds of PCMs: one suitable for space cooling and other suitable for space heating. Therefore, it is convenient to modify the design of the referential roof, by inserting two layers of materials as the phase change material (PCM) can store the both hot and cold energy in order to restore it later for minimizing the transmission of outside climate energy in indoor.

Indeed for a conventional roof without PCM, the amount of consumption energy is obviously influenced by the evolving external thermal conditions but also there is physical properties effect of roof components. Certainly, the thermal inertia of traditional building compounds used in this study, cheek an important role in delaying the transition of heat flow fluctuations. From **Table 1**, it is found that the values of thermal diffusivity and thermal conductivity of concrete which possesses thicker part of the roof are higher compared to same physical properties of cement rendering and plaster. Hence lower thermal conductivities and diffusivities of plaster and cement rendering lead to protection around concrete layer against outside climate conditions. This protection can be improved by replacing part of the Cement Rendering and Plaster by a same size of PCMs layers with requirement to keep the same total thickness of the roof. According to **Table 3**, the thermo physical properties of PCMs provide generally a low thermal conductivity and a less thermal diffusivity which allows to improvement prediction of the thermal behavior roof.

Table 1. Physical properties of different building materials used in the study.

Materials	Thermal conductivity K (W/m·K)	Specific heat C (kJ/kg·K)	Density ρ (kg/m^3)	Thermal diffusivity α (m^2/s)
Cement rendering	0.230	835	770	3.58×10^{-7}
Concrete	1.730	840	2400	8.58×10^{-7}
Plaster	0.76	1050	1500	4.82×10^{-7}

Table 2. Temperature evolution for two extreme periods; one is hottest, while other is coldest.

Location designed to insertion of PCM	Temperature in hottest period (˚C)		Temperature in coldest period (˚C)	
	Min	Max	Min	Max
Concrete/Plaster	24.24	27.06	18.89	21.18
Concrete/Cement rendering	25.04	33.84	17.42	22.57

3.2. Comparative Study of the Annual Internal Flux of Different Combinations of Phase Change Materials (PCM)

Numerical study focuses on roof thermal behavior simulation for each combination of selected PCMs. The aim is to determine that which be able to reduce the effect of outside climate energy fluctuations during all year on the air-conditioning energy consumption. The configurations schematized in **Figure 2** are carried out in periodic regime laid, under same external environmental conditions. This study includes the various configurations analyzed and processed for all selected PCMs, such as one PCM selected for hottest period is combined with all PCMs selected for coldest period, with swapping of PCMs. The same procedure is repeated for all PCMs of coldest period, where each PCM is combined with others of hottest period. Thus, there are three different configurations with

Table 3. Physical properties of selected PCMs.

PCMs	Melting temperature T_m(°C)	Latent heat λ_f(kJ/kg)	Thermal Conductivity K(W/m·K)	Specific heat C(kJ/kg·K)	Density ρ(kg/m^3)	Thermal Diffusivity α (m^2/s)
			PCM for Coldest Period			
S19 [10]	19	160	0.430	1.90	1520	1.49×10^{-7}
Clim Sel C21 [10]	21	144	0.7(liquid) 0.5(solid)	3.6	1380	1.41×10^{-7} (liquid) 1.01×10^{-7} (solid)
S23 [10]	23	175	0.540	2.2	1530	1.60×10^{-7}
			PCM for Hottest Period			
Clim Sel C24 [10]	24	126	0.7 (liquid) 0.5 (solid)	3.6	1380	1.41×10^{-7} (liquid) 1.01×10^{-7} (solid)
RT25-RT30 [9]	26.6	232	0.18 (liquid) 0.19 (solid)	1.8 (liquid) 1.41 (solid)	749 (liquid) 785 (solid)	1.34×10^{-7} (liquid) 1.72×10^{-7} (solid)
Polyethelene glycol 900 [9]	34	150.5	0.19 (liquid) 0.19 (solid)	2.26 (liquid) 2.26 (solid)	1100 (liquid) 1200 (solid)	0.76×10^{-7} (liquid) 0.69×10^{-8} (solid)

swapping of PCM selected, such as:

Configuration 1: Both PCMs are separated by concrete layer.

Configuration 2: Both PCMs are above concrete layer.

Configuration 3: Both PCMs are below concrete layer.

Further, these last configurations are compared with referential roof without PCM under the same external conditions and for the same thickness of concrete layer.

The annual energy consumption, defined by equation 10, is calculated for all PCMs combination formed in each configuration and also for roof without PCM. The annual accumulation of the monthly average of daily heat flow is given in (kWh/m^2), at the inner face of the roof that is mean in contact with the room at $x = 0$, where the temperature is kept at 20°C. Hence its data summarizes Φ_{Annual} as total of daily energy Φ_{Daily} calculated for each months by Equation (9), at the inner surface for different cases.

The first observation is that the combination of the same PCM with other PCMs gives different results in terms of energy consumption, to ensure internal temperature fixed at 20°C. This is also remarkable when going from one configuration to another for the same combination. Generally the annual energy consumption is often shows that the integration of PCMs, improves the residential buildings roof by reducing the energy consumption. This decrease can reach until 0.61 kWh/m^2, when PCM of melting temperature 34°C is up and PCM of melting temperature 21°C is lower in configuration 2 such as the both PCMs are over concrete, while the referential roof without PCM uses about 3.686 kWh/m^2, *i.e.* reduction of 83%. The importance of studying these combinations in details is summarized in the next section.

3.3. Combination Analysis

In order to assess the study mentioned previously, the results achieved are presented in **Figures 5-10**. These Figures enable analysis of annual energy consumption

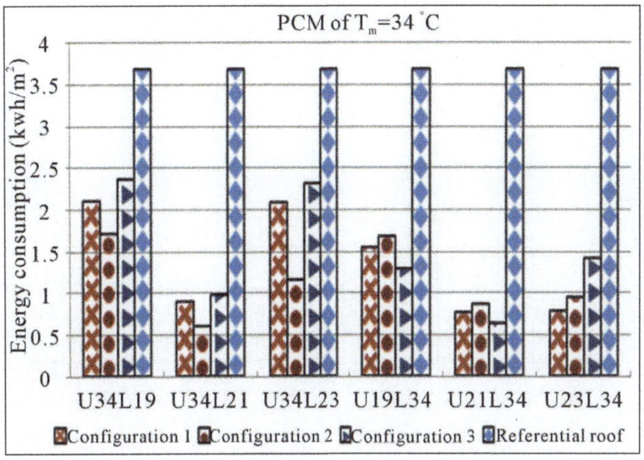

Figure 5. Combinaisons of PCM of T_m= 34°C with PCMs of coldest period.

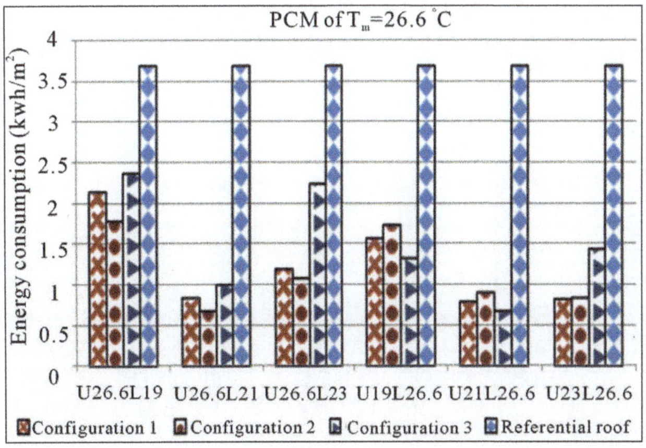

Figure 6. Combinaisons of PCM of T_m= 26.6°C with PCMs of coldest period.

Φ_{Annual} development for the three configurations and for referential roof.

Generally, the obviously observation as shown in **Figures 5-10** is that for roof without PCM, energy amount is greater than roof with two layer of different PCMs. The

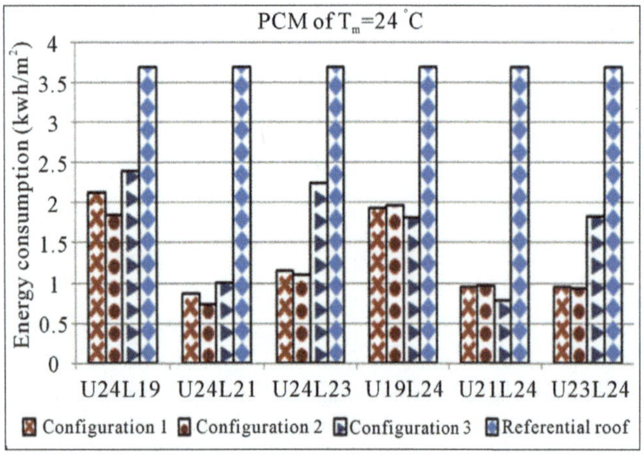

Figure 7. Combinaisons of PCM of T_m = 24°C with PCMs of coldest period.

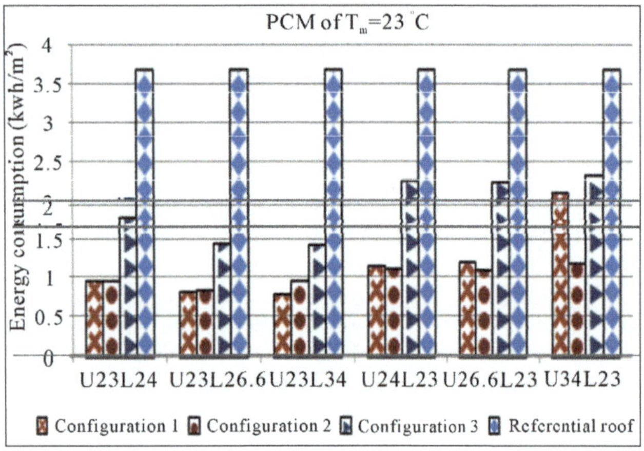

Figure 8. Combinaisons of PCM of T_m = 23°C with PCMs of hottest period.

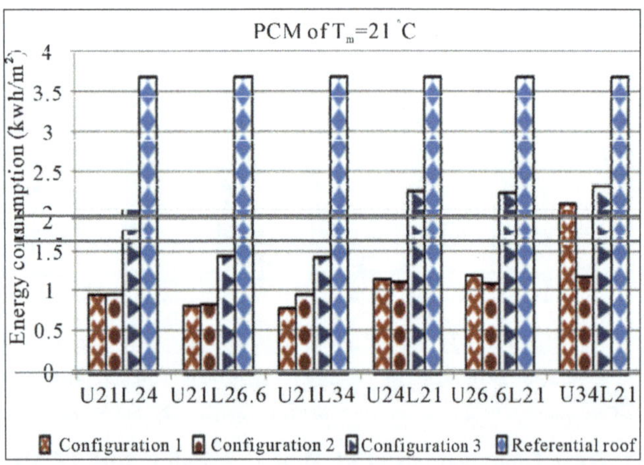

Figure 9. Combinaisons of PCM of T_m = 21°C with PCMs of hottest period.

integration of PCMs in the roof is always suitable and may even decrease the energy necessary to maintain constant local temperature.

The facts are confirmed for combination of PCMs with

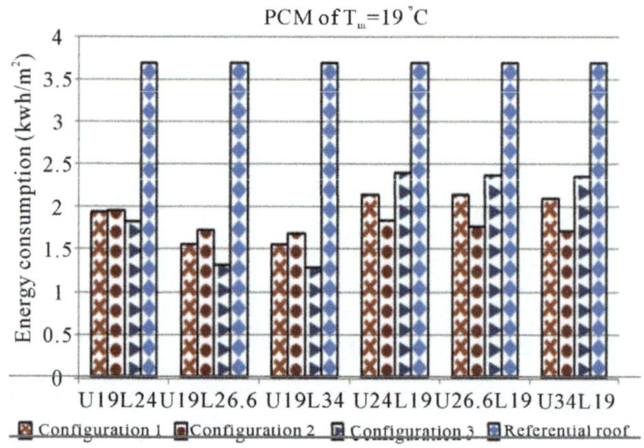

Figure 10. Combinaisons of PCM of T_m = 19°C with PCMs of hottest period.

significant advantage compared with the roof without PCM regarding the annual energy consumption Φ_{Annual}. Even that is true, but it is not always possible to have a significant reduction as shown results obtained when the PCM of melting temperature 19°C is combined with the selected PCM for the hottest period. Whereas the insertion of two PCMs with different melting temperature is very important, it is necessary to realize a study corresponding to thermal behavior of roof undergoes incorporation of PCMs. This thermal behavior evolution is influenced by other physical properties besides the melting temperature as latent heat and thermal diffusivity.

Because PCMs are selected according extreme periods, results will be discussed as two sub-paragraphs: one for the hottest period and other for the coldest period. Such a similar notation as U34L19 ~~is~~ used in abscissa axis, means: Upper PCM of melting temperature 34°C and Lower PCM of melting temperature 19°C.

3.3.1. Combinations of PCMs Selected for Hottest Period

Figures 5-7 illustrated annual consumption energy according to different combinations of PCMs selected for hottest period with the other PCMs, allow to comparison of three studied configurations.

This shows that location of PCMs having a higher melting temperature (T_m = 34°C, T_m = 26.6°C, T_m = 24°C), above PCMs with lower melting temperature (T_m = 23°C, T_m = 21°C, T_m = 19°C) is preferred for configuration 2 followed by the configuration 1, while configuration 3 is placed thirdly.

However, when the location of PCMs with lower melting temperature is above PCMs having a high melting temperature, configuration 1 is better than configuration 2, with an even evolution. The results obtained become nearly the same, as well gradually the melting temperature decreases for PCMs hottest period, especially for PCM of T_m = 24°C. Regarding the configuration 3, which behaves differently, it is obvious that the combination of PCMs selected for hottest period with PCMs of T_m = 19°C and T_m = 21°C, leads a saving of energy better than configuration 1 and 2. For PCM of T_m = 23°C results by these PCMs mentioned previously show that configuration 3 is third in front of configurations 1 and 2.

Generally, for the three configurations, when PCMs selected for hottest period are above of all PCMs of coldest period, the energy consumption is rising compared to the inverse permutation. So the location of PCMs of the upper melting temperature, below to PCMs of lower melting temperature, leads to save the energy designed to provide thermal comfort in room interior.

3.3.2. Combinations of PCMs Selected for Coldest Period

Evolution of the annual energy consumption of air conditioners providing T_{int} = 20°C, is evaluated by testing the PCMs selected to reduce this energy in the coldest period. Firstly the **Figures 8-10** show that location of PCMs having a higher melting temperature, above PCMs with lower melting temperature is preferred for configuration 2 followed by the configuration 1, whose configuration 3 is placed thirdly. Accordance this order, due to melting temperature decrease of PCMs hottest period. The results of configurations 1 and 2 begin to approach until to have almost a same value, specially for the PCM T_m = 23°C.

Secondly, the location of PCMs having a higher melting temperature below PCMs with lower melting temperature (T_m = 21°C and T_m = 19°C) is preferred for configuration 3 followed by the configuration 1, whose configuration 2 is placed thirdly. Regarding PCM of melting temperature T_m = 23°C, the classification of configurations is respectively 1, 2 and 3. Accordance this classification, due to melting temperature decrease of PCMs of hottest period, the results of configurations 1 and 2 start approximated until to have almost a same value, exept for the PCM of T_m = 34°C.

The results mentioned in **Figures 8-10** have confirmed for PCMs selected for coldest period the same remark previously concluded, but with one exeption. The location of PCMs having low melting temperature above to PCMs with high melting temperature is relevant, except PCM of melting temperature 21°C which has evolved inversely, but only for configuration 2. The location below of concerning PCM, relative to other PCMs for the hottest period is slightly important.

3.4. Criteria to Choose the Most Suitable PCMs

To determine the criteria for selecting the most suitable combination of PCMs reducing annual energy consumption through the roof of an air-conditioned room, the results obtained previously should be recapitulated. From the above **Figures 5-10**, it is obvious that the PCM with melting temperature T_m = 21°C is most able for reducing energy consumption, whatever configuration of PCMs insertion. This PCM combined with any PCM selected for hottest period, allows save up around three quarters 3/4 of the annual energy consumed by the roof without PCM.

Thus, it appears that the comfort temperature affects the choice of coldest PCMs period whose melting temperature should be close to the desired temperature in the room. This is also true for the PCM with fusion temperature T_m = 19°C, but its contribution with hottest PCMs period lead to energy supplied greater compared to all other coldest PCMs.

So when the comfort temperature is fixed at specific value, it seems that the most suitable coldest PCM to be combined with the PCMs of the hottest period is the one whose melting temperature is slightly greater, about 1°C, than the comfort temperature.

The checking of this summary needs to repeat the same study carried out for the comfort temperature at T_{int} = 20°C with another comfort temperatures such for example T_{int} = 18°C. Knowing that operating intervals for the hot and cold periods of the air-conditioned room at the new comfort temperature T_{int} = 18°C lead to the same selection of PCMs listed above in **Table 3**, the results of this further study summarized in **Figure 11** shows a minimum energy consumption for all combinations made by the PCM of T_m = 19°C whatever the integration configuration in the roof.

Regarding the selected of hottest period PCM, the choice should be based on the offset of excess heat caused by the increase in temperature up to a maximum value. In this effect PCM of temperature T_m = 34°C is most likely to be convenient for both comfort temperatures T_{int} = 18°C and 20°C. For other comfort temperatures tested, the results obtained confirm the same criteria previously mentioned.

Hence according to this remark, must be selected the most appropriate configuration for combinations containing suitable PCMs. Moreover the most favorable combinations of phase change materials for each configuration will be selected according to the top yields achieved regardless combinations of PCMs, such as the results will be discussed in the next section.

3.5. The Most Suitable of Phase Change Materials Combinations for Each Configuration

As already mentioned the top yields achieved by each configuration will be discussed for both comfort temperatures T_{int} = 20°C and 18°C. For this purpose, a comparative study of the relative deviation of energy con-

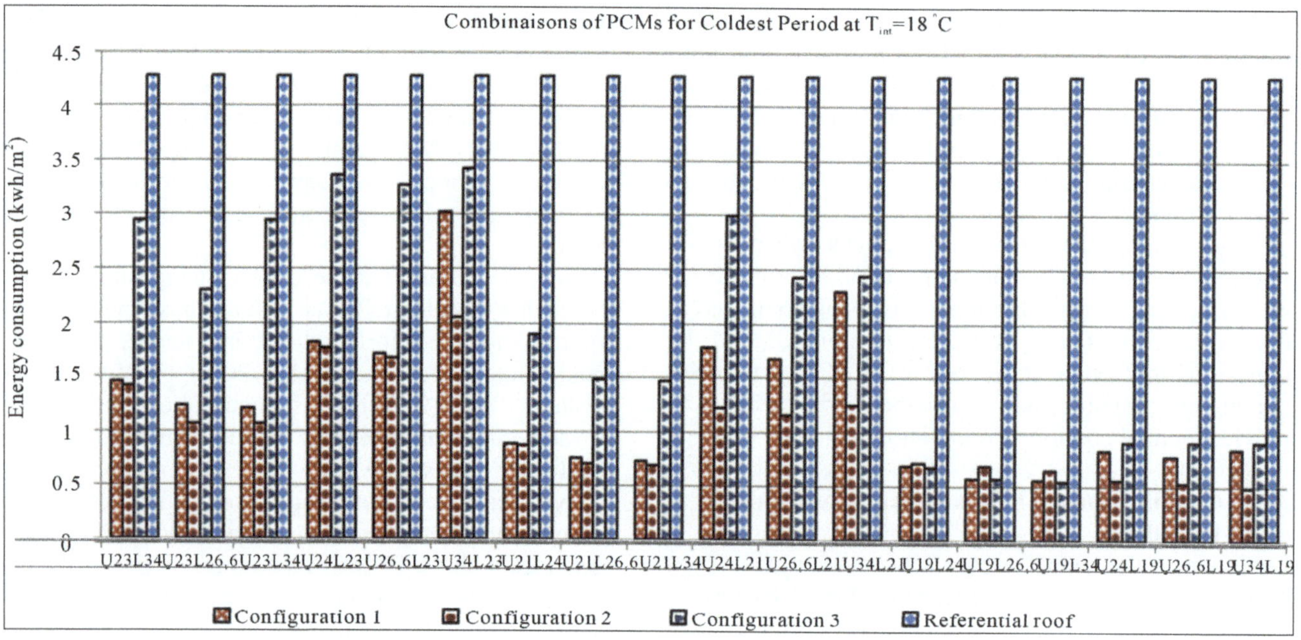

Figure 11. Combinaisons of PCMs for coldest period at T_{int} = 18°C.

sumed is conducted for each combination relative to the referential roof calculated as:

$$\sigma = \left[\left(\Phi^{\text{UXLY}}_{\text{Annual}} - \Phi^{\text{without PCM}}_{\text{Annual}}\right) / \Phi^{\text{without PCM}}_{\text{Annual}}\right] \times 100 \quad (11)$$

where:

- UXLY is such X and Y are two different melting temperatures PCMs, while U and L means that PCMs X and Y are respectively Upper and Lower.
- $\Phi^{\text{without PCM}}_{\text{Annual}}$ is the annual energy consumption at the inner surface of roof without PCM.
- $\Phi^{\text{UXLY}}_{\text{Annual}}$ is the annual energy consumption at the inner surface of roof with PCMs of any combination of PCM X and Y.

Numerical values of this parameter, defined by Equation (11), are reported respectively below in **Tables 4** and **5** for comfort temperatures T_{int} = 20°C and 18°C, according to top yields for the three configurations of PCMs.

According to **Tables 4** and **5**, performances can reach respectively over than (79% and 86%) in the energy economy compared to roof without PCM for three configurations 1, 2 and 3, respectively for comfort temperatures T_{int}=20°C and 18°C. For the three configurations (1, 2 and 3), the best performance is associated to the presence of PCMs of T_m = 21°C and T_m = 34°C for comfort temperatures T_{int} = 20°C, while for comfort temperatures T_{int} = 18°C the best performance is associated to the presence of PCMs of T_m = 19°C and T_m = 34°C.

Since in configuration 1 the both PCMs are separated by concrete, always a part of energy is absorbed by concrete layer. Then, it appears that the latent heat storage in configuration 2 and configuration 3 is more important

Table 4. Top yields for three configurations for comfort temperatures T_{int} = 20°C.

	Configuration (1)	Configuration (2)	Configuration (3)
Combination	H21B34	H34B21	H21B34
Performance (%)	−79.30	−83.44	−82.78

Table 5. Top yields for three configurations for comfort temperatures T_{int} = 18°C.

	Configuration (1)	Configuration (2)	Configuration (3)
Combinations	H19B34	H34B19	H19B34
Performance (%)	−86.82	−88.48	−87.28

compared to configuration 1; therefore its performance is lower compared with other configurations. While for configurations 2 and 3, these yields can reach maximum values shown in **Tables 4** and **5**. In configuration 2, the both PCMs are above of concrete; this means that the highest level of heat transfer is received by PCMs. Therefore, when the melting temperature of the upper PCM is higher than the lower PCM, the performance of the energy consumed is important

However, third configuration where both PCM are below of concrete, this means that the highest level of heat transfer is received by concrete layer. Thus melting temperature of the upper PCM is lower than the below PCM and provide the important performance of the energy consumed.

4. Conclusions

Thermal behavior of two roofing systems is studied, under climate conditions of Casablanca city during all year. One roof is constituted just by conventional building materials. In the other hand, two phase change materials (PCM) are introduced in different locations within traditional roof.

The parametric study produces the roof thermal behavior was carried out using a numerical program developed and successfully validated. Three multilayer roofing configurations provided with two phase change materials have been treated at different comfort temperatures; such results of two comfort temperatures 18˚C and 20˚C are detailed.

In general, combining two PCMs leads to lower annual energy consumption, when the PCM for coldest period is placed above, that is to say, close to the external environment. This is due to the fact that when it is moved down, it contributes less by a phase change as it does not undergo to external excitation.

Even the integration of PCMs in the roof is suitable, it is not always possible to have a significant reduction. Hence it is necessary to realize a study of thermal behavior under internal and external conditions imposed.

In addition, the criteria for choosing the most suitable PCMs, according to the results obtained show that each PCM must correspond to one temperature interval (hottest or coldest) with a choice of a PCM able to compensate the maximum consumption of energy imposed by the highest temperature of the hottest period, however, the other PCM suitable for the cold period should be that its melting temperature is slightly above that of comfort around +1˚C. Eventually it seems that the thermal comfort level dictates the choice of PCM for cold period while the PCM selected for hot period should be able to compensate the highest temperature transferred to roof studied.

Finally, the three configurations lead to different results, depending on the combination of PCMs. This difference becomes less important when the criteria to select the PCMs are respected.

REFERENCES

[1] G. Barrios, G. Huelsz and J. Rojas, "Thermal Performance of Envelope Wall/Roofs of Intermittent Air-Conditioned Rooms," *Applied Thermal Engineering*, Vol. 40, 2012, pp. 1-7.

[2] A. D. Granja and L. C. Labakiny, "Influence of External Surface Color on the Periodic Heat flow through a Flat Solid Roof with Variable Thermal Resistance," *International Journal of Energy Research*, Vol. 27, No. 8, 2003, pp. 771-779.

[3] M. Amir, M. Lacroix and N. Galanis, "Comportement Thermique de Dalles Chauffantes Electriques pour le Stockage Quotidien," *International Journal of Thermal Sciences*, Vol. 38, No. 2, 1999, pp. 121-131.

[4] B. Abouelkhayrat, J. Lahjomri and A. Oubarra, "Study of Latent Accumulator Storage by Heat Transfer Fluid (HTF) and External Heat Source (EHS)," *Physical and Chemical News*, Vol. 66, 2012, pp. 56-68.

[5] E. M. Alawadhi, "Temperature Regulator Unit for Fluid Flow in a Channel Using Phase Change Material," *Applied Thermal Engineering*, Vol. 25, No. 2-3, 2005, pp. 435-449.

[6] J. Kośny, K. Biswas, W. Miller and S. Kriner, "Field Thermal Performance of Naturally Ventilated Solar Roof with PCM Heat Sink," *Solar Energy*, Vol. 86, No. 9, 2012, pp. 2504-2514.

[7] A. Pasupathy, L. Athanasius, R. Velraj and R. V. Seeniraj, "Experimental Investigation and Numerical Simulation Analysis on the Thermal Performance of a Building Roof Incorporating Phase Change Material (PCM) for Thermal Management," *Applied Thermal Engineering*, Vol. 28, No. 5-6, 2008, pp. 556-565.

[8] P. Brousseau and M. Lacroix, "Study of the Thermal Performance of a Multi-Layer PCM Storage Unit," *Energy Conversion Management*, Vol. 37, No. 5, 1996, pp. 599-609.

[9] F. Agyenim, N. Hewitt, P. Eames and M. Smyth, "A Review of Materials, Heat Transfer and Phase Change Problem Formulation for Latent Heat Thermal Energy Storage Systems (LHTESS)," *Renewable and Sustainable Energy Reviews*, Vol. 14, No. 2, 2010, pp. 615-628.

[10] P. Tatsidjodoung, N. L. Pierres and L. Luo, "A Review of Potential Materials for Thermal Energy Storage in Building Applications," *Renewable and Sustainable Energy Reviews*, Vol. 18, 2013, pp. 327-349.

[11] A. P. Simard and M. Lacroix, "Study of the Thermal Behavior of a Latent Heat Cold Storage Unit Operating under Frosting Conditions," *Energy Conversion and Management*, Vol. 44, No. 10, 2003, pp. 1605-1624.

[12] K. Ouyang and F. Haghighat, "A Procedure for Calculating Thermal Response Factors of Multi-Layered Walls-State Space Method," *Building and Environment*, Vol. 26, No. 2, 1991, pp. 173-177.

[13] A. K. Athienitis, "Thermal Analysis of Buildings in a Mathematical Programming Environment and Applications," *Building and Environment*, Vol. 34, No.4, 1999, pp. 401-415.

[14] R. Yumrutas, M. Nsal and M. Kanoglu, "Periodic Solution of Transient Heat Flow through Multilayer Walls and Flat Roofs by Complex Finite Fourier Transform Technique," *Building and Environment*, Vol. 40, No. 8, 2005, pp. 1117-1125.

[15] A. Pasupathy and R. Velraj, "Effect of Double Layer

Phase Change Material in Building Roof for Year Round Thermal Management," *Energy and Buildings*, Vol. 40, No. 3, 2008, pp. 193-203.

[16] F. Kuznik and J. Virgone, "Experimental Investigation of Wallboard Containing Phase Change Material: Data for Validation of Numerical Modeling," *Energy and Buildings*, Vol. 41, No. 5, 2009, pp. 561-570.

[17] V. Alexiades and A. D. Solomon, "Mathematical Modeling of Melting and Freezing Processes," Hemisphere Publishing Corporation, Washington DC, 1993, p. 305.

[18] G. Zhou, Y. Yang and H. Xu, "Performance of Shape-Stabilized Phase Change Material Wallboard with Periodical outside Heat Flux Waves," *Applied Energy*, Vol. 88, No. 6, 2011, pp. 2113-2121.

[19] Z. A. Hammou and M. Lacroix, "Anew PCM Storage System for Managing Simultaneously Solar and Electric Energy," *Energy and Building*, Vol. 38, No. 3, 2006, pp. 258-265.

[20] R. Velraj, K. Anbudurai, N. Nallusamy and M. Cheralathan, "PCM Based Thermal Storage System for Building Air Conditioning at Tidel Park, Chennai," *Proceeding of WREC Cologne*, Germany, 2002.

[21] C. Stetiu and H. E. Feustel, "Phase Change Wallboard and Mechanical Night Ventilation in Commercial Buildings," Lawrence Berkeley National Laboratory, 1998.

[22] G. Barrios and G. Huelsz, R. Rechtman and J. Rojas, "Wall/Roof Thermal Performance Differences between Air-Conditioned and Non Air-Conditioned Rooms," *Energy and Building*, Vol. 43, No.1, 2011, pp. 219-223.

[23] N. A. Yahaya and H. Ahmed, "Numerical Investigation of Indoor Air Temperature with the Application of PCM Gypsum Board as Ceiling Panels in Buildings," *Procedia Engineering*, Vol. 20, 2011, pp. 238-248.

[24] D. Pal and Y. K. Joshi, "Thermal Management of an Avionics Module Using Solid-Liquid Phase-Change Materials," *Thermophysics and Heat Transfer*, Vol. 12, No. 2, 1998, pp. 256-262.

[25] "ASHRAE Handbook of Fundamentals Chapter 26," American Society of Heating, Refrigerating and Air-Conditioning Engineers (ASHRAE), New York, 1997.

[26] S. Barbaro, G. cannistraro, C. Giaconia and A. Orioli, "The Ashrae Clean Sky Model, an Evaluation in the Mediterranean Zone," *Solar and Wind Technology*, Vol. 1, No. 1, 1988, pp. 111-116.

7

Analysis and Modeling of Buck Converter in Discontinuous-Output-Inductor-Current Mode Operation*

Jianbo Yang[1], Weiping Zhang[2], Faris Al-Naemi[1], Xiaoping Chen[2],

[1]Materials and Engineering Research Institute (MERI), Sheffield Hallam University (SHU), Sheffield, UK

[2]Lab of Green Power & Energy System (GPES), North China University of Technology (NCUT), Beijing, China

ABSTRACT

The Buck converter with LC input filter operating in discontinuous output current mode has a high power factor with a constant duty cycle. A Buck converter in this operation mode can reduce the reverse recovery loss of the freewheeling diode thus increase the efficiency. The operation, power factor analysis and modeling of the converter are studied in this paper. Experimental results are presented to verify the theoretical predictions.

Keywords: Buck Converter; Power Factor Correction; Modeling

1. Introduction

AC/DC converter has been studied as a high-power-factor rectifier. The most popular power stage circuit of the AC/DC converter is a Boost converter [1,2]. The Boost converter has an input inductor which naturally makes the input current continuous and less of harmonics. However, the disadvantage is the output has to be higher than the peak input voltage. Unlike the Boost AC/DC converters, the Buck AC/DC converters have step-down characteristics as the output voltage is lower than the peak input voltage. The buck converter with LC input filter operating in discontinuous output inductor current mode has resistive and constant input impedance with a constant duty cycle D and a constant switching cycle T_s. Thus, the average input current of the converter follows the input voltage with a constant duty ratio control.

The low-frequency behavior modeling of the Buck converter with LC input filter operating in discontinuous output inductor mode is presented in this paper. Based on the model, the characteristic of the converter and the conditions for power factor correction is studied. The experimental verification is also given.

2. Power Stage

2.1. Circuit Configuration

The buck converter with LC input filter is shown in **Figure 1**. The discontinuous output inductor current mode means the current through the inductor L_2 is zero during part of one switching cycle. The converter can still operate in discontinuous capacitor voltage mode which means the voltage across C_1 is zero during part of a switching cycle. These two operation modes can be considered as dual of each other [3]. However, the voltage stress of the switch and diode will be very high when the converter operates in discontinuous capacitor voltage mode [4].

2.2. Operation Principles

The operation principles of the converter and the characteristic waveforms are presented in **Figure 2**.

There are three phases for the Buck converter with LC input filter operating in discontinuous inductor current mode.

Phase 1: 0~DTs. switch S is turned on. D is reverse-biased. When the operation is in steady state, the average voltage across L_1, over one switching cycle, is zero. The average voltage across C_1 over one switching cycle is

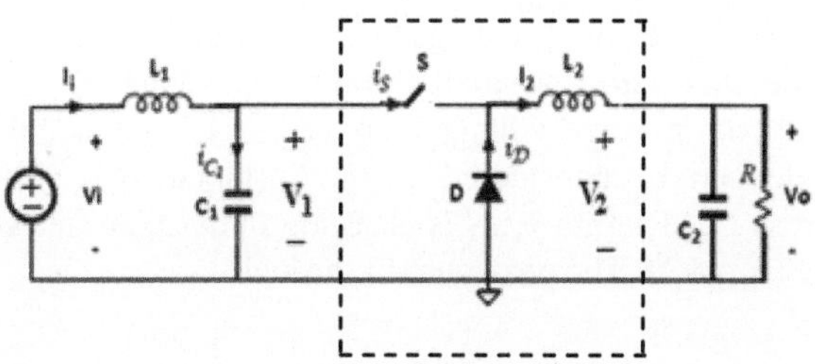

Figure 1. Buck AC/DC converter with LC input filter.

*Project supported by Natural Science foundation of China (NO. 51277004). The Importation and Development of High-Caliber Talents Project of Beijing Municipal Institutions (No.IDHT20130501)

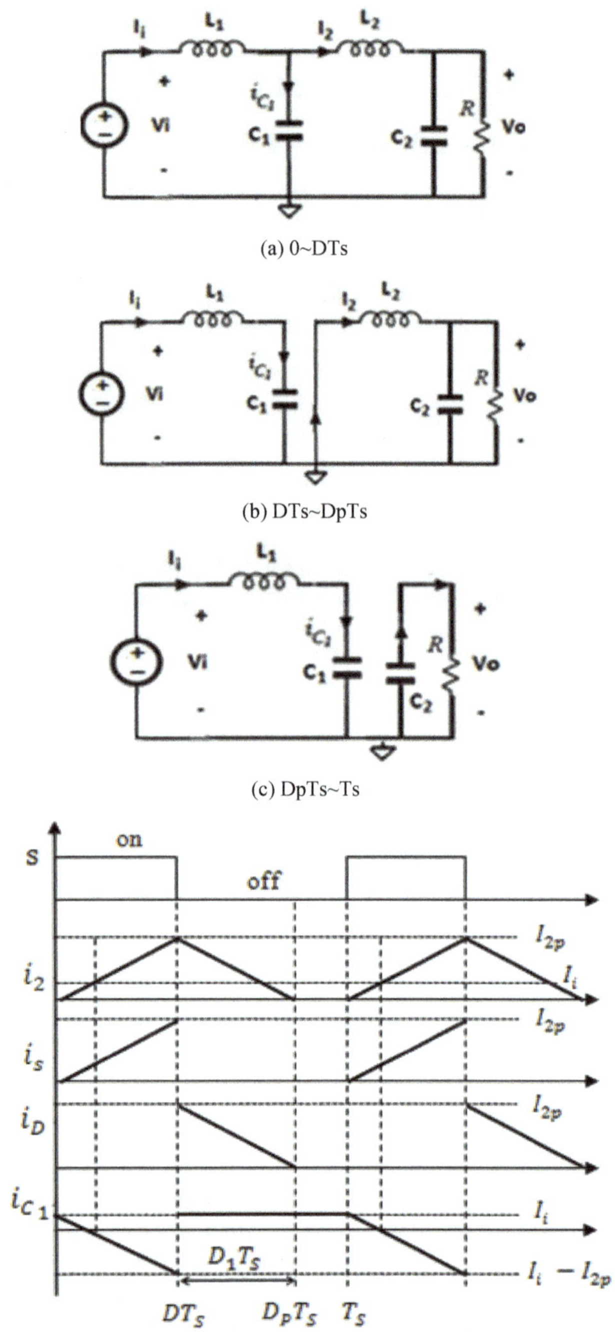

(a) 0~DTs

(b) DTs~DpTs

(c) DpTs~Ts

(d) Swiching waveforms

Figure 2. Discontinuous inductor current operation and waveforms.

equal to the input voltage. C_1 operates in continuous mode that Δu_{c1} is small during one switching cycle. Thus, the average voltage across C_1 over one switching cycle can be considered as the instantaneous voltage across C_1. Therefore, L_2 is charging under constant voltage ($V_i - V_o$). The peak current through L_2 is:

$$I_{2p} = \frac{(V_i - V_o)}{L_2} DT_s \tag{1}$$

The current through C1 is $I_i - I_2$. When $I_2 < I_i$, i_{c1} is positive, C_1 is charging. When $I_2 > I_i$, C_1 is discharging.

Phase 2: DTs~DpTs. *S* is closed *D* is forward-biased. L_2 is discharging and the current through L2 falls to zero at DpTs. C_1 is charging. The peak current through L_2 can also be obtained as:

$$I_{2p} = \frac{V_o}{L_2}\left(D_p - D\right)T_s \tag{2}$$

Phase 3: DpTs~Ts; the switch *S* is still turned off. Current through L_2 falls to zero. Output is supported by C_2 individually. Combination of (1) and (2), the following equation can be obtained.

$$\frac{V_o}{V_i} = \frac{D}{D_p} > D \tag{3}$$

If the switching period Ts is much smaller than the input cycle Ti, the input voltage can be considered constant over one switching cycle [3]. Then, the input voltage V_i in (1), (3) can be directly replaced by $V_i \sin w_i t$ during half of line cycle T_i.

$$I_{2p} = \frac{(V_i \sin w_i t - V_o)}{L_2} DT_s \tag{4}$$

$$\frac{V_o}{V_i \sin w_i t} = \frac{D}{D_p} > D \tag{5}$$

V_o/V_i can be defined as η. then from equation (5), the relation between η and *D* is

$$\frac{\eta}{D} > \sin w_i t \tag{6}$$

Therefore, the converter will operate in discontinuous inductor current mode throughout half line cycle T_i when $\eta/D > 1$.

3. Power Analysis

As shown in **Figure 2**, the input current I_i is the sum of the switch current i_s and current through $C_1\,i_{c1}$. The average current through C_1 is zero during one switching cycle in the steady state. Thus, the average input current equals to the average switching current over one switching cycle. According to (2) and (4), the average input current, over one switching cycle, can be obtained as,

$$I_{iavg} = \frac{(V \sin w_i t - V_o)}{2L_2} D^2 T_s \tag{7}$$

where I_{iavg} means the average input current over one switching cycle. As the switching cycle T_s is much smaller than the input cycle T_i, the input current can be considered constant over one switching cycle. Buck converter operates only when the input voltage is higher than the output voltage. Therefore, (7) is valid only when $V\sin w_i t > V_o$. When the input rectified voltage equals to the output voltage,

$$t_1 = \frac{1}{w_i}\arcsin\frac{V_o}{V_i} = \frac{1}{w_i}\arcsin\alpha \qquad (8)$$

α is defined as the ratio between the output voltage and the peak of the rectified input voltage. Thus, over half input line cycle, operation is possible only for

$$t \in (t_1, \frac{T_i}{2} - t_1),$$

as shown in **Figure 3**. The average input power with constant duty ratio and constant switching cycle T_s is provided in (9). As referred to (5), the maximum duty cycle D is η for maintaining the converter operate in discontinuous inductor current mode. Substitution of D for η in (9), the following relation can be obtained in (10).

$$P_i = \frac{2}{T_i}\int_{t_1}^{\frac{T_i}{2}-t_1} V_i \sin w_i t \cdot I_{i,avg}\,dt = \frac{V_i^2 D^2 T_s}{4L_2}\left(1 - \frac{2}{\pi}\arcsin\eta - \frac{2\eta\sqrt{1-\eta^2}}{\pi}\right) \qquad (9)$$

$$P_i = \frac{V_i^2 \eta^2 T_s}{4L_2}\left(1 - \frac{2}{\pi}\arcsin\eta - \frac{2\eta\sqrt{1-\eta^2}}{\pi}\right) \qquad (10)$$

The relation between active input power and conversion ratio η with maximum duty ratio is plotted in **Figure 4**. As shown, with $D = \eta$, the converter operates in the boundary of the discontinuous inductor current mode and the maximum active input power will be obtained when η is near 0.58.

The root mean square value of the input current can be obtained from (7) as

$$I_{irms} = \sqrt{\frac{2}{T_i}\int_{t_1}^{\frac{T_i}{2}-t_1} \frac{(V\sin w_i t - V_o)}{2L_2} D^2 T_s\,dt} = \frac{V_i D^2 T_s}{2\sqrt{2}L_2}\sqrt{(1+2\eta^2)(1-\frac{2}{\pi}\arcsin\eta) - \frac{6\eta\sqrt{1-\eta^2}}{\pi}} \qquad (11)$$

the power factor is obtained from (10) and (11) as,

$$P.F = \frac{P_i}{V_{irms}\cdot I_{irms}} = \frac{1 - \frac{2}{\pi}\arcsin\eta - \frac{2\eta\sqrt{1-\eta^2}}{\pi}}{\sqrt{(1+2\eta^2)(1-\frac{2}{\pi}\arcsin\eta) - \frac{6\eta\sqrt{1-\eta^2}}{\pi}}} \qquad (12)$$

where P.F is the power factor. It is clear that the power factor is affected only by conversion ratio η when the duty cycle D is constant. The relation is plotted in **Figure 5**.

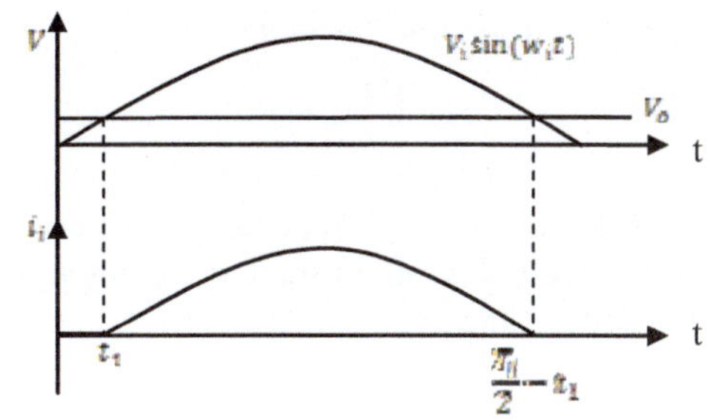

Figure 3. Operation waveforms during half input cycle.

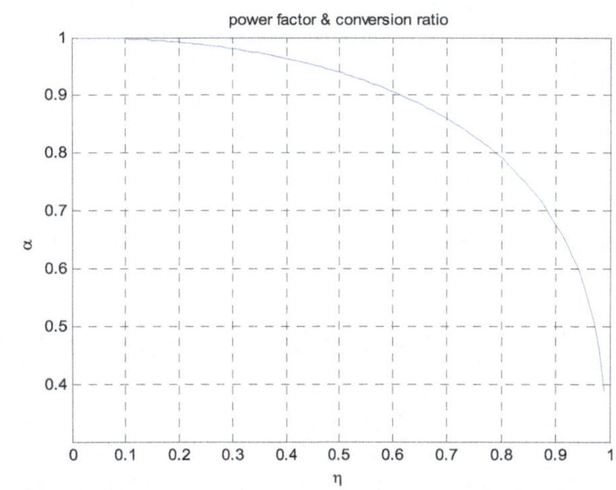

Figure 4. Active Input Power vs Conversion Ratio η (from equation (10)).

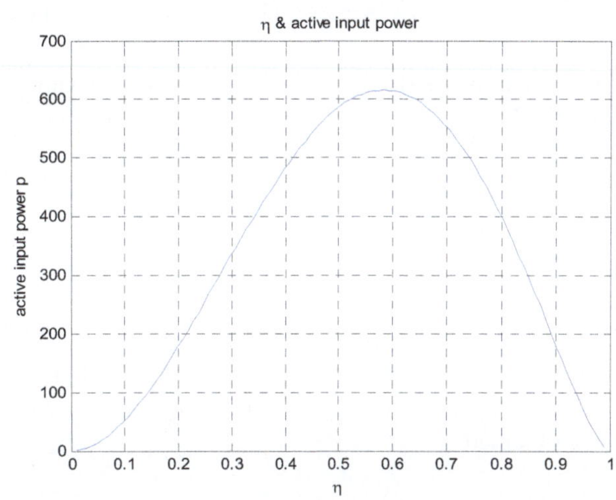

Figure 5. Power Factor vs Conversion Ratio η (from equation (12)).

It is shown that the power factor is reversely proportional to the conversion ratio η.

When $\eta < 0.6$ the power factor will be higher than 0.9. The results are exact the same as [5]. When η is very small, $V_o \ll V_i \sin w_i t$. Then the input current will be

$$I_{iavg} \approx \frac{V_i \sin w_i t}{2L_2} D^2 T_s \qquad (13)$$

The equivalent input impendence is resistive,

$$R_i = \frac{2L_2}{D^2 T_s} \tag{14}$$

The power factor turns out to be unity. Therefore, with constant duty cycle D and constant switching cycle Ts, the lower output voltage, the higher power factor.

4. Modeling

The low-frequency model is obtained by considering the relative voltage or current will not change over one switching cycle since the switching cycle is much higher than the input line frequency. Therefore, the average value, over one switching cycle, is considered as an instant value during one switching cycle [4,6]. By assuming that the voltage across C1 is continuous and constant over a switching cycle, the average switch current i_s and average current through L_2 over one switching cycle can be obtained from **Figure 2**.

$$I_s = \frac{V_2}{2L_2} D_1 D T_s \tag{15}$$

$$I_2 = \frac{V_2}{2L_2}(D_1 + D) D_1 T_s \tag{16}$$

As the average voltage across L2 is zero over one switching cycle.

$$D_1 = \frac{V_1 - V_2}{V_2} D \tag{17}$$

As shown in **Figure 1**, the following equations can be obtained.

$$I_s(s) = \frac{1}{sL_1}(V_i(s) - V_1(s)) - V_1(s) * sC_1 \tag{18}$$

$$I_2(s) = V_2(s)(sC_2 + \frac{1}{R}) \tag{19}$$

$$V_o(s) = V_2(s) \tag{20}$$

Derivation and Laplace transform of (15) ~ (17) and substitution of the results into (18) ~ (19), the following can be obtained.

$$\begin{cases} \frac{V_o(s)}{V_i(s)} = \frac{(2D_1 + D)D}{\alpha_3 s^3 + \alpha_2 s^2 + \alpha_1 s + \alpha_0} \\ \frac{V_o(s)}{D(s)} = \frac{b_2 s^2 + b_1 s + b_0}{\alpha_3 s^3 + \alpha_2 s^2 + \alpha_1 s + \alpha_0} \end{cases} \tag{21}$$

where,

$$\begin{cases} \alpha_3 = \frac{2L_1L_2C_1C_2}{T_s} \\ \alpha_2 = L_1C_1(D_1 + D)^2 + \frac{2L_1L_2C_1}{RT_s} + L_1C_2D^2 \\ \alpha_1 = \frac{2L_2C_2}{T_s} + \frac{L_1D^2}{R} + \frac{T_sL_1D_1^2D^2}{2L_2} \\ \alpha_0 = (D_1 + D)^2 + \frac{2L_2}{RT_s} \end{cases}$$

$$\begin{cases} b_2 = 2L_1C_1V_2D_1(\frac{D_1 + D}{D}) \\ b_1 = -\frac{V_2L_1T_sDD_1^2}{L_2} \\ b_0 = 2V_2D_1\left(\frac{D_1 + D}{D}\right) \end{cases}$$

$V_o(s)/V_i(s)$ is output-input transfer function and $V_o(s)/D(s)$ is the output-control transfer function. The transfer function between output voltage and duty cycle given by (21) has three poles and two zeros. This can be approximate to have a single pole which is exactly the same as the traditional Boost converters [6,7].

5. Results

5.1. Simulations

Simulations were carried out to verify the discontinuous inductor current operation of the circuit. The components used are: L_1 = 500u. L_2 = 20u, C_1 = 220n. C_2 = 2000u, D = 0.1, R = 13. The input voltage is 220VAC and the output voltage is 36 V. The results are illustrated as follows (**Figures 6 - 9**), which show that input current will follow the input voltage automatically when Buck converter with an input LC filter operates in discontinuous inductor current with a constant the duty ratio.

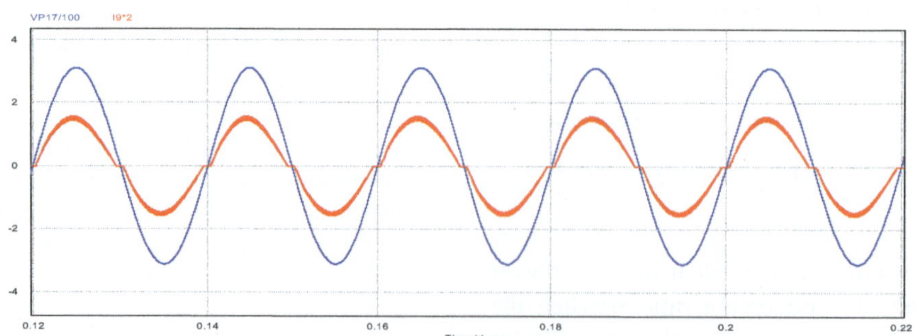

Figure 6. Input voltage and input current (220Vac; 0.52A).

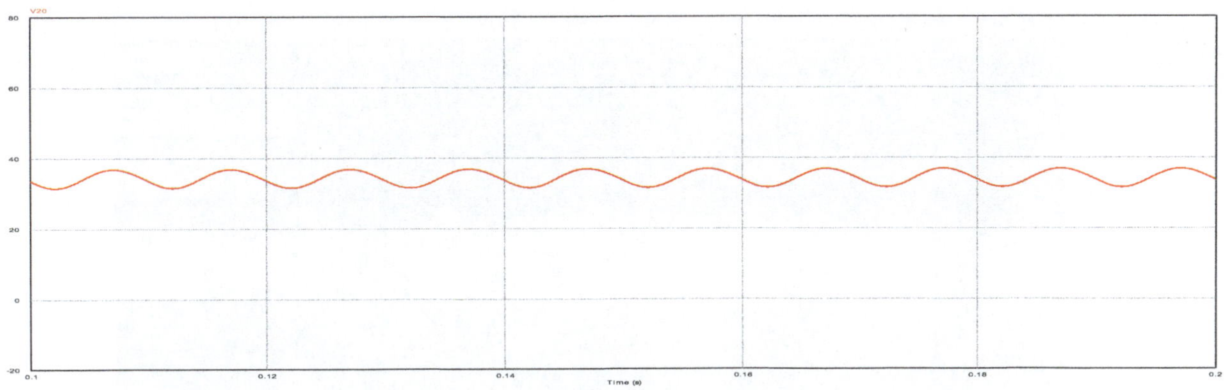

Figure 7. Output voltage: 36v (ripple: 5v),

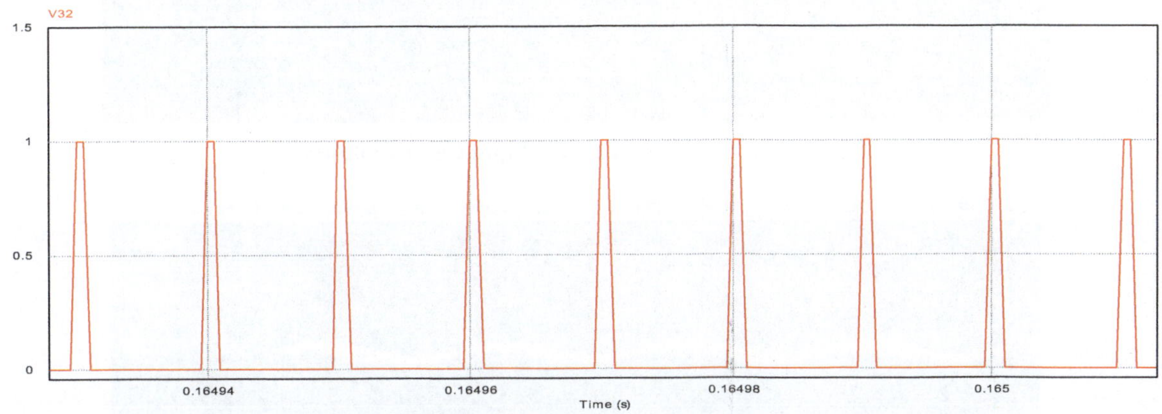

Figure 8. Constant duty cycle (0.15).

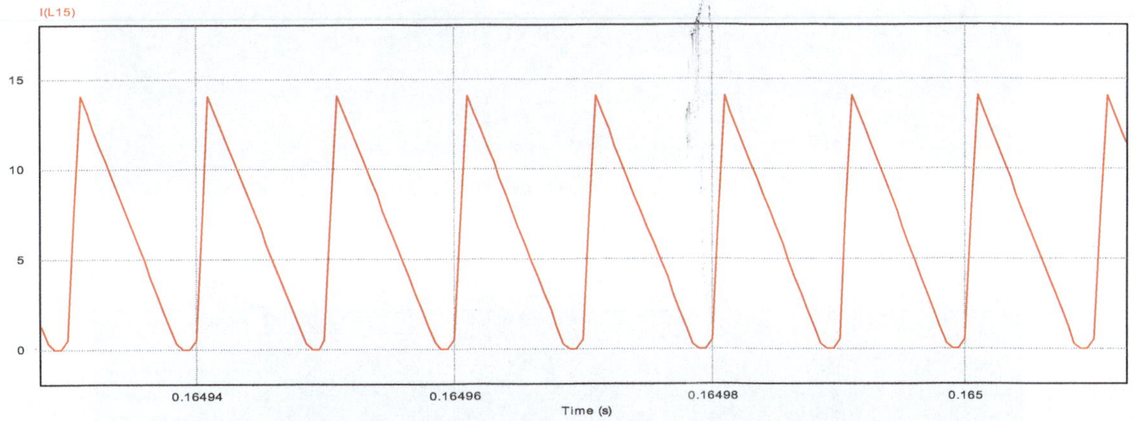

Figure 9. Output inductor current (L2 in Figure 1: DCM).

5.2. Experiments

An experiment circuit was also built and the parameters and the components used is the same as the simulations. The control signal was generated using an UC3854. The results are as follows.

The experimental results are in accordance with the simulations. The input power is 125w and the output power is 100w. Thus the efficiency of the system is 80%. The power factor is 0.98. As shown in **Figures 10 - 13**, the duty cycle D is a simple constant value to gain a high power factor.

6. Conclusions

The operation and characteristics of the Buck converter with LC input filter operating in discontinuous inductor current mode has been studied.

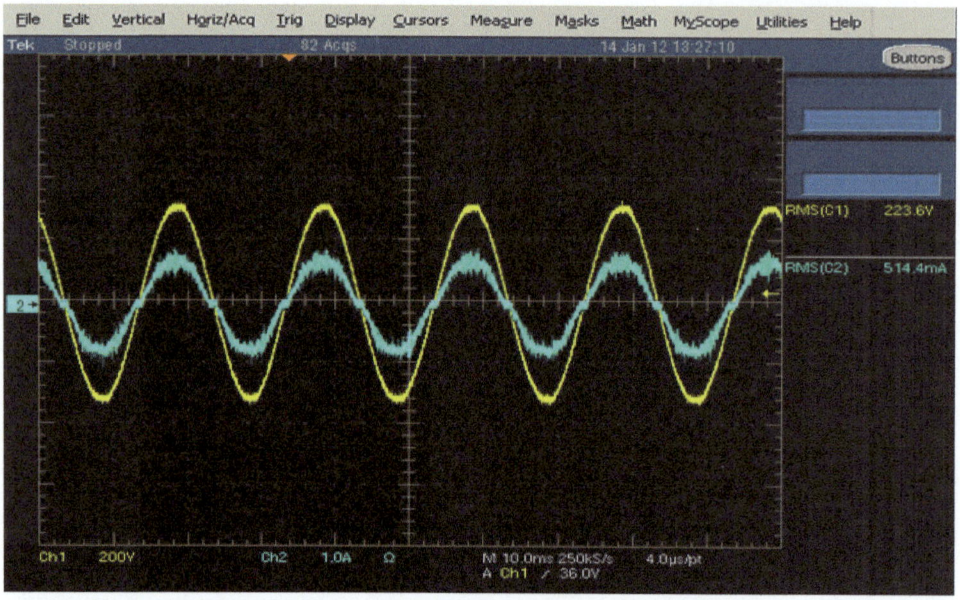

Figure 10. Input Voltage: 220vac Input Current: 0.54mA.

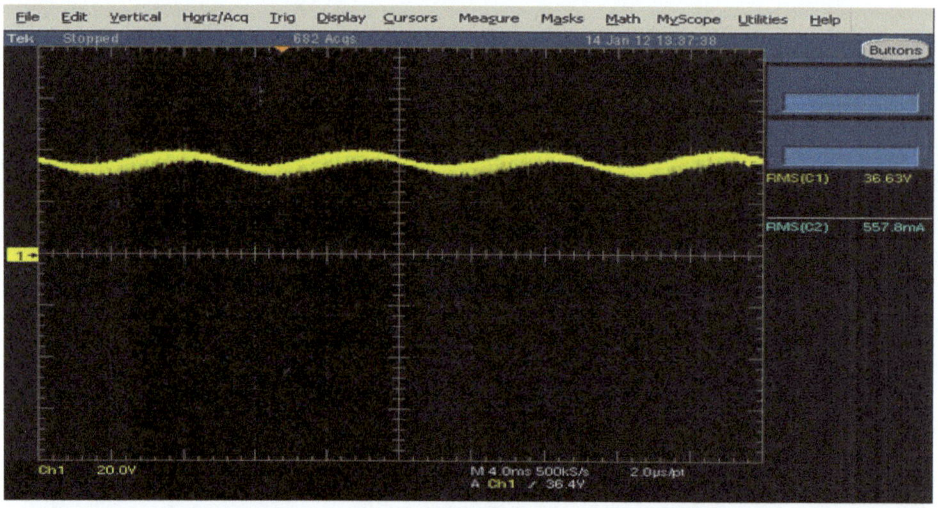

Figure 11. Output voltage: 36VDC (ripple: 5v).

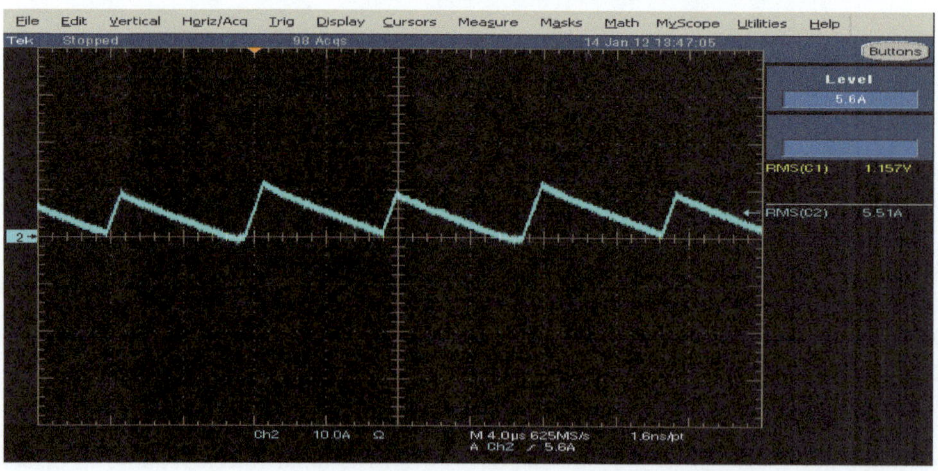

Figure 12. Output inductor current (L2 in Figure 1) DCM.

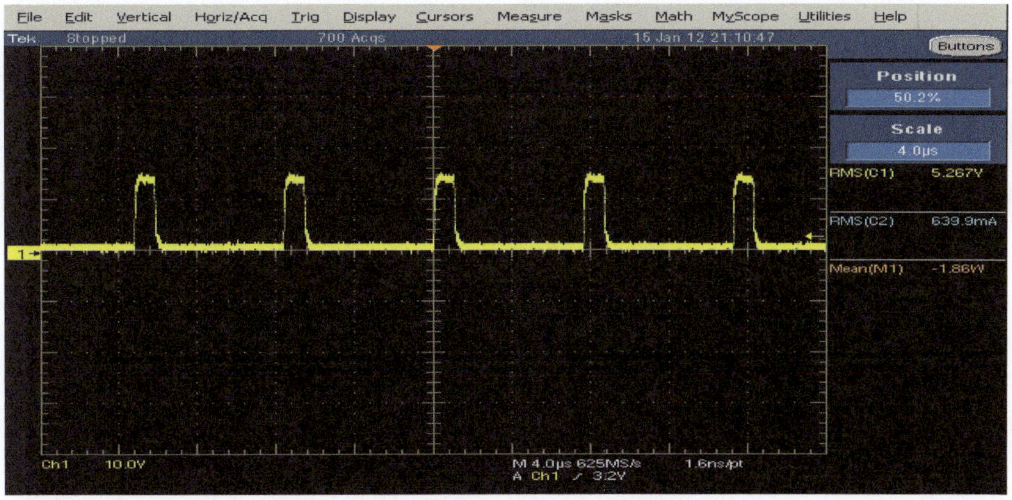

Figure 13. Constant duty cycle.

It is found that the converter can gain a high power factor when the duty ratio maintain constant. A low frequency model of the converter has been developed. A 100w experimental circuit has been built to confirm the theoretical analysis.

REFERENCES

[1] R. Hammano and R. Neidorff, "Improving Input Power Factor - A New Active Control Simplifies the Task," *in Proceedings of the 19th International PCIM Conference*, 1989.

[2] R. Keller and G. Baker, "Unity Power Factor Off Line Switching Power Supplies," *in IEEE INTELEC Record.* 1984, pp. 332-339.

[3] V. Grigore and J. Kyyra, "High Power Factor Rectifier Based on Buck Converter Operating in Discontinuous Capacitor Voltage Mode," *in Proc. IEEE Appl. Power Electron. Conf. Expo.*, Mar. 1999, pp. 612-618.

[4] Y. S. Lee, "Modeling, Analysis, and Application of Buck Converters in Discontinuous-Input-Voltage Mode Operation," *IEEE Trans. Ind. Electron*, Vol. 12, No. 2, 1992.

[5] H. Endo, T. Yamashim and T. Sugiura, "A High-Power-Factor Buck Converter," *Proc. of the IEEE Power Electronics Speciulists Conference*, pp. 1071-1076. R. W. Erickson, "Fundamentals of Power Electronics," M, second edition, 2001

[6] PHILIP C. TODD, "UC3854 Controlled Power Factor Correction Circuit Design," UNITRODE application note, U134.

Real-Time Implementation of Solar Inverter with Novel MPPT Control Algorithm for Residential Applications

Ayaz Ahmad[1], Rajaji Loganathan[2]
[1]Singhania University, Pacheri Bari, India
[2]P.B. College of Engineering, Chennai, India

ABSTRACT

Solar energy is a fast growing energy resource among the renewable energy resources in the market and potential for solar power is huge to contribute towards the power demand almost in all the countries. To capture the maximum power from the sun light in order to generate maximum power from the inverter, control system must be an equally efficient with the well designed power electronic circuits. Maximum power point tracking (MPPT) control system in general is taking care of extraction of maximum power from the sun light whereas current controller is mainly designed to optimize the inverter power to feed to power grid. In this paper, a novel MPPT algorithm using neuro fuzzy system is presented to ensure the maximum MPPT efficiency in order to ensure the maximum power across the inverter terminals. Simulation and experimental results for residential solar system with power electronic converters and analysis have been presented in this paper in order to prove the proposed algorithm.

Keywords: PV Cells; Solar Inverter; MPPT Control; DCDC Converter and Current Controller

1. Introduction

Solar energy is the fastest growing renewable energy market which will meet the energy demand across the globe. Photovoltaic (PV) panels are feeding the power either to battery or to the power grid through the power converters and they are termed as off-grid solar converters and on-grid solar converters respectively. In off grid solar system, PV panel output power is converted into DC power by means of boost converter and stored in a battery. For low power applications, off grid solar energy system is very much used. A grid-connected PV system is made up of an array of panels mounted on rack-type and these panels are connected in series or parallel to achieve optimal voltage and current, and feed into an inverter transforming direct current into alternating current at a phase and at the same voltage as the grid. Operating voltage of an array of panels is varied based on the rating of inverter and the maximum value of this voltage even up to 700 V. **Figure 1** shows the block diagram of solar inverter system which comprises of array of solar panels, DCDC converter (DCDC power conversion), DCAC converter (DCAC power conversion) and both are coupled with the help DC link capacitor. This solar inverter system is connected to Power utility system known as "Power Grid". The solar inverter is equipped with a maximum power point tracking (MPPT) system that constantly adjusts the entry voltage to the characteristics of the PV modules, which vary according to temperature and solar radiation. MPPT system is normally available in DCDC converter and based on the power point tracking DCDC converter will deliver the DC power to DCAC converter for the AC power conversion. DC link voltage between DCDC and DCAC converter is maintained within the specified value by control systems [1].

In general, DCDC converter converts the photovoltaic power (DC) to controllable DC power and DCAC converter converts controlled DC power into controlled AC power which will be transferred to power grid. This research work designs and develops novel control strategies for better MPPT tracking at any kind of sun light fluctuations and also for optimum current control in DCAC converter in order to achieve lower harmonics level when inverter gets connected to power grid. Unlike wind energy system, solar energy system does not need any mechanical or thermal interlink and produces electrical power from sun light. Photovoltaic (PV) cells are interconnected together and formed as solar panel which

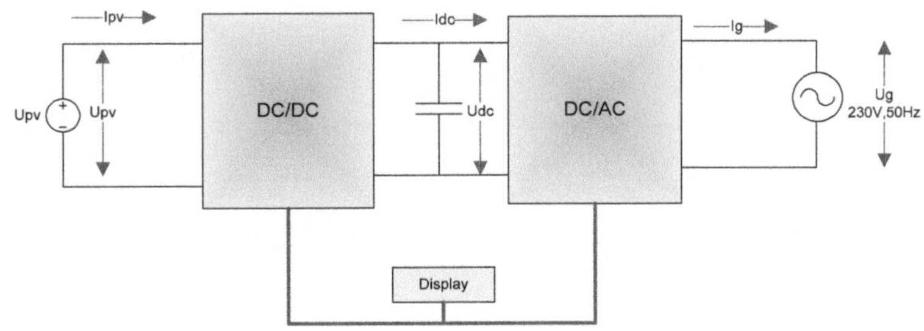

Figure 1. System Overview.

captures sun light directly and DC voltage is produced based on the number of cells connected. Normally, each PV cell produces 0.3 V to 0.6 V DC voltage from the captured sun light [2]. To get the maximum DC voltage of 400 V, minimum 667 PV cells are required to capture the sunlight. Although, PV panels are capable of generating the required amount of DC voltage, DC voltage output from the PV cells is also depending on the solar irradiation. To capture the maximum voltage from the available irradiation level, an efficient maximum power point tracking algorithm should be deployed so that it will track the irradiation level periodically. **Figure 2** shows the Voltage (Upv) − Power (Ppv) characteristics of a typical PV Panel.

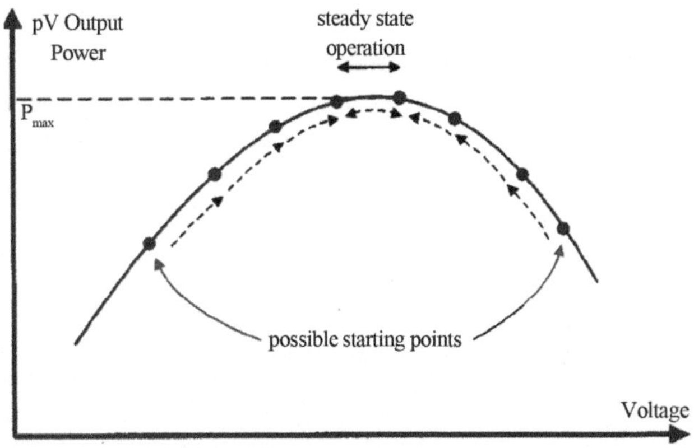

Figure 2. P-V characteristics of a PV module.

2. MPPT Control Algorithms

With reference to different types of MPPT control algorithms, Perturb and Observe (P & O) algorithm, constant voltage (CV) algorithm and Incremental Conductance (INC) algorithms are commonly used algorithms in MPPT control systems [3-6]. Generally, MPPT controller gives the reference voltage from the inputs of PV voltage, PV current and the DC link voltage as shown in **Figure 3**. Adaptive MPPT approach has also been presented by many authors [7-11] which is more suitable to non-linear irradiation conditions. INC algorithm is also used by many researchers and presented in [12,13]. One of the artificial intelligence techniques such as Fuzzy control is proposed in [14] and in this reference author has implemented the proposed methodology using field programmable gate array (FPGA). Based on the understanding from the previous work and current scenario, it is well understood that efficient and dynamic MPPT control is very much required to get the maximum output across the input terminals. In this paper hence author proposed novel MPPT approach using adaptive neuro fuzzy approach to get the optimized reference voltage from the MPPT controller.

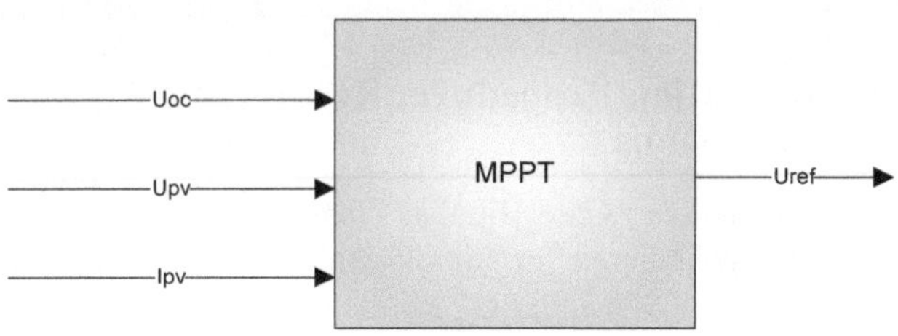

Figure 3. Input/Output Block for MPPT subsystem in DC/DC module.

3. Adaptive Neuro Fuzzy MPPT Algorithm

Hybrid learning control algorithm in adaptive neuro fuzzy system is used to identify the membership functions parameters of single output. **Figure 4** shows the block diagram of the PV model with MPPT algorithm. The control block takes input from the PV panel and computes the duty cycle for the DC/DC converter. Inside the control block, MPPT block computes the reference voltage and proportional controller computes the duty cycle of PWM based on the Panel Voltage (Upv) and reference voltage calculated from MPPT block. Neuro fuzzy technique is incorporated in MPPT block which gives the optimized reference voltage to proportional controller. **Figure 5** shows the simulink model of proposed controller with three inputs and single output. Combination of least squares and back propagation descent methods are used to train the membership function parameters [15].

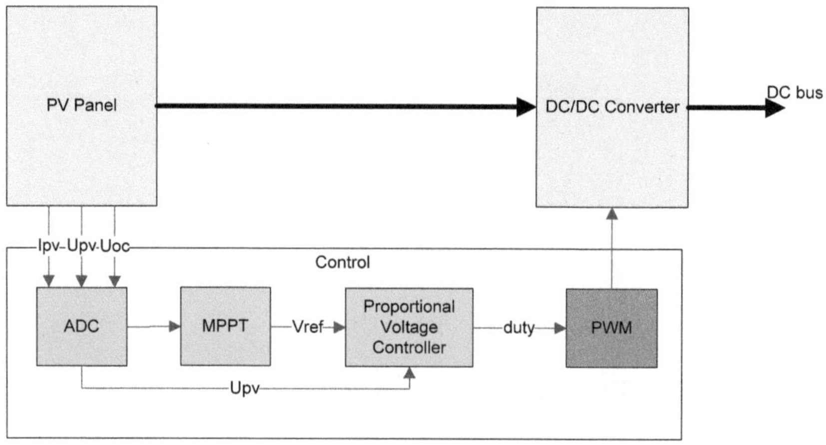

Figure 4. Block diagram of MPPT implementation.

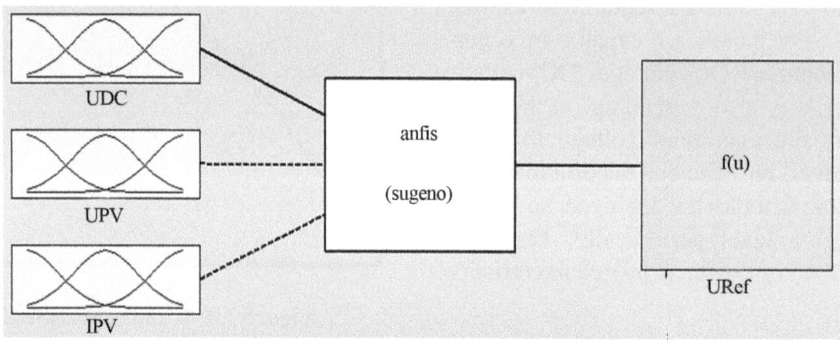

Figure 5. ANFIS based MPPT controller.

4. Simulation Procedure, Results and Discussions

IGBT (Insulated Gate Bipolar Transistor) based PV inverter is considered for Simulation of proposed control algorithm. In this inverter, 3 nos. of DCDC converters which are connected in parallel and single DCAC module rated 6 kW are considered for the simulation. Simulation model of complete system is shown in **Figure 6**. The intensity of the sunlight (W/m^2) is given as input in the form of Square wave or Sine wave to the PV Module. PV Module is duplicated with Ipv = 0, to measure the open circuit voltage of the PV module for the given light intensity. The Output of the PV module is connected to the DC/DC converter. The reference voltage of DC link is also given as input to the DC/DC Converter. The voltage and Current from PV Module are sampled and given as input to the MPPT block and voltage of PV module is given to the Proportional controller. The MPPT block calculates the U_{ref} and it is fed to Voltage controller. The proportional voltage controller block calculates the duty cycle for PWM signal based on the PV Panel voltage and Uref. The inputs to the MPPT system in the simulation model are PV Panel Voltage (Ppv), Panel Current (Ipv) and Open Circuit Voltage of the PV Panel (Uoc).

Tracking Error and Efficiency are calculated based on the Equations (1) and (2).

$$\text{Tracking Efficiency}(\%) = \frac{\text{Ppv}}{\text{Pmpp}} \times 100 \quad (1)$$

$$\text{Tracking Error}(\%) = \frac{\text{Pmpp-Ppv}}{\text{Pmpp}} \times 100 \quad (2)$$

The Matlab function block implements the combined two method hybrid algorithm for MPP Tracking. The output of the MATLAB function would be the voltage at which MPP occurs (Vmpp), which is given as reference to the proportional voltage controller. For DCDC converter circuit, an average model of DCDC converter is taken for simulation. The irradiance of 500 W/m^2 was considered for static condition and variation of 300 W/m^2 is considered over 500 W/m^2. In **Figure 7**, Square wave of 0.5 Hz varying between 500 W/m^2 and 800 W/m^2 irradiation is shown. It is assumed that intensity is varying between 500 W/m^2 and 800 W/m^2 every second. Though this case is practically not possible as huge variation of intensity in shorter duration, it has been considered for measuring the effectiveness of MPPT algorithm to track the actual Maximum Power Points in the worst case scenario. In **Figure 8**, Power was dropped to 250 W during transient state and attained the steady state value of 700 W with 1 second. Though there exists a fluctuation in

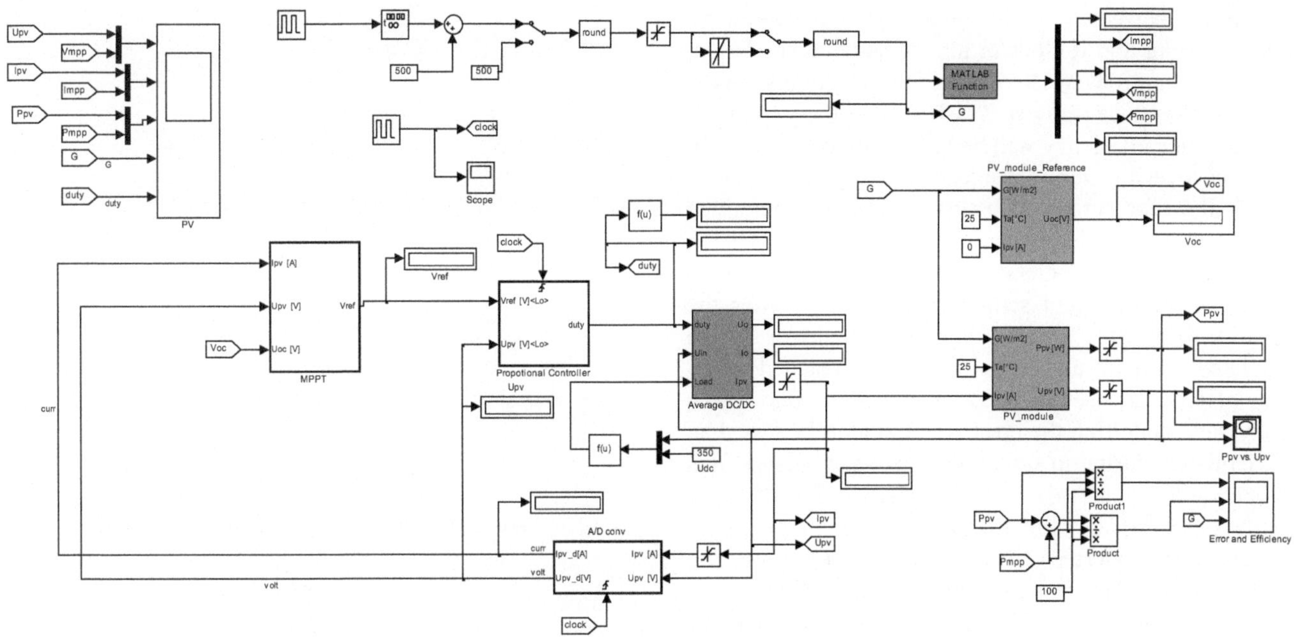

Figure 6. Simulink Model of PV panel, MPPT and DC/DC Converter.

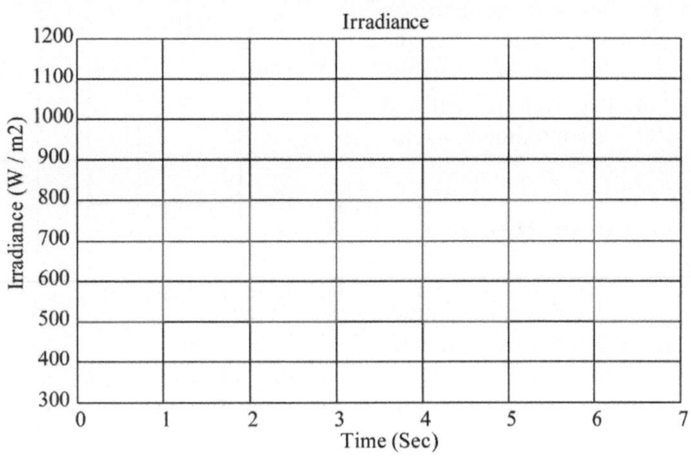

Figure 7. Solar Light Intensity with varying irradiance.

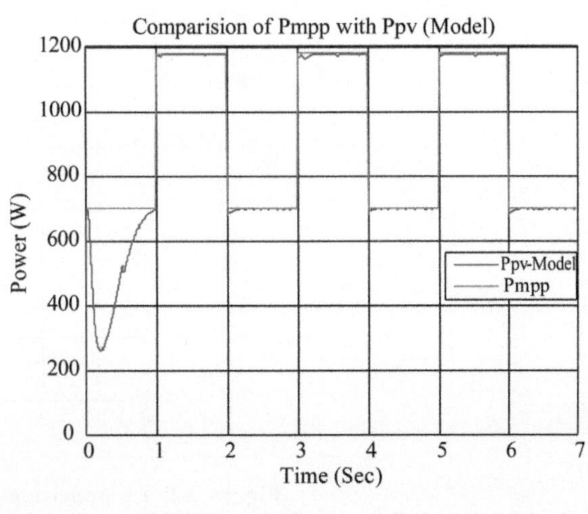

Figure 8. Comparison of actual PV Power (*Ppv*) and Calculated Maximum power (*Pmpp*).

voltage and current, the amount of power delivered remains constant and it close or almost same as the calculated maximum power. But during the transition state the amount of produced power was less than the maximum power. Hence efficiency will be less during the transition state whenever the sunlight intensity changes. **Figures 9** and **10** show percentage of Error and Efficiency obtained from the model under dynamic conditions. It has been observed that the efficiency was dropped to the 37% during the start up and settled around 99.9% after the transition time of 1 second. Similarly the error in the calculated set point and actual PV power reached maximum of 63% and settled with in 1 second. During the steady state condition the error was found to be less than 1%. However during in sunlight intensity, efficiency was dropped to 97% and Error was increased to 3%. It's very clearly observed that the frequent change in Set point will affect the inverter efficiency. **Figure 11** shows the comparison of maximum power tracked and operating power at the 50% of maximum irradiance level. **Figure 12** shows the comparison of voltage at maximum power tracked and the operating voltage at the 50% of maximum irradiance level. **Figure 13** shows the comparison of Maximum Power tracked and operating power at the 100% of maximum irradiance level. **Figure 14** shows the comparison of voltage at maximum power tracked and the operating voltage at the 100% of maximum irradiance level.

5. Experimental Implementation, Results and Discussion

Specification of solar inverter considered for the simulation has also been considered to design the experimental system. **Figure 15** shows the experimental circuit of DCDC converter which has the H-bridge with MOSFETs. Solar inverter hardware consists of MOSFET based

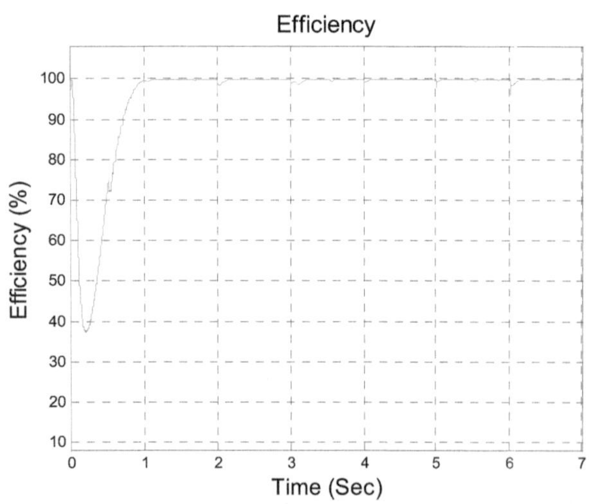

Figure 9. Efficiency under dynamic condition.

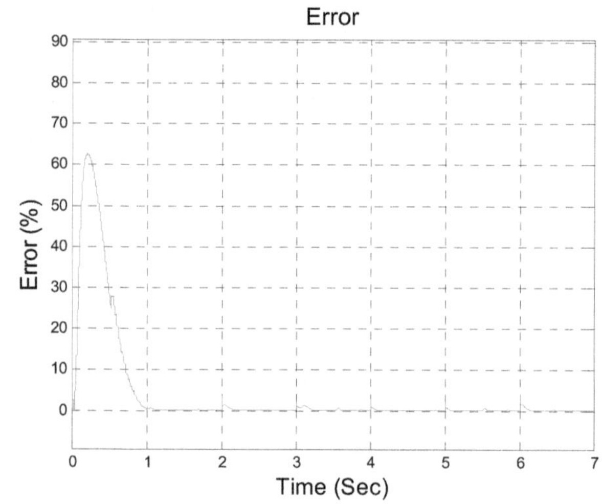

Figure 10. Error under dynamic condition.

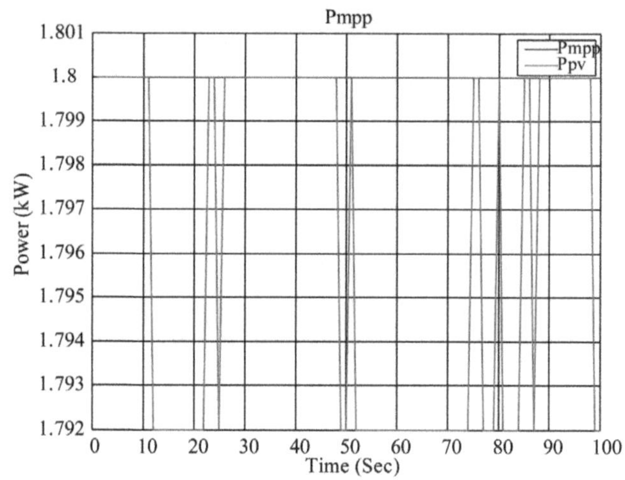

Figure 11. Comparison of MPP Power and Operating Power at 50% of irradiance.

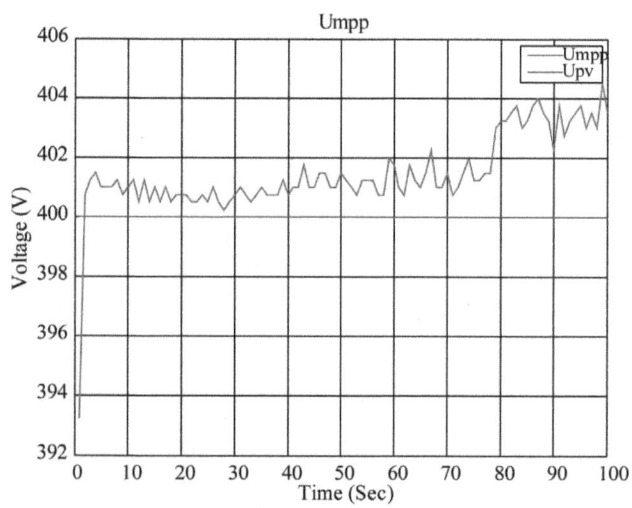

Figure 12. Comparision of MPP Voltage and Operating Voltage at 50% of irradiance.

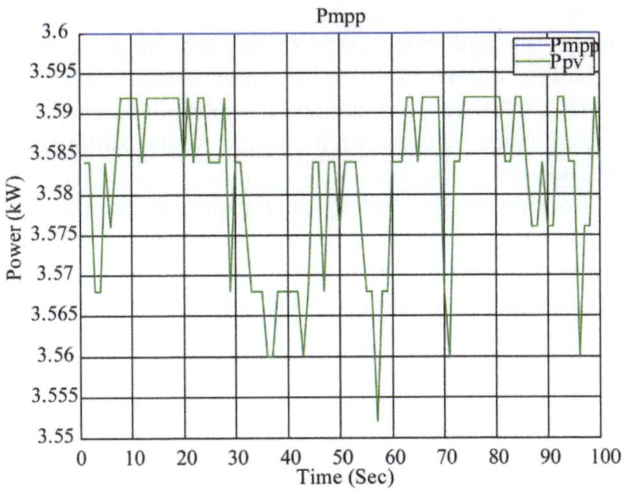

Figure 13. Comparision of MPP Power (Pmpp) and Operating Power (Ppv) at 100% of irradiance.

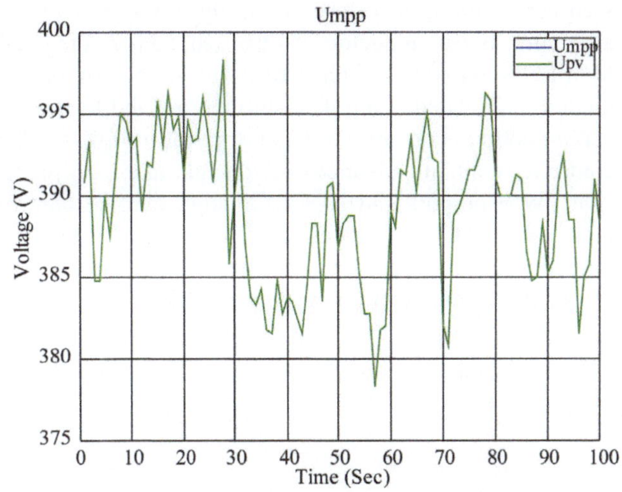

Figure 14. Comparision of MPP Voltage (Umpp) and Operating Voltage (Upv) at 100% of irradiance.

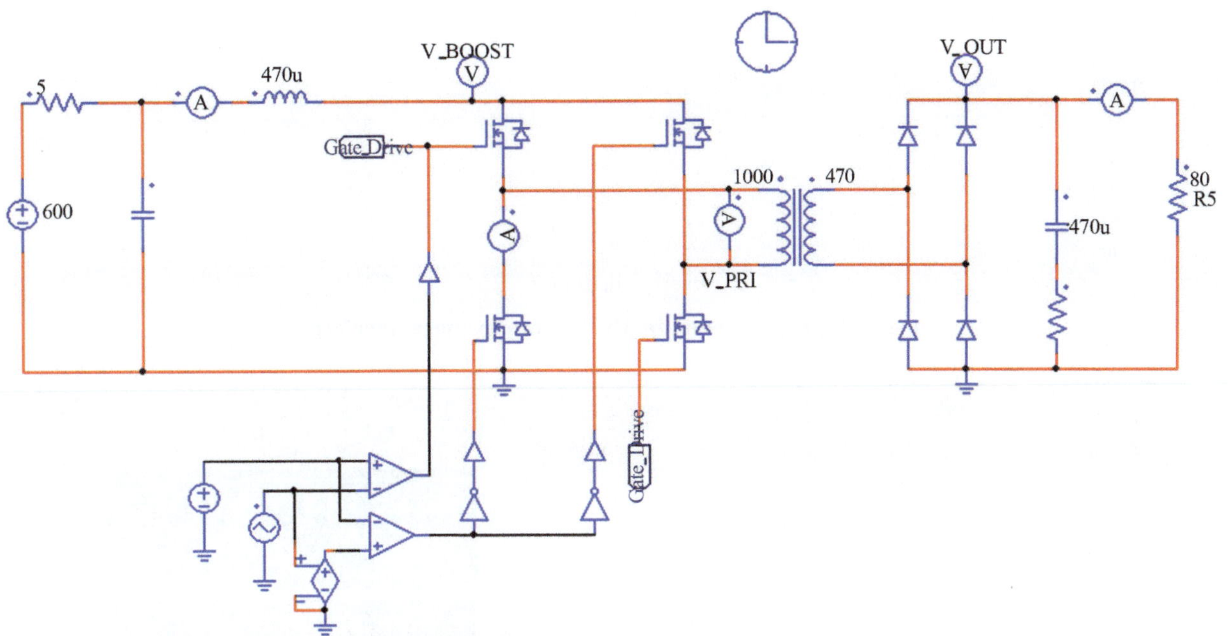

Figure 15. Experimental layout of DCDC converter.

DCDC converter, DC link capacitor, IGBT based DCAC inverter and LCL filter. This power electronic circuit combination is integrated and input to DCDC converter is captured from the PV cell and output of inverter is connected to the power grid. Major part of experimental set up consists of DCDC board, DC Link capacitor, DCAC board and LCL filter. 6 kW DCDC board consists of three 2 kW DCDC boards connected in parallel in order to optimize the operation of the board. Based on the power availability at the solar panel, no. of boards in DCDC board would be selected and maximum power captured from DCDC board could be 6 kW. PV voltage range is varied from 180 V to 400 V and DCDC converter is boosting the voltage whenever it is required and this voltage is appeared across the DC link capacitor as DC link voltage. DCDC converter acts as boost converter which operates at 90 KHz and the full bridge operates at 45 KHz. In the DC converter, the input Photovoltaic (PV) voltage is fed to H-bridge power module with a front end filter and boost section. The input voltage is boosted with the boost inductor. The Power module consists of an H bridge MOSFET, a current sense resistor and diode bridge rectifier. In the first cycle initially all the MOSFETs in the H-bridge are switched on to charge the boost Inductor and the voltage is boosted. In the off time, diagonal MOSFETs will be switched off and the voltage is seen across the transformer. Similarly in the next cycle, after the boost inductor is charged, other diagonal MOSFETs will be switched off. The output of power module is isolated using Transformer and fed to the inverter section. Transformer has a turn ratio of 0.75. The controller contains advanced peripherals like high precision PWM outputs and ADCs for implementing control loops. The ADC measures variables, such as the PV output voltage and current, and then adjusts the DC/DC or DC/AC converter by changing the PWM duty cycle. C2000 is used to read the ADC and adjust the PWM within a single clock cycle so that real time control is achieved. **Figure 16** shows the output voltage waveforms of DCDC converter circuit and for convenience, captured experimental data has been considered and various output waveforms drawn.

Booster voltage, voltage at transformer primary and voltage at transformer secondary are shown in **Figure 16**. **Figure 17** shows the PV and IV curves captured from the photovoltaic simulator for the given irradiance level

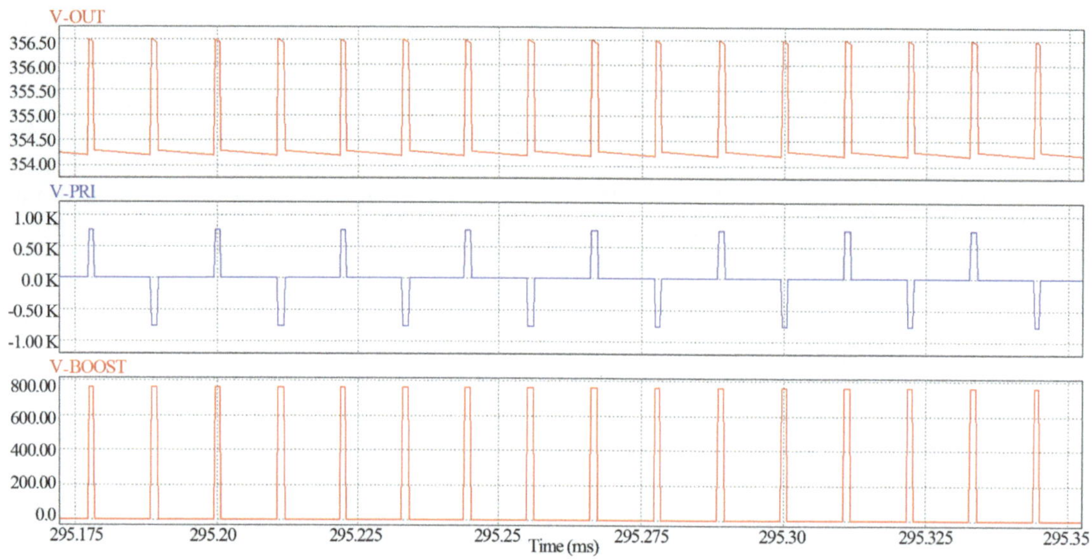

Figure 16. Experimental DCDC converter voltage waveforms.

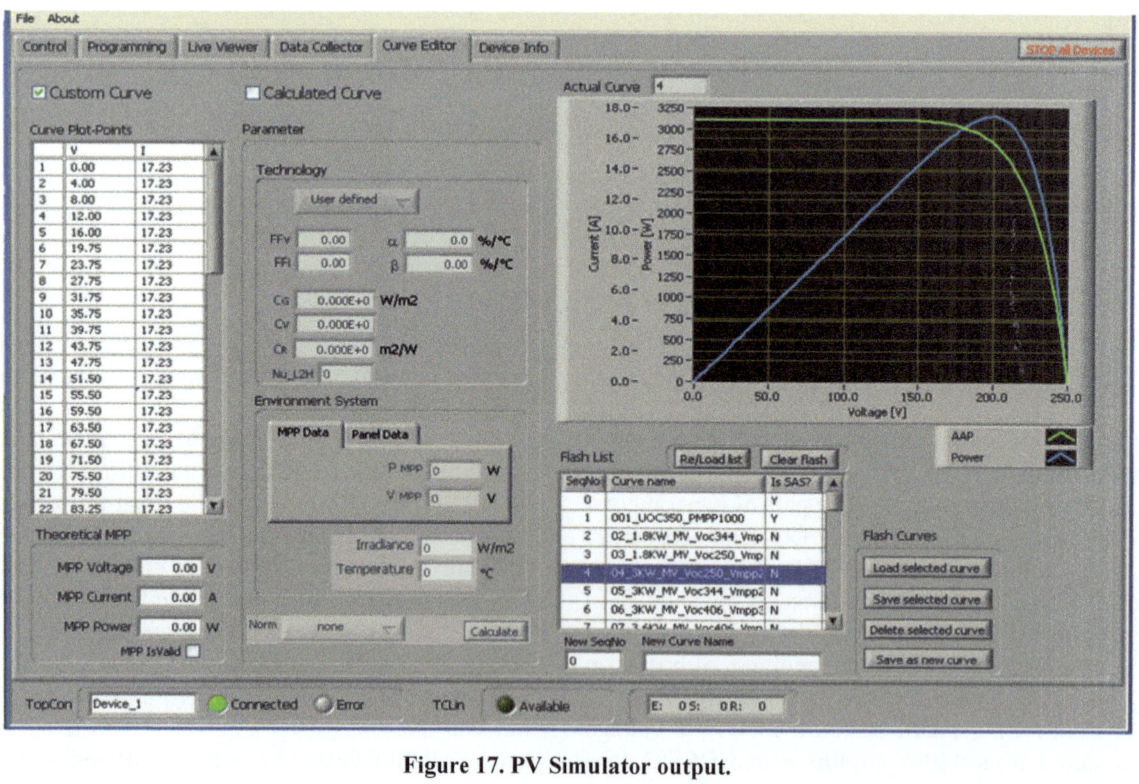

Figure 17. PV Simulator output.

of 500 W/m^2. **Figure 18** shows the voltage and current output at the PV inverter. From this figure, it is clearly understood that current is lagging voltage at 0.8 PF and this lagging power is fed to power grid.

Table 1 shows the duty cycle ranges for various input and output voltage levels and based on this duty cycle values, DCDC converter output voltage values will be delivered from the given input voltage. **Table 2** shows the MPP tracking efficiency without changing irradiance value at different power levels. Values are given in **Table 2** shows the capability of proposed MPPT algorithm.

6. Conclusion

An efficient Maximum power point tracking control algorithm using adaptive neuro fuzzy technique is presented in this paper. For the proposed control algorithm, 6 kW rated single phase inverter is considered for modeling and simulation in Matlab/Simulink environment. Dynamic variation of irradiation level is considered as input and tracking efficiency is computed. Based on the simulation and experimental results, it is clearly observed

Table 1. Duty cycle for various input and output voltage ranges

		Input Voltage					
	D	250	290	330	370	410	450
Output Voltage	350	0.46	0.38	0.29	0.21	0.12	0.04
	355	0.47	0.39	0.30	0.22	0.13	0.05
	360	0.48	0.40	0.31	0.23	0.15	0.06
	365	0.49	0.40	0.32	0.24	0.16	0.08
	370	0.49	0.41	0.33	0.25	0.17	0.09
	375	0.50	0.42	0.34	0.26	0.18	0.10
	380	0.51	0.43	0.35	0.27	0.19	0.11
	385	0.51	0.44	0.36	0.28	0.20	0.12
	390	0.52	0.44	0.37	0.29	0.21	0.13
	395	0.53	0.45	0.37	0.30	0.22	0.15
	400	0.53	0.46	0.38	0.31	0.23	0.16

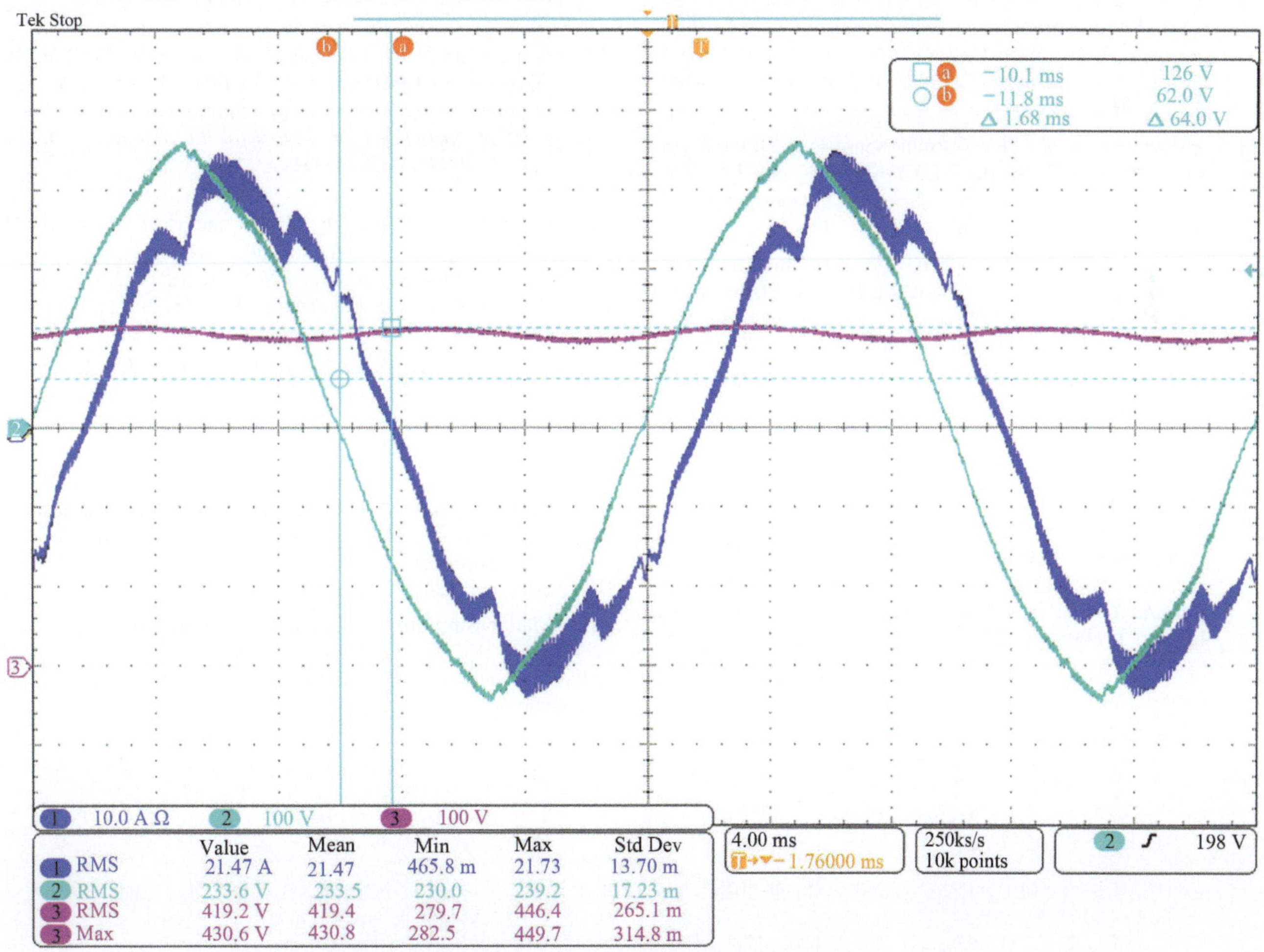

Figure 18. Voltage and current waveforms of DCAC inverter circuit.

Table 2. MPP Tracking Efficiency without change in Irradiance @ different Power Levels.

S. No	% of Power	Efficiency (%)
1	5	100
2	10	99.58
3	20	99.72
4	25	99.32
5	30	99.75
6	50	99.75
7	75	99.21
8	100	99.43

that proposed control technique would be the best one to capture the maximum power across the input terminals since the tracking efficiency is achieved as 99.9%.

REFERENCES

[1] H. Patel and V. Agarwal, "MPPT Scheme for a PV-Fed Single-Phase Single-Stage Grid-Connected Inverter Operating in CCM With Only One Current Sensor," *IEEE Transactions on Energy Conversion*, Vol. 24, No. 1, 2009, pp. 256-263.

[2] G. Escobar, *et al.*, "PV Current Sensorless MPPT for a Single-Phase PV Inverter," *IECON* 2011—*The* 37*th Annual Conference on IEEE Industrial Electronics Society*, Melbourne, 7-10 November 2011, pp. 3906-3911.

[3] N. Kasa, T. Iida and L. Chen, "Fly Back Inverter Controlled by Sensorless Current MPPT for Photovoltaic Power System," *IEEE Transactions on Industrial Electronics*, Vol. 52, No. 4, 2005, pp. 1145-1152.

[4] O. Lopez-Lapena, Penella, M.T. and M. Gasulla, "A New MPPT Method for Low-Power Solar Energy Harvesting," *IEEE Transactions on Industrial Electronics*, Vol. 57, No. 9, 2010, pp. 3129-3138.

[5] R. Leyva, C. Alonso, I. Queinnec, A. Cid-Pastor, D. Lagrange and L. Martinez-Salamero, "MPPT of Photovoltaic Systems Using Extremum-Seeking Control," *IEEE Transactions on Aerospace and Electronic Systems*, Vol. 42 , No. 1, 2006, pp. 249-258.

[6] N. Mutoh, M. Ohno and T. A. Inoue, "Method for MPPT Control While Searching for Parameters Corresponding to Weather Conditions for PV Generation Systems," *IEEE Transactions on Industrial Electronics*, Vol. 53, No. 4, 2006, pp. 1055-1065.

[7] N. Fermia, D. Granozio, G. Petrone and M. Vitelli, "Predictive & Adaptive MPPT Perturb and Observe Method," *IEEE Transactions on Aerospace and Electronic Systems*, Vol. 43, No. 3, 2007, pp. 934-950.

[8] N. Femia, G. Petrone, G. Spagnuolo and M. Vitelli, "A Technique for Improving P&O MPPT Performances of Double-Stage Grid-Connected Photovoltaic Systems," *IEEE Transactions on Industrial Electronics*, Vol. 56, No. 11, 2009, pp. 4473-4482.

[9] A. K. Abdelsalam, A. M. Massoud, S. Ahmed, P. N. Enjeti, "High-Performance Adaptive Perturb and Observe MPPT Technique for Photovoltaic-Based Microgrids," *IEEE Transactions on Power Electronics*, Vol. 26, No. 4, 2011, pp. 1010-1021.

[10] D. Sera, R. Teodorescu and J. Hantschel, "Optimized Maximum Power Point Tracker for Fast-Changing Environmental Conditions," *IEEE Transactions on Industrial Electronics*, Vol. 55, No. 7, 2008, pp. 2629-2637.

[11] G. Carannante, C. Fraddanno, M. Pagano and L. Piegari, "Experimental Performance of MPPT Algorithm for Photovoltaic Sources Subject to Inhomogeneous Isolation," *IEEE Transactions on Industrial Electronics*, Vol. 56, No. 11, 2009, pp. 4374-4380.

[12] Q. Mei, M. W. Shan, L. Y. Liu and J. M. Guerrero, "A Novel Improved Variable Step-Size Incremental-Resistance MPPT Method for PV Systems," *IEEE Transactions on Industrial Electronics*, Vol. 58, No. 6, 2011, pp. 2427-2434.

[13] F. R. Liu, S. X. Duan, F. Liu, B. Y. Liu and Y. Kang, "A Variable Step Size INC MPPT Method for PV Systems," *IEEE Transactions on Industrial Electronics*, Vol. 55, No. 7, 2008, pp. 2622-2628.

[14] Z. Cheng, H. Z. Yang and Y. Sun, "FPGA-Based PV Systems Fuzzy MPPT Control Algorithm," 2010 *Seventh International Conference on Fuzzy Systems and Knowledge Discovery* (*FSKD*), Vol. 3, Yantai, 10-12 August 2010, pp. 1244-1248.

[15] Matlab/Simulink—Tool and User Guide, 201.

Characteristic Analysis of DS18B20 Temperature Sensor in the High-voltage Transmission Lines' Dynamic Capacity Increase

Song Nie[1], Yang-chun Cheng[1,2], Yuan Dai[3]
[1]Beijing Key Laboratory of High Voltage & EMC, North China Electric Power University, Beijing, China
[2]State Key Laboratory of Alternate Electrical Power System with Renewable Energy Sources, North China Electric Power University, Beijing, China
[3]Electric Power Research Institute of Guangdong Power Grid Corporation, Guangzhou, China

ABSTRACT

Dynamic capacity increase in high voltage electric power transmission line is currently the most economical method for solving electric power transmission bottleneck nowadays. DS18B20 temperature sensor is applied to the dynamic capacity increase of high voltage transmission lines to measure the conductor temperature and ambient temperature. The paper is focused on the experiment of DS18B20 both in the laboratory and outside. From the result of the lab temperature measurement data analysis, using 4 DS18B20's is the most suitable plan, considering both accuracy and economical efficiency. In the experiment outside, we get four groups of conductor (uncharged) temperature and four groups of ambient temperature. The data proved that DS18B20 has good stability, and small measurement error. It is suitable for measuring the temperature of conductor and ambient in dynamic capacity increase, and helpful to improve the accuracy of the calculation of capacity increasing.

Keywords: DS18B20 Temperature Sensor; Measurement Error; Dynamic Capacity Increase; Data Analysis

1. Introduction

In recent years, with the development of China's sustained and rapid, power consumption will also continue to increase. In economically developed areas, due to the partial slow speed of grid construction, the bottleneck problem of the power system transmission capacity of transmission lines has become increasingly prominent. However, because of the limited line level of thermal stability, transient stability level and the level of dynamic stability, about a quarter of the transmission line transmission capacity was significantly lower than the level of foreign. In order to improve the transmission capacity of transmission lines, the comment measures are UHV technology, flexible AC transmission technology, the series compensation technology, dynamic reactive power compensation with the rod back and compact transmission, heat-resistant wires of large cross-section and other technologies,etc.[1]. Among various researches, increasing the heat capacity of transmission line has been carefully studied by most electrical departments and has been proven to be an effective and the most economical way to enhance the current capacity of transmission lines.

The current capacity of transmission line is depend on its heat-balanced equation. The line's current capacity is related to various factors like sunlight intensity, wind velocity and direction, characteristics of line (such as diameter, aging and AC resistance , etc.) and ambient temperature. The maximum current capacity of transmission line is changing with the line's temperature. However, the maximum current capacity can only be calculated by the conductor temperature model. It can't be monitored online. It is important to get conductor temperature exactly by online monitoring. The error of conductor temperature is mainly from two aspects. One is measurement error, sensor error and sensor acquisition device error; the other is the algorithm error, reduce from the regularity of distribution of the sample to. Therefore, this paper is focused on analyzing the validation and precision of DS18B20 temperature sensor which we decided to use on measuring temperature, by doing experiments in laboratory and outside.

2. Temperature Sensor Experiments

2.1. Laboratory Experiments

1) Experimental equipments

DS18B20 temperature sensor: measuring range-55℃

-125℃, accuracy±0.5℃; two standard mercury thermometer: measuring range-5℃-150℃, accuracy±0.1℃, it is the standard value; GDWJS-250 alternating wet heat test box: temperature range -40℃-150℃,temperature volatility≤±0.5℃, temperature uniformity≤±2℃.[2]

2) Experimental methods

In the laboratory, eight DS18B20's and a mercury thermometer are put into an alternating wet heat test box, and the data collection equipments are put outside the box. According to the measuring range of each temperature sensor, We choose 0℃,20℃,30℃,40℃,60℃ and 80℃ as test points. When it reaches to the test temperature and stipulated time, we record the temperature of mercury thermometer and data collection equipments in each test points for four or five times, every minutes. We use SPSS software to process data, and do the Levene-test and t-test to the sample[3].

2.2. Field Experiments

1) Experimental equipments

We Beijing Key Laboratory of High Voltage & EMC designed the temperature measurement device. As it is showed in **Figure 1**, there are four sensors in the device. Sensor 1 is for measuring the temperature of transmission line. Sensor 2 is for measuring the reference wire. Sensor 3 and 4 are for ambient temperature.

2) Experimental methods

In the experiment of measuring real transmission line, we hang the four temperature measuring devices on wire between two towers, which span is 187.8 meters, **Figure 2** shows how the devices are hanged on the wire, we record a data every 5 minutes, testing for 24 hours. We use SPSS software and Excel to process data, and do Time series analysis and Paired-samples T-test to the sample.

3. Analysis on Experimental Data of Temperature Sensor

3.1. Analysis on Experimental Data of Laboratory Experiments

Because of different measuring points and the differences among temperature measurements, it is difficult to analysis overall. We unify the experiment data at first. Let the measured values minus the standard to get the measurement error and then analyze the measurement error. The obtained experimental data were histogram, interval estimation, normal distribution hypotheses test. From **Figure 3**, the sample of single sensor of measurement error is not entirely belonging to the normal distribution.

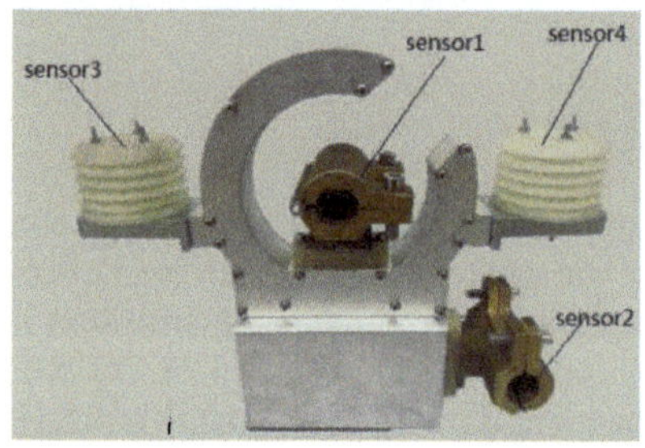

Figure 1. Temperature measurement device.

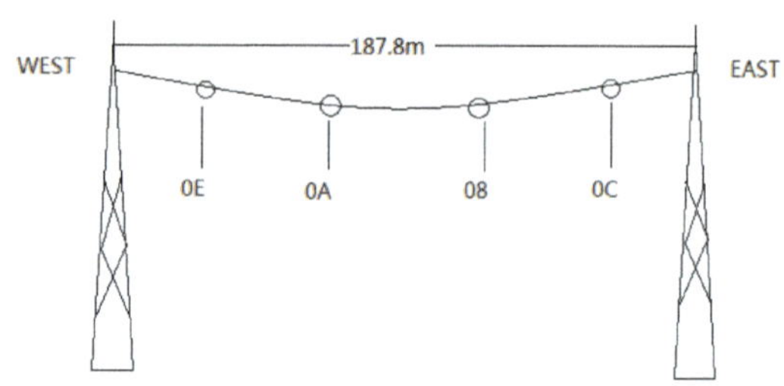

Figure 2. Device suspension schemes.

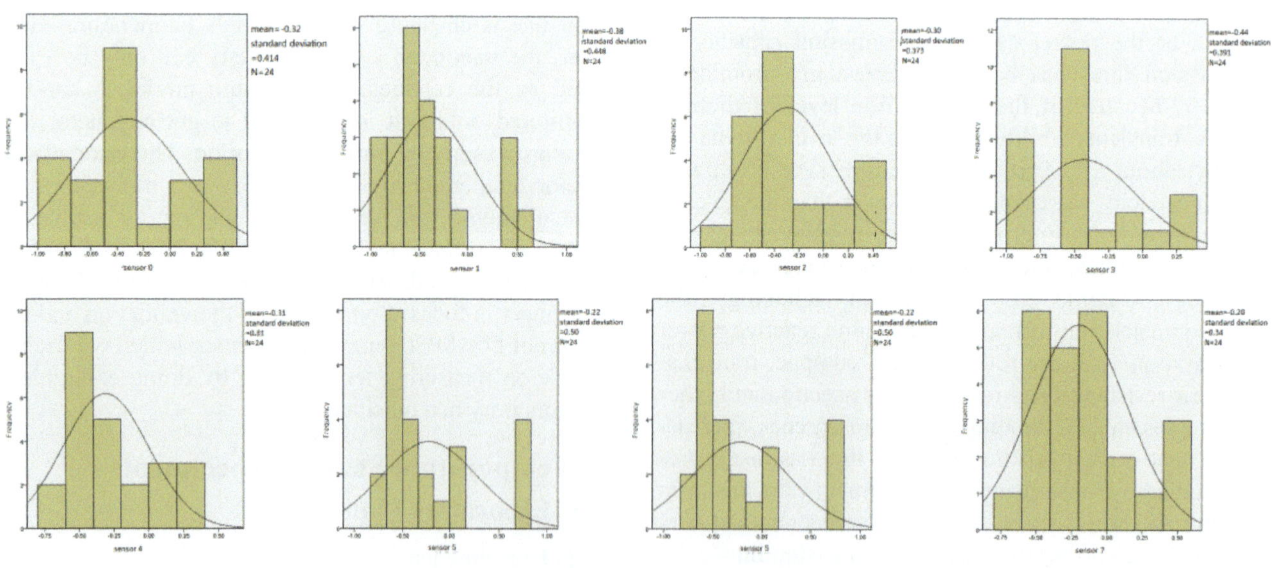

Figure 3. DS18B20's data histograms.

Table 1 [3] shows that the columns asymptotically significant (both sides) values greater than 0.05, based on hypothesis testing knowledge to know, the original assumption that the sample (each sensor measurement deviation) from a normally distributed population.

Single DS18B20 has no significant impact on measuring error, and multiple DS18B20's 95% upper and lower limits of the confidence interval of the difference are significantly lower than single DS18B20. So that multiple sensor measurement value of mean values can significantly reduce error. Using four DS18B20's is the most suitable plan, Considering both accuracy and economical efficiency.

3.2. Analysis on Experimental Data of Field Experiments

According to the curve of real transmission line measuring data [4], when DS18B20 enter the stable operation, data are changing smoothly, have no mutation, and hard real time. According to weather conditions, from the trend of the curve can be seen, the sensor mounted just needs a stable period of time, in order to accurately reflect the conductor and the ambient temperature; data for temperature measurement devices 0E is selected 11:13 as the start value, 0A is selected 11:16, 08 is selected 10:58, and 0C is selected 10:55s the start value. The standard of selecting the start value is the inflection point of the temperature started to climb. Each group extracts 253 data according to time sequence. **Figure 4** is the curve of real transmission line measuring data.

We do paired sample T-test on the difference between sensor 3 and 4 of these four measured devices. The significance probability of measurement device 0E, 0A and 08 are greater than 0.05, so that there are no significant differences between paired sample data.[3] However measurement device 0C is different from others, the reason maybe anthropic factor. The difference between the sensor 3, 4 were less than 0.2°C, in line with project requirements error which is lower than 0.3℃.[4]

4. Conclusions

We could make conclusions from the analyzing above. Single DS18B20 has no significant impact on measuring error, and multiple DS18B20's 95% upper and lower limits of the confidence interval of the difference are significantly lower than single DS18B20. So that multiple sensor measurement value of mean values can significantly reduce error. Using four DS18B20's is the most suitable plan, considering both accuracy and economical efficiency.

Table 1. DS18B20 statistics.

NO.		0	1	2	3	4	5	6	7
Mean value		-0.3192	-0.3817	-0.2983	-0.4358	-0.3108	-0.2192	-0.2942	-0.1983
Standard deviation		0.41444	0.44767	0.37261	0.39113	0.30974	0.49959	0.42007	0.34016
Mean 95% confidence interval	Lower limit	-0.4942	-0.5707	-0.4557	-0.601	-0.4416	-0.4301	-0.4715	-0.342
	Upper limit	-0.1442	-0.1926	-0.141	-0.2707	-0.18	-0.0082	-0.1168	-0.0547
Asymptotically significant (both sides)		0.667	0.367	0.353	0.478	0.131	0.155	0.916	0.502

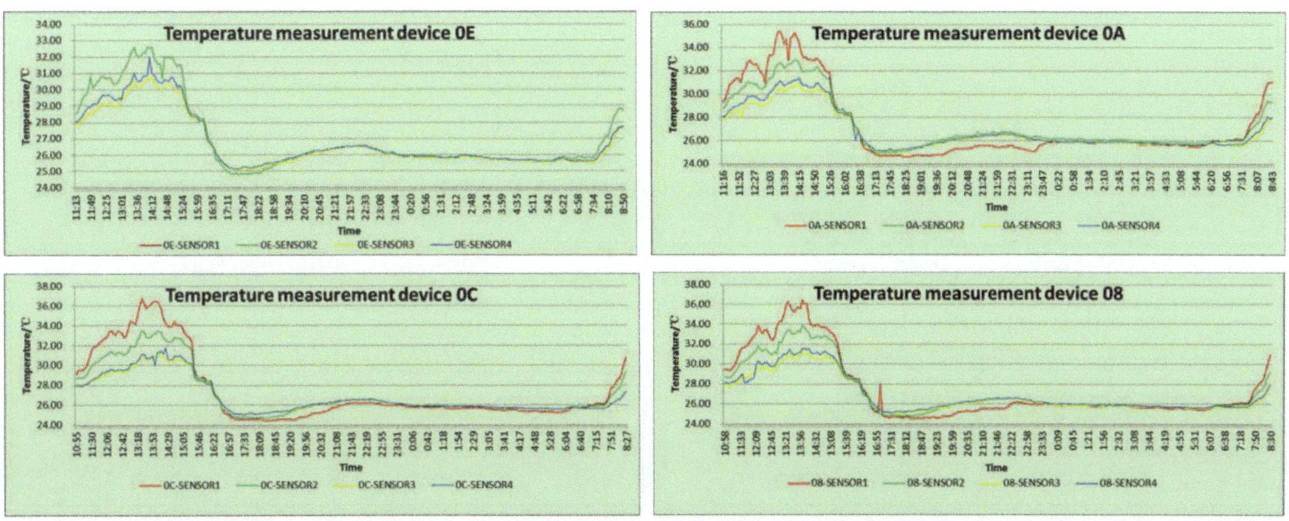

Figure 4. Curve of real transmission line measuring data.

According to the curve of real transmission line measuring data, when DS18B20 enter the stable operation, data are changing smoothly, have no mutation, and hard real time. The significance probability of measurement device 0E, 0A and 08 are greater than 0.05, so that there are no significant differences between paired sample data. However measurement device 0C is different from others, the reason maybe anthropic factor.

According to the analysis above, DS18B20 is suitable for measuring the temperature of conductor and ambient in dynamic capacity increase, and helpful to improve the accuracy of the calculation of capacity-increasing.

5. Acknowledgement

This research is supported by Guangdong Power Grid Corporation, key project "Research on forecasting and monitoring method of real-time dynamic capacity expansion in high-voltage transmission line".

REFERENCES

[1] L. J. Ren, "Dynamic Overhead Transmission Line Rating Based on Tension Measurement," *Doctor Degree thesis*, Shanghai, China, 2008.

[2] A. Q. Song, Y. C. Cheng and Y. Dai, "Characteristics of Temperature Sensor Using in the Dynamic Capacity of the High Voltage Line," *The Chinese Society of Electrical Engineering Professional Committee of the high voltage*, Jinan, China, 2011.

[3] J. M. Su, "SPSS12.0 for Windows Application and Development Guide," Publishing House of Electronics Industry, Beijing, 2004.

[4] M. K. Du, "Excel the data processing and preliminary statistics," The Electronic Industry Press, Beijing, 2011.

Studying the Role Played by Evaporative Cooler on the Performance of GE Gas Turbine Existed in Shuaiba North Electric Generator Power Plant

O. R. AL-Hamdan, A. A. Saker
Water Resources Dept., High Institute of Energy, Public Authority of Applied Education & Training, Kuwait

ABSTRACT

It's well known that the performance of a gas turbine (efficiency, heat rate and power generated) is largely dependent on mass flow rate of air, inlet air temperature and turbine inlet temperature (TIT). As turbine inlet temperature is dependent on quantity of burned fuel so that this factor is dropped out from this paper. It's also known that gas turbines are constant volume machines *i.e.* at a given shaft speed they always move the same volume of air, but the power output of a turbine depends on the flow of mass through it. This is precisely the reason why on hot days, when air is less dense, power output falls off. A rise of one degree Centigrade temperature of inlet air decreases the power output by 1% and at the same time heat rate of the turbine also goes up. This is a matter of great concern to power producers. Many techniques have been developed to cool the inlet air to gas turbine. Some of these techniques to decrease the inlet air temperature are discussed here. The evaporative cooling technique is taken as a case study in this paper. A comparative studying is carried out between a unit using this technique and the same unit when the evaporative cooler is idle. The results advert to an increase in power output by 11.07% and a decrease in heat rate by approximately 4% when inlet air temperature drops from 50˚C to 26˚C.

Keywords: Gas Turbine; Shaft Speed; Power Output; The Evaporative Cooling Technique

1. Introduction

Electric generator power plants play a great role in developing countries. The modern techniques in generating electricity from fossil fuel are by using the so-called combined units through combined cycle. The combined cycle in Shuaiba North power plant consists of three unit gas turbine and one unit steam turbine. The performance characteristics of these units are the most important judging element. The main factors affecting unit performance characteristics are the inlet air temperature to the compressors of the gas turbine units. The research will focus on the system of inlet air cooling to a compressor of GE Gas turbine unit, 250 MW capacities as a case study.

2. Literature Survey

An inlet cooling system is a beneficial option for applications where significant operations occur during times of warmer temperatures combined with low relative humidity. Cooler air is denser, resulting in a higher mass flow rate to the compressor. This results in an increase in turbine output and efficiency [1].

Evaporation of water is one of the simplest and oldest methods of cooling air. Even with the sophisticated technology available today, including mechanical chillers, absorption chillers and thermal energy storage systems, evaporative cooling remains a most cost-efficient method for temperature control of the gas turbine inlet air supply [2].

Evaporative cooling can be achieved by many methods. In practical two forms are typically used. Media type evaporative cooling and spray type evaporative cooling.

The most famous for media type evaporative coolers consist of re-circulated water sprayed over an extended surface media mounted downstream from the inlet air filters. As inlet air passes through the water soaked evaporative cooler media, evaporation occurs performing a double function:

1) Energy in the form of heat is removed from the air

and in the same time that energy is used to evaporate water in the media; and

2) Water vapor content of the air increases due to evaporated water approaching saturation along constant wet bulb lines.

These two processes increase the density of the air, which in turn increases the mass flow and output of the gas turbine.

3. Cycle Thermodynamic Consideration

3.1. Gas Turbine Cycle Thermodynamics

The compression process consumes as much as 66% of the total work produced by the gas turbine and therefore any means of reducing the work of compression will enhance the power output of the gas turbine. The compressor work per unit mass of air is given by:

$$\frac{W}{m} = \frac{h_2' - h_1}{\eta_c} = \frac{C_p\left(T_2' - T_1\right)}{\eta_c} = \frac{C_p T_1\left[\left(\frac{P_2}{P_1}\right)^{\frac{\gamma-1}{\gamma}} - 1\right]}{\eta_c} \quad (1)$$

In examining this equation, it can be seen that increasing T_1 would increase the compression work.

Assuming no pressure losses, and equal specific heats in the cycle gas turbine, thermal efficiency can be reduced to the equation below (Sorenson, 1950) [3].

$$\eta_{th} = \frac{\frac{\eta_t T_3}{B} - \frac{T_1}{\eta_c}}{\frac{T_3 - T_1}{B-1} - \frac{T_1}{\eta_c}} \quad (2)$$

where:

$$B = \left[\left(\frac{P_2}{P_1}\right)^{\frac{\gamma-1}{\gamma}}\right]$$

η_c = Compressor efficiency, η_t = Turbine efficiency, T_3 = Turbine inlet temp., p_2, p_1 = Cycle pressure limits.

Examination of this equation shows that the cycle efficiency decreases with an increase in compression inlet temperature. When the inlet temperature goes up the compressor discharge pressure and temperature drops and more fuel is required to attain the same turbine inlet temperature (TIT).

An increase in ambient temperature T_1 causes the numerator, representing net power output to decrease at a faster rate than the denominator, representing the heat delivered. So the cycle efficiency decreases with increasing inlet compressor temperature.

Figure 1 shows a T-S diagram on a hot day the drop in pressure ratio that occurs on the hot day can be seen on this diagram. The cycle peak temperature is still limited as before and so the expansion ratio also drops meaning that less work is extracted from the turbine. As the compressor work increases and the turbine work decreases, the output power drops [3].

Further, as the ambient temperature increases, the pressure ratio of the compressor decreases and more fuel has to be added to attain the same turbine inlet temperature (TIT) this further decreases the output of the gas turbine unit and the cycle efficiency drops [3].

3.2. Effect of Inlet Air Temp. on GE Gas Turbine Performance Characteristics

GE offers a table explains the effect of inlet air temperature on the power output and heat rate of many models of gas turbine. **Table 1** provides a complete listing of the available outputs and heat rates of the GE heavy-duty gas turbines [4].

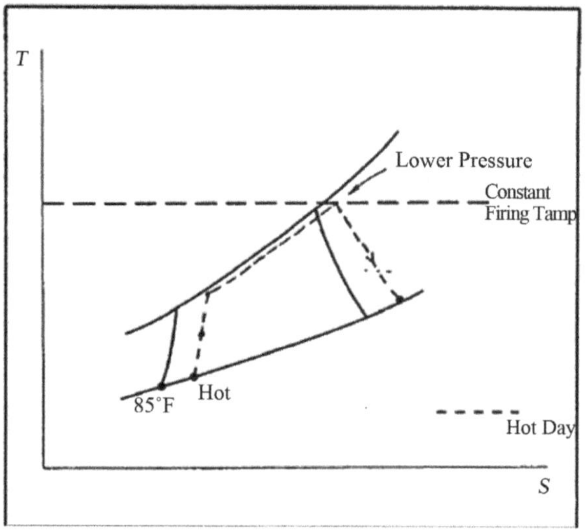

Figure 1. T-S Diagram of Simple Brayton Cycle on Hot Day.

Table 1. GE gas turbine performance characteristics—Mechanical drive gas turbine ratings.

Model	ISO rating continuous (KW)	Heat rate (KJ/KWh)	Mass flow (Kg/s)	Exhaust temp. (degree °C)
M3142 (J)	11,290	13,440	53	542
M3142R (J)	10830	10,450	53	370
M5261 (RA)	19,690	13,270	92	531
M5322R (B)	23,870	10,000	114	352
M5352 (B)	26,110	12,490	123	491
M5352R (C)	26,550	9,890	121	367
M5382 (C)	28,340	12,310	126	515
M6581 (B)	38,290	11,060	134	545

3.3. Factors Affecting Performance of Gas Turbine

The performance of Gas Turbine viz. the power output and the heat rate (measure of efficiency, *i.e.* the amount of energy consumed per kWh of electricity produced) depends on the following major parameters [5].

1) Site altitude *i.e.* atmospheric pressure.

2) Inlet pressure drops in the filters and intake system.

3) Outlet pressure drops in the HRSG (Heat Recovery Steam Generator).

4) Site design temperature.

5) Site design relative humidity corresponding to the site design temperature.

3.4. Impact of Higher Inlet Air Temperature on Gas Turbine Performance

A gas turbine is a constant volume machine *i.e.* the volume of air compressed is fixed, irrespective of ambient temperature. Hence, as the temperature of air raises the density of air decreases and the mass flow rate of compressed air gets reduced. The power output of the gas turbine is proportional to the mass flow rate of air. Thus as the ambient temperature increases, the power output decreases. Further, the efficiency of the gas turbine also falls, as more power is required to compress warmer air, from the above paragraph (factors affecting performance of gas turbine). For a given site and the configuration, the first three parameters are constant and cannot be changed. However, it is possible to alter the other two parameters and obtain a higher output and improved efficiency by cooling the air before it is admitted in the gas turbine compressor section. Impact of higher ambient air temperature is very well spelled in a given **Figure 2** [5], and **Table 2** [6].

4. Methods Currently Adopted by Industry

Industry has implemented various techniques to reduce ambient temp. They are as follows [5]:

1) Vapor Compression:

Various refrigerant type air-chilling systems are available in the market. The biggest advantage of this is that one can reduce temp of inlet air up to 15˚C. The biggest disadvantage in this system is higher power consumption to cool inlet air and pressure drop in the air,

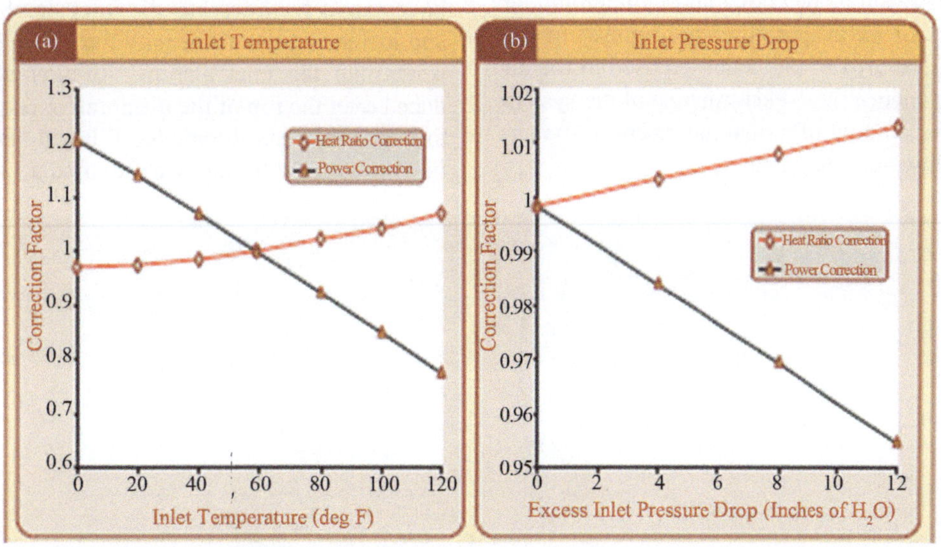

Figure 2. Changes in Power Output and Heat Rate for a 40 MW Gas Turbine Output to: (a) Inlet Air Temperature. (b) Inlet Pressure Drop.

Table 2. Performance of various gas turbine models with respect to inlet air temperature.

Model		GT 10 B			Frame 5 (5371 PA)			Frame 6 (6561 B)			*LM 6000PC		
Inlet temp.	˚C	15	35	40	15	35	40	15	35	40	15	35	40
GT power	MW	23.1	20.1	18.9	25.2	21.4	20.4	38.6	33.5	32.3	40.9	31.9	27.4
GTG efficiency	%	33	32.11	31.8	28	26.4	25.9	31.5	30.3	30	40.6	37.8	35.4
Decrease in power w.r.t 15˚C	%	0	13	18	0	15.1	19.2	0	13.1	16.3	0	22	32.9
Decrease in efficiency w.r.t 15˚C	%	0	3	4	0	5.82	7.64	0	3.8	5.04	0	7.04	12.9

thus adversely affecting the performance of compressor.

2) Vapor Absorption Chillers:

Steam is used to give the chilling effect and one can achieve 15°C inlet air temperature.

This system consumes steam which reduces output of the steam turbine .It also reduces the inlet air pressure which adversely affects the performance of compressor.

3) High Pressure Fog system:

It's one of the recent technologies employed for inlet air cooling. It's similar to evaporative cooling, but instead of using water as an evaporative medium, the water is atomized into billions of supper small droplets thereby creating a large evaporative surface area. With fog over-cooling arrangement, one can generate more power. The biggest disadvantages of this technique are high cost of high pressure nozzles, high vibration, prices of high pressure pumps and the ability of erosion in compressor blades.

4) Evaporative Cooler:

Evaporative cooling works on the principle of reduce-ing the temperature of an air stream through water evaporation. The process of converting water from liquid to a vapor state requires energy. This energy is drawn from the air stream, the result being cooler and more humid air. The effectiveness of an evaporative cooling system depends on the surface area of the water exposed to the air stream and the residence time. Performance of the system is restricted by the amount of moisture present in the air. It works well in low humidity area.

5. The Present Work (Case Study)

For better choice between the cooling methods it's necessary to take into consideration the following factors:

1) Capital cost.
2) Maximum achievable cooling range.
3) Auxiliary power/steam consumption.
4) Reliability of the system.

5.1. Evaporative Cooler in Shuaiba North Gas Turbine Power Plant

The system existed in Shuaiba North gas turbine power plant for decreasing the inlet air temperature to the compressor consists of two centrifugal water pumps with capacity 20 m^3/hr and 120 m head. The two pumps are parallel arrangement. **Figure 3** shows the line diagram of the system arrangement.

The evaporative cooling system arrangement is also shown in **Figure 4** [7].

5.2. Thermodynamic Relationships of Evaporative Cooler System

The function of the evaporative cooler is to increase turbine output by lowering the inlet dry bulb temperature and increasing the air density through the evaporation of water into the inlet airflow. Re-circulated water introduced over the top of the evaporative cooler media drains through and wets the media. Filtered ambient air comes into contact with the wetted media where the air is

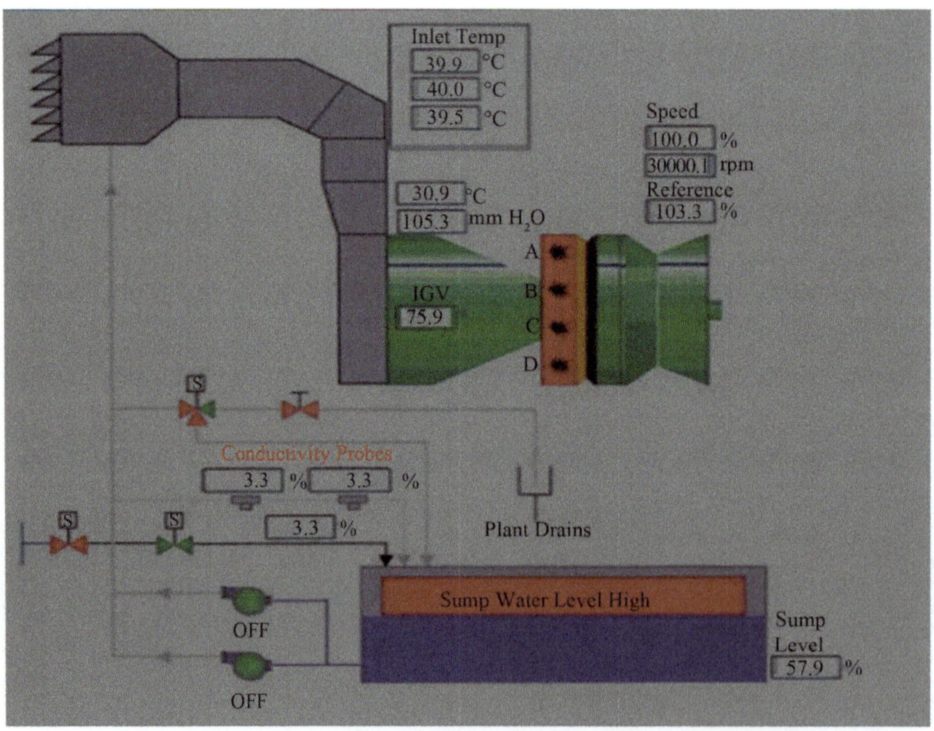

Figure 3. Schematic Diagram of Evaporative Cooling Components.

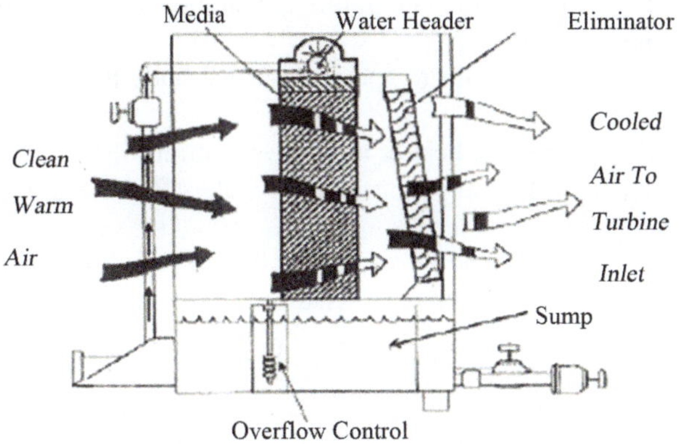

Figure 4. Schematic Diagram of an Evaporative Cooling.

cooled by the heat of evaporation of water. Gas turbine mass flow rate increases due to the lower air temperature and increased water content. Higher mass flow results in increased power output from the gas turbine. When discussing media type evaporative coolers, a couple of key parameters are essential to the proper operation and maintenance of evaporative coolers. These evaporative cooler parameters are: Saturation Efficiency, Evaporation Rate, Blow down Rate, Makeup Water Rate, Water Carryover, and Water Bypass

5.2.1. Saturation Efficiency (Effectiveness)

The performance of an evaporative cooler is based on the ratio of the number of degrees it can cool the air compared to the wet bulb temperature depression. As air passes through an evaporative cooler, the dry bulb temperature is lowered, but the wet bulb temperature remains approximately the same.

The dry bulb temperature will then approach the wet bulb temperature as the air is cooled and the relative humidity will rise (due to moisture added to the airflow through evaporation). The entering wet bulb temperature is a function of the entering conditions of the airflow (air dry bulb temperature, ambient pressure or altitude with respect to sea water level, and relative humidity). Since an evaporative cooler operates along constant wet bulb temperature lines, the entering wet bulb temperature determines the maximum amount of cooling that can be achieved. We can define the saturation efficiency of an evaporative cooler as the ratio between the actual amounts of cooling achieved (difference between entering and exiting airflow temperatures) and the maximum allowable cooling per the entering conditions (Wet Bulb Depression). The effectiveness of an evaporative cooler may be calculated using the following formula and **Figure 5** [7].

$$\eta = \frac{T_{DBE} - T_{DBL}}{WBD} = \frac{T_{DBE} - T_{DBL}}{T_{DBE} - T_{WBE}} \quad (3)$$

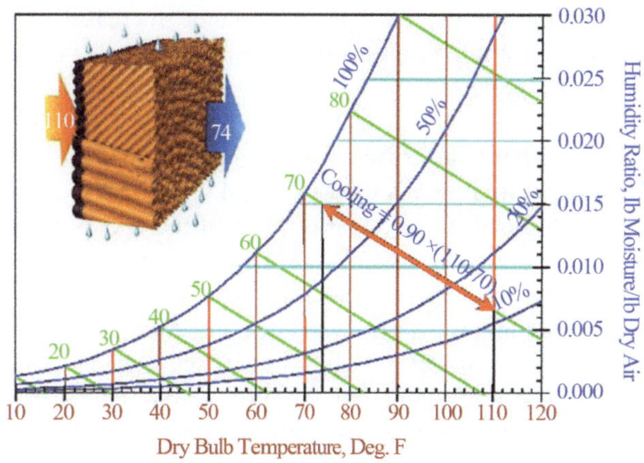

Figure 5. Evaporative Cooler Process on Psychometric Chart.

where:

η = Efficiency in percent.

T_{DBE} = Entering Dry Bulb Temp

T_{WBE} = Entering Wet Bulb Temp

T_{DBL} = Leaving Dry Bulb Temp

WBD = Wet Bulb Depression, or $=T_{DBE} - T_{WBE}$

1) Dry Bulb Temperature

The temperature as measured by a standard thermometer.

2) Wet Bulb Temperature

If the air is at a temperature above its dew point, evaporation of water will occur in the wick causing cooling and the reading of a temperature below the dry bulb temperature, the wet bulb and dry bulb temperatures are equal at 100% humidity.

The Wet Bulb Depression is the difference between the ambient dry bulb and the wet bulb temperature. The Wet Bulb Depression is the maximum amount of cooling achievable with a 100% efficient evaporative cooler system.

3) Humidity

Humidity, which is expressed by the water vapor pressure in air, is controlled by the ambient temperature. Relative Humidity is the ratio of the vapor pressure of water in air compared the saturated water vapor pressure at the same temperature. Dew Point Temperature of the air is the temperature at which condensation of liquid water may occur.

4) Ambient Pressure

The temperature and barometric pressure of atmospheric air vary considerably with altitude as well as with local geographic and weather conditions. The standard atmosphere gives a standard of reference for estimating properties of air at various altitudes. At sea level, the standard barometric pressure is 14.696 Ib/in^2 [29.921 in. Hg] and the standard temperature is 59°F [15°C]. The temperature is assumed to decrease linearly with increasing altitude while the specific volume increases. The lower atmosphere is assumed to consist of dry air that behaves as a perfect gas.

5.2.2. Evaporation Rate

The evaporation rate is the amount of water that is added to the gas turbine airflow as a result of the energy transfer between airflow and water, and the subsequent evaporation of water into the air. The amount of water evaporated into the air or evaporation rate will depend on the entering ambient air conditions. Factors such as dry bulb temperature, relative humidity, and air ambient pressure determine the maximum amount of water that may be evaporated before air becomes "saturated" with water.

The amount of water evaporated in gallons per minute as the air passes through the cooler can be calculated using the psychometric chart using the following formula:

$$\text{Evaporation} = \frac{V_{\text{air}}\left(W_L - W_E\right)}{\rho_{\text{water}} v_{\text{air}}} = \frac{\rho_{\text{air}} V_{\text{air}}\left(W_L - W_E\right)}{\rho_{\text{water}}} = \frac{m_{\text{air}}\left(W_L - W_E\right)}{\rho_{\text{water}}} \quad (4)$$

where:

V_{air} = Actual volumetric flow rate of air (cubic feet of air per minute or CFM)

ρ_{water} = Density of water (lb water/gal) at T_{WBE}

ρ_{air} = Density of air (lb air/ft^3) at entering airflow conditions

W_E = Moisture content of entering air (lbs. water/lb. dry air)

W_L = Moisture content of leaving air (lbs. of water/lb. dry air)

m_{air} = Mass airflow rate (lb. dry air/min) at entering conditions

V_{air} = Specific volume of air (ft^3/lb. dry air) at entering conditions

Similarly, since the evaporation rate is a function of the temperature of the ambient air and airflow, the amount of water evaporated may be obtained from the energy balance equation between air entering and leaving the evaporative cooler media:

$$\text{Evaporation} = \frac{m_{\text{air}} C_{P\text{air}}\left(T_{DBE} - T_{DBL}\right)}{\rho_{\text{water}} h_{fg\,\text{water}}} = \frac{\rho_{\text{air}} V_{\text{air}} C_{P\text{air}} \eta\left(T_{DBE} - T_{WBE}\right)}{\rho_{\text{water}} h_{fg\,\text{water}}} \quad (5)$$

where:

T_{DBL} = Dry bulb temperature at leaving from evaporative cooler.

T_{DBE} = Dry bulb temperature at entrance to evaporative cooler.

T_{WBE} = Wet bulb temperature at entrance to evaporative cooler.

$C_{P\text{air}}$ = Specific heat of air at T_{DBE} (Btu/lb. °F)

$h_{fg\text{water}}$ = Heat of evaporation of water (Btu/lb water) at T_{WBE}

For practical purposes, the following version of the energy balance equation is used:

$$\text{Evaporation} = \frac{V_{\text{air}}\left(T_{DBL} - T_{DBE}\right)}{500000} = \frac{V_{\text{air}}\eta\left(T_{DBE} - T_{WBE}\right)}{500000} \quad (6)$$

5.2.3. Blow down (Bleed-Off) Rate

Water always contains a certain amount of dissolved minerals. The process of evaporative cooling removes pure water as vapor from the re-circulating flow and leaves behind the solids that had been dissolved in the water when it was added as makeup. Accordingly, enough water must be blown down from the re-circulating flow to control the level of these solids and to avoid build-up of insoluble minerals on the pad surface (commonly referred to as "scaling"), which results in an increase in pressure drop, and a loss of evaporation area and efficiency.

Blow down is the flow of water that must be continuously removed from the cooler in order to maintain the chemistry of the re-circulating water at the design value. It is a function of evaporation rate and the cycles of concentration that can be achieved with a given water quality in the system.

$$\text{Blow down} = \text{Evaporation rate}/\left(\text{No. of cycle-1}\right) \quad (7)$$

The chemistry of the sump is established by determining the maximum cycles of concentration that the makeup water can go through. By definition, the number of cycles of concentration is equal to the number of times that incoming (makeup) water may be re-circulated before it is removed (bled) from the system. Two methods are commonly used for controlling blow down and maintaining the desired chemistry in the sump of the evaporative cooler: Constant flow and Conductivity control blow down.

5.2.4. Makeup Water Rate

Makeup water is the water added to the sump of the evaporative cooler to replace the water lost by evaporation into the airflow and the water removed from the sump through blow down or bleed-off as required.

$$\text{Makeup Rate}(\text{gpm}) = \text{Evaporation Rate}(\text{gpm}) + \text{Blow down Rate}(\text{gpm}) \quad (8)$$

When conductivity control is used to control blow down, the makeup water required while blow down is on will be higher than the average identified by the makeup rate formula. When the blow down is off (Blow down rate = 0 gpm), the required make up rate will be equal to the current evaporation rate.

A high makeup water rate is typically required during startup of the evaporative cooler system in order to ensure enough water is available out of the sump (recirculation) tank during initial wetting of media.

5.3. Role Played by Evaporative Cooler on Power Output and, Heat Rate of Shuaiba North Gas Turbine Power Plant

From **Table 3** which show the properties of air at entering and leaving from evaporative cooler, When using the evaporative cooling techniques the power output = 224.9 MW and heat rate = 9882.2 kj/kw hr - when the ambient conditions are: dry bulb inlet temperature T_{DB} = 42.48 C, inlet pressure = 0.99 bar and relative humidity = 12%, the leaving conditions from the evaporative cooling are as follows:

Exit dry bulb temperature T_{DBL} = 22.43 C,

Exit pressure= 0.98 bar, and the relative humidity = 85.11%.

At entrance to evaporative cooler from **Figure 5.**

$W_E = 0.0054$ Ib/Ib dry air.

At leaving from evaporative cooler from Figure(5)

$W_L = 0.015$ Ib/Ib dry air.

$$\because m_{\text{air}} = 2141.2 \text{ Ton/hr.} \tag{9}$$

$$\because m_W = m_a\left(W_L - W_E\right)$$

$$\therefore m_W = \frac{2414.2 \times 1000}{3600} \times 2.204 \times (0.015 - 0.0054)$$

$$= 12.58 \text{ Kg/s}$$

6. Calculation of Evaporative Cooling Efficiency

Using the data in **Table 4** which shows the humidity ratio at entering and leaving from evaporative cooler:

$$\eta_{evapo.cooler} = \frac{T_{DBE} - T_{DBL}}{W_{BD}} = \frac{42.48 - 22.43}{42.48 - 2.1} = 93.34\%$$

Table 3. Properties of air at entering and leaving from evaporative cooler.

State	Entering to cooling evaporator	Leaving from cooling evaporator
Dry bulb temperature T_{DB}	42.48	22.43
Wet bulb temperature T_{WB}	21	21
Relative humidity %	12	85.11

6.1. Humidity

Amount of air entered when power output = 224.9 MW, m_{air} = 2141.2 Ton/hr.

Amount of water used when power output = 224.9 Mw is

$$m_w = m_{\text{air}} \times \left(W_L - W_E\right) = 12.58 \text{ kg/s}$$

6.2. Power Saved at Various Inlet Temperatures

Table 5 determines the amount of power saved as a result of using evaporative cooler in services in Shuaiba North gas turbine power plant at different inlet air temperature.

7. Differences in Heat Rate When Using the Evaporative Cooling Techniques

Table 6 below shows the quantity of gas oil and natural gas fuels consumed in producing 225 Mw/unit from Shuaiba North gas turbine power plant when evaporative cooler in service and when it was out of services. It's clear that the quantity of fuel consumed increases when evaporative cooler is out of services, especially during hot weathers.

8. Money Saved When Using the Evaporative Cooler Techniques

The unit price of gas oil and natural gas fuels in 23 January 23, 2012 are as follows [8].

Table 4. Humidity Ratio at Entering and Leaving from Evaporative Cooler.

State	Entering	Leaving	Entering temp. °C
Absolute humidity W (Ib/Ib dry air)	0.0054	0.015	42.48
Absolute humidity W (kg/kg dry air)	2.45×10^{-3}	6.8×10^{-3}	

Table 5. Output power saved as a result of using an evaporative cooler.

Output power Mw when		Power saved in Mw	Inlet air temperature
Evaporative cooler in service	Evaporative cooler idle		
215	190	25	50
220	196	24	44
224	198	26	40
225	199	26	35
225	204	21	30
225	211	14	25
225	220	5	20

Table 6. Quantity of fuel consumed at different inlet air temperature.

Power output MW	Inlet air temperature °C		Quantity of gas oil kg/s when		Quantity of natural gas kg/s when	
	Before cooler	After cooler	Evaporative cooler in service	Evaporative cooler idle	Evaporative cooler in service	Evaporative cooler idle
225	50	26	12.21	13.38	12.2	14.91
225	44	23	11.75	13.1	11.5	14.3
225	40	39.9	12.85	12.85	14	14
225	42.48	22.43	11.5	12.93	11.2	13
225	42.5	23	11.62	12.93	11.2	13
225	42.1	22.42	11.75	12.89	11.2	13

Table 7. Money saved due to using evaporative cooler in Shuaiba North gas turbine power plant.

Power output MW	Inlet air tem. °C		Quantity of gas oil (kg/s) saved due to operation of evaporative cooler	Unit price $/kg	Total Money saved from gas oil $/s	Total money saved in 4 month
	Before cooler	After cooler				
225	50	26	1.17	1.1911	1.3935	14,442,624.00
225	44	23	1.35		1.608	16,671,744.00
225	40	39.9	0		0	0
225	42.48	22.43	1.43		1.703	17,656,704.00
225	42.5	23	1.31		1.5603	16,174,080.00
225	42.1	22.42	1.14		1.3578	14,077,670.40

Table 8. Decrease in heat rate in (kj/kw·hr) When Evaporative Cooling in Service.

Power output MW	Inlet air temp. °C		Quantity of gas oil kg/s consumed when		C.V of gas oil kj/kg	Heat rate (kj/kw·hr) when		Decrease in heat rate (kj/kw·hr)	
	Before cooler	After cooler	Evaporative cooler in service	Evaporative cooler idle		Evaporative cooler in service	Cooling evaporation idle	(kj/kw·hr)	%
225	50	26	12.85	13.38	40000	8224	8563.2	339.2	3.96
225	44	23	12.63	13.1		8083.2	8384	300.8	3.59
225	40	39.9	12.85	12.85		8224	8224	0	0
225	42.48	22.43	12.49	12.93		7993.6	8275.2	281.6	3.40
225	42.5	23	12.52	12.93		8012.8	8275.2	226.4	2.74
225	42.1	22.42	12.34	12.89		7997.6	8249.6	352	4.27

Diesel (Gas oil) $3.945 per Gallon (US).
Natural Gas $1.0324 per Thermo.

When using evaporative cooler during hot weathers a lot of money can be saved as a result of decreasing the amount of fuel burned. **Table 7** determine a sample of money saved when power output = 225 MW. When inlet air temperature ranging from 40°C - 50°C.

Heat Rate

Another important factor in determining the role played of evaporative cooler in Shuaiba North gas turbine power plant is the heat rate in (Kj/Kw·hr). The decrease of heat rate means the more efficient the gas turbine unit.

Table 8 determines the heat rate for Shuaiba North gas turbine power plant when power output 225 Mw. and inlet air temperature range 40°C - 50°C.

9. Discussions

It's clear from **Figure 6** that as the inlet air temperature decreases, the output power increases, which is due to increase in inlet air density which retails increases of inlet mass flow rate which leads to increases of gas turbine power output.

From **Figure 7** there is a considerable amount of

money saved as a result of using evaporative cooling techniques because less fuel consumed at constant turbine inlet temperature (TIT = constant).

From **Figure 8** and **Figure 9** as the inlet air temperature decreases, a significant decreases in heat rate is also detected.

From **Figure 10** it can be noticed that there is a significant needs to use evaporative cooler especially at high temperature and low humidity.

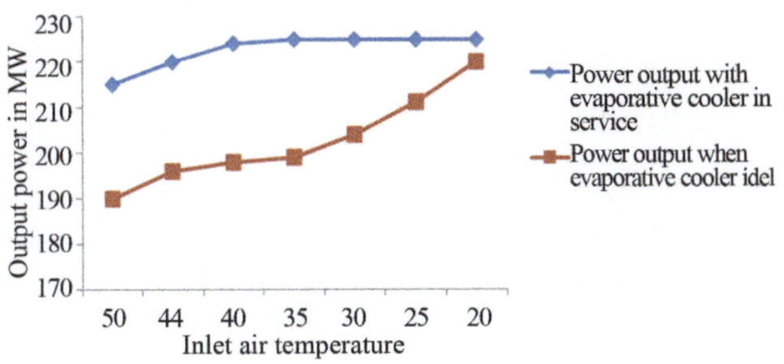

Figure 6. Effect of Inlet Air Tem. On the Power Output from Shuaiba North Gas Turbine.

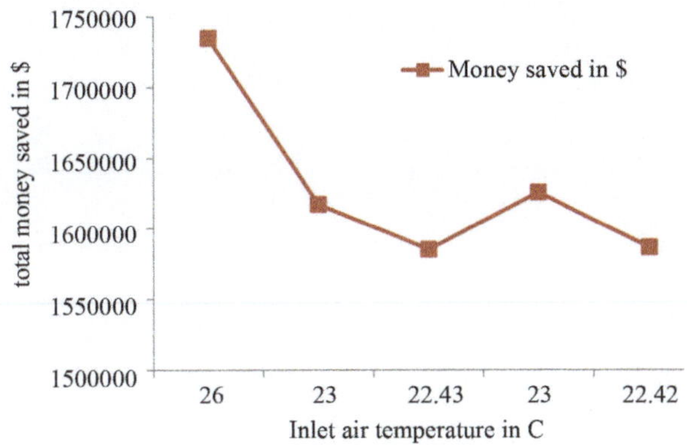

Figure 7. Money Saved When Using Evaporative Cooler in Shuaiba North Gas Turbine in $.

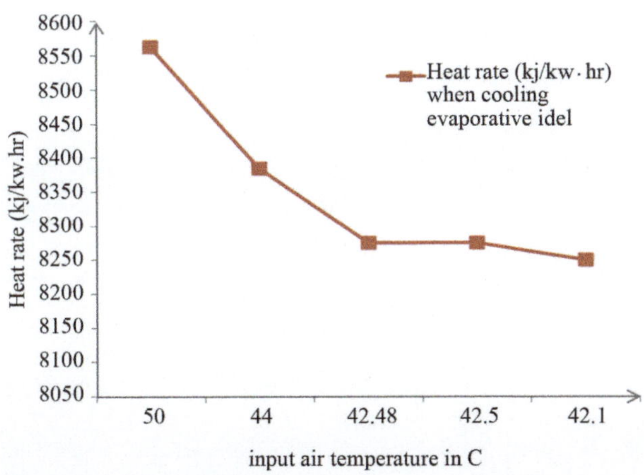

Figure 8. Effect of Input Air Tem. On Heat Rate (Kj/Kw·hr) When Cooling Evaporative Idle.

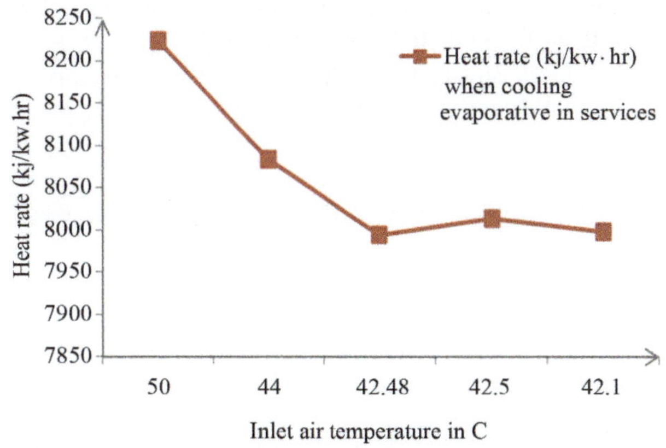

Figure 9. Effect of Input Air Tem. On Heat Rate (Kj/Kw·hr) When Cooling Evaporative in Service.

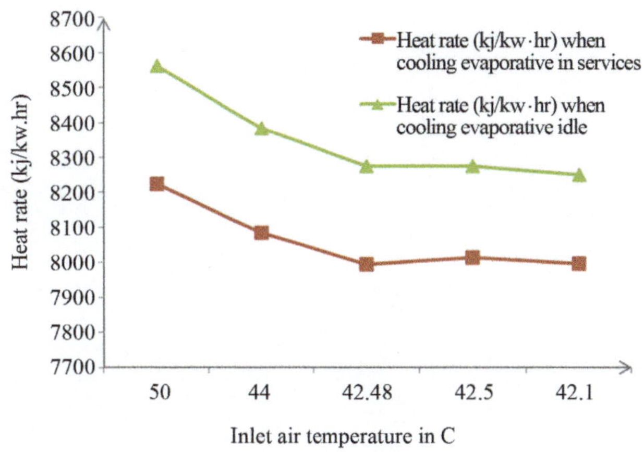

Figure 10. Comparison between Heat Rate in (Kj/Kw·hr) Versus Inlet Air Temp.

10. Suggestions and Recommendations

For more power output and lower heat rate from a gas turbine, a combination of two methods of air cooling techniques can be used but, this techniques needs more analysis in future works.

REFERENCES

[1] M. Chaker, C. B. M. Homji and Thomas Mee III, "Inlet Fogging of Gas Turbine Engines-Part A: Fog Droplet, Thermodynamics, Heat Transfere and Practical Considerations," *Proceedings of ASME Turbo Expo* 2002, 3-6 June 2002, Amsterdam.

[2] M. Chaker, C. B. M. Homji and Thomas Mee III, "Inlet Fogging of Gas Turbine Engines," *Proceedings of ASME Turbo Expo* 2001, 4-7 June 2001, New Orleans.

[3] C. B. M. Homji and Thomas Mee III, "Inlet Fogging of Gas Turbine Engines—Part B: Practical Considerations, Control, and O&M Aspects," *Proceedings of ASME Turbo Expo* 2000, 8-11 May 2000, Munich.

[4] F. J. Brooks, "GE Gas Turbine Performance Characteristics," GE Power Systems, Schenectady.

[5] P. K. Patel, "Better Power Generation from Gas Turbine along With Improved Heat Rate," 2003.

[6] J. Petek and P. Hamilton, "Performance Monitoring of Gas Turbine," *GE Energy*, Vol. 25, No. 1, 2005.

[7] Shuaiba North Power Plant Operation Manual, "Evaporative Cooler Operation & Maintenance".

[8] http://oilprice.com/Energy/Oil-Prices/Natural-Gas-Analysis-for-the-Week-of-January-23-2012.html

DSP Based Simulator for Speed Control of the Synchronous Reluctance Motor Using Hysteresis Current Controller

Abdel-Karim Daud[1], Basim Alsayid[2]
[1]Electrical and Computer Engineering Department, Palestine Polytechnic University (PPU), Hebron, Palestine
[2]Electrical Engineering Department, Palestine Technical University (PTU), Tulkarm, Palestine

ABSTRACT

This paper presents the field oriented vector control scheme for synchronous reluctance motor (SRM) drives, where current controller followed by hysteresis comparator is used. The test motor has a star-connected wound stator and a segmental rotor of the multiple barrier type with an external incremental encoder to sense rotor position. The magnetic characteristics of this motor are described using 2D finite element method, which is used firstly for rotor design of SRM. The field oriented vector control, that regulates the speed of the SRM, is provided by a quadrature axis current command developed by the speed controller. The simulation includes all realistic components of the system. This enables the calculation of currents and voltages in different parts of the voltage source inverter (VSI) and motor under transient and steady state conditions. Implementation has been done in MATLAB/Simulink. A study of hysteresis control scheme associated with current controllers has been made. Experimental results of the SRM control using TMS320F24X DSP board are presented. The speed of the SRM is successfully controlled in the constant torque region. Experimental results of closed loop speed control of the SRM are given to verify the proposed scheme.

Keywords: Field Oriented Control; 2D Finite Element Method; SRM; Hysteresis Current Controller; DSP; MATLAB/Simulink

1. Introduction

Synchronous reluctance motor (SRM) has recently attracted the efforts of a number of researchers and is gaining favor as a possible alternative for ac drives [1-4]. Since the converter fed SRM does not need a starting cage, an optimized rotor for synchronous performance can be designed. The simple and rugged structure allows high speed drives to be readily attained while the copper and iron losses confined into the stator allow the motor to be operated with high electric and magnetic loads with advantage on the torque/weight ratio. A drawback of these drives resides in the need of synchronization with the rotor position [5-7]. The complicated coupled nonlinear dynamic performance of SRM can be significantly improved using vector control theory [4,8-16] where torque and flux can be controlled separately. Simulations have helped the process of developing new systems including motor drives, by reducing cost and time.

Simulation tools have the capabilities of performing dynamic simulations of motor drives in a visual environment so as to facilitate the development of new systems [14,17,18].

In this work, the simulation of a field oriented controlled SRM drive system is developed using MATLAB/Simulink.

The motor has a segmental rotor of the multiple barrier type with an external incremental encoder to sense rotor position. The simulation circuit will include all realistic components of the drive system. This enables the calculation of currents and voltages in different parts of the inverter and motor under transient and steady state conditions. A closed loop control system with a PI controller in the speed loop has been designed to operate in constant torque region. A study of hysteresis control scheme associated with current controller has been made. Simulation results are given for the speed range in constant torque region of motor operation. Finally, the experimental verification obtained by using the DSP based vector control is presented.

2. Prototype Rotor Design

2.1. Rotor Types

Almost all of the important parameters of the synchronous reluctance motor depend on the synchronous inductance ratio or saliency ratio, $\xi = L_q/L_d$. The main classes of rotor design aimed at maximizing ξ are transversely and axially laminated multiple barrier rotors as shown in **Figures 1** and **2**, respectively. In all cases the objective is to achieve a high L_q by providing, essentially, flux guides for q-axis flux; and low L_d by providing flux barriers to d-axis flux.

In respect to axially laminated rotor, a transverse laminated one has many advantages such as simplicity in the mechanical construction, lower manufacturing cost and rotor skewing possibility. On the contrary it has a quite large level of torque ripple due to the stator slots and rotor flux barriers that produce a non sinusoidal air-gap permeance variation.

2.2. Direct Finite Element Design

A two-dimensional finite element method is used to design the rotor and to analyze the magnetic characteristics of the motor by taking into account the non-linearity and saturation phenomena. The solutions of the magnetic field with different rotor displacements and phase excitations are obtained. The magnetic field characteristics are described over the entire area of the motor in terms of magnetic vector potentials.

The result of magnetic flux calculation is shown in **Figure 3** for only 1/6 of the machine due to symmetry considerations. The achievable saliency ratio is limited by two factors: the d-axis permeance cannot be zero and the laminations are subject to saturation in the q-axis.

Figure 4 shows the cross-sectional view of the proposed reluctance motor. It is a three-phase six-pole motor, with 4 barriers per pole. The stator is the same as that of the PM synchronous motor. The rotor cores are laminated in the rotor shaft direction. Magnetic reluctance in the d-axis is large and in the q-axis is small. There is a narrow connection ring in the fringe of the rotor and narrow connection bars from the centre to the fringe in the d-axis. They are necessary for mechanical stability of the rotor.

The rotor design is based on maximising torque. **Figure 4** shows more detail of the rotor construction indicating two key variables, W_{iron} and W_{air}, corresponding to the width of each rotor segment and of the width of the air gap between segments respectively. The sum of the $n^*(W_{iron} + W_{air})$ is chosen so as to always equal the width

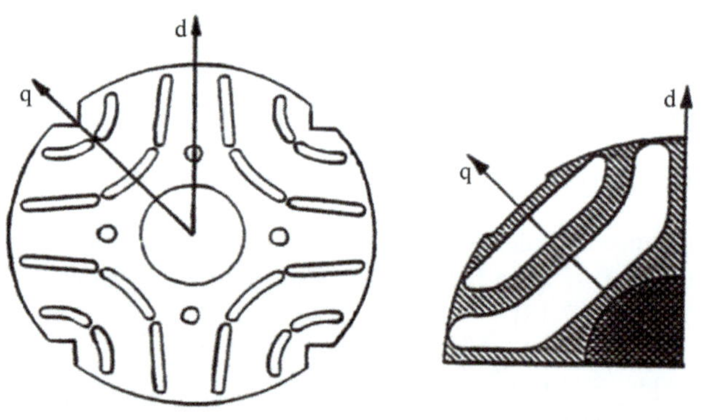

Figure 1. Transversely laminated multiple-barrier rotors.

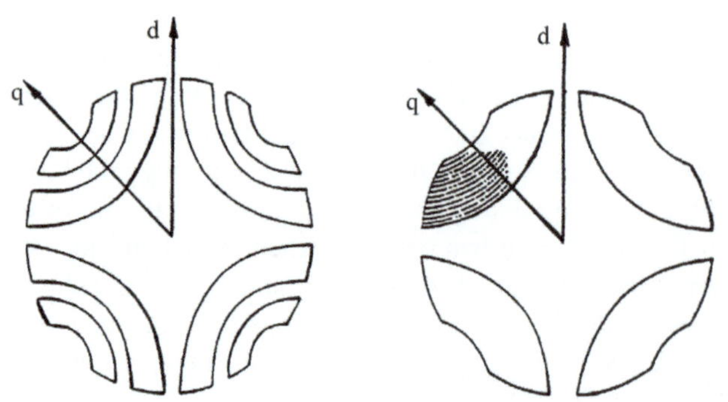

Figure 2. Axially laminated multiple-barrier rotor.

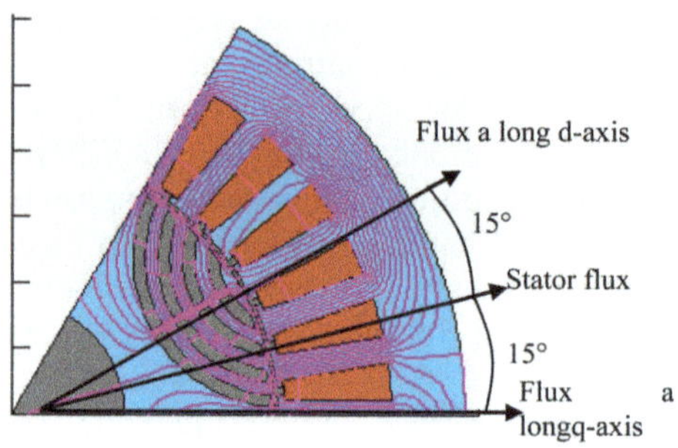

Figure 3. Stator and rotor fluxes (MTC45).

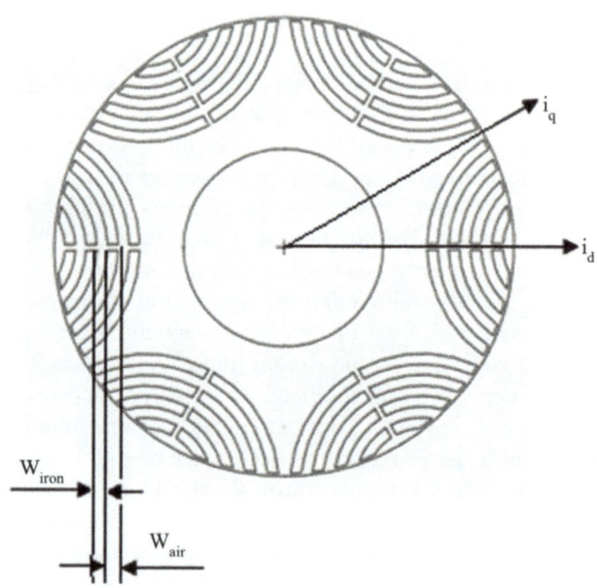

Figure 4. Cross-sectional view of the proposed SRM.

of one stator tooth (W_t) in order to limit, as much as possible, pulsating fluxes (torque ripple)in the stator teeth. In practice, values of $n = 1, 2$ and 3 were investigated.

For purposes of comparison between different geometries it is useful to define the ratio $K_w = W_{air}/W_{iron}$. Clearly, when $K_w = 0$, the rotor is assumed to be completely made of iron (no saliency). When $K_w = 1$ the rotor is constructed of lamination segments in which the air space and lamination segments are equal. FEA for $n = 1$, 2 and 3 gave maximum torque for $K_w = 1$ and $n = 3$, in which $W_{air} = W_{iron} = 1.86$ mm.

2.3. Torque and Torque Ripple

Three-phase motor currents have been considered to supply the stator expressed by Equation(1) and the resulting stator flux forms an electrical angle of 45° (15° mechanical) with q-axis, which gives a maximum torque curve (MTC45) (see **Figure 3**).

$$\begin{aligned} I_a &= 122\sin(\theta) = 106 \text{ A} \\ I_b &= 122\sin(\theta - 120^\circ) = 0 \text{ A} \\ I_c &= 122\sin(\theta - 240^\circ) = -106 \text{ A} \end{aligned} \quad (1)$$

where $\theta = 120°$ is the electrical angle corresponding to stator flux at 45° electric degrees from rotor flux, 122 A is maximum current for half stator slot. The same normal B-H characteristic is computed for stator and rotor magnetic material. With FEA a value of 4.2 N·m of Torque has been calculated, it was the best value for different values of electrical angle between stator and rotor fluxes, which was verified experimentally. The torque ripple results from the non-sinusoidal air-gap permeance function for the stator slots and rotor flux barriers; its estimation requires an accurate FEA with different stator-rotor relative positions. A few significant rotor positions, among infinite ones, have been considered in order to predict the torque ripple with a good accuracy.

Five rotations have been considered, each rotation of 1 mechanical degree or 3 electrical degrees, which correspond to a half stator tooth pitch.

The results of calculated torque for each position are obtained in **Figure 5**. It shows the torque ripple for a rotation of an electrical angle of ≅15° (equal to a mechanical angle of 5°, six-pole motor). This is the angle of skewing of the rotor adopted to reduce torque ripple.

The main dimensions of the test motor are shown in **Table 1**. It is a 6-pole sinusoidal synchronous reluctance motor (SRM). It has a star-connected wound stator, a segmental rotor of the multiple barrier type (see **Figure 6**) and external incremental encoder to sense rotor position.

3. SRM Drive System

The well-known d-q model of AC machines is widely

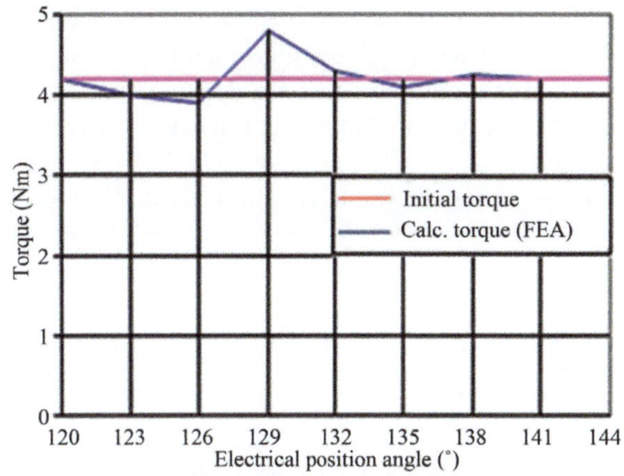

Figure 5. Torque ripple.

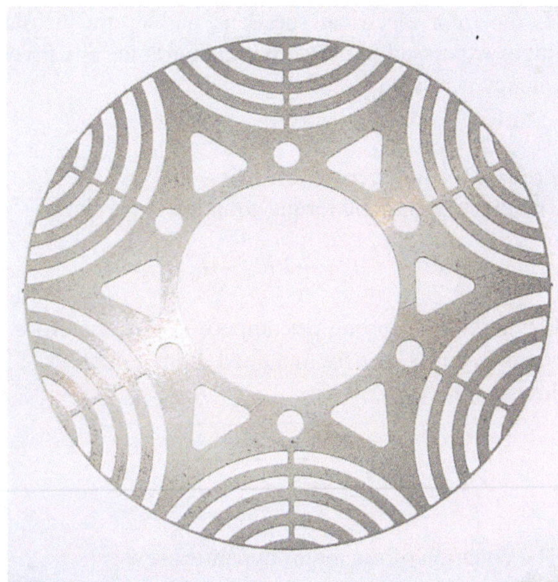

Figure 6. Cross section of the synchronous reluctance machine.

Table 1. SRM parameters.

Name	Symbol	Value
Rated voltage	V_n	220 V
Rated torque	Tn	5 N·m
Rated current	I_n	14.9 A
Rated speed	n_n	1000 RPM
Maximum speed	n_{max}	6000 RPM
Number of pole pairs	P	3
Stator resistance	R_s	0.3 Ω
q-axis Inductance	L_q	0.009 H
d-axis Inductance	L_d	0.004 H
Motor Inertia	J	0.0755 kg·m²

used for simulation purposes. Based on the assumption that the stator windings are sinusoidally distributed, the d-q model is a powerful tool for dynamic simulation of most AC machines including the reluctance synchronous machine.

The d-q reference frame equations describing the motor are:

$$\frac{\mathrm{d}}{\mathrm{d}t}\begin{bmatrix} i_q \\ i_d \end{bmatrix} = \begin{bmatrix} -\frac{R}{L_q} & -\omega_r \frac{L_d}{L_q} \\ \omega_r \frac{L_q}{L_d} & -\frac{R}{L_d} \end{bmatrix} \begin{bmatrix} i_q \\ i_d \end{bmatrix} + \begin{bmatrix} \frac{1}{L_q} & 0 \\ 0 & \frac{1}{L_d} \end{bmatrix} \begin{bmatrix} v_q \\ v_d \end{bmatrix} \quad (2)$$

where i_d and i_q are the d and q axis stator currents, L_d and L_q are the stator phase d-axis and q-axis winding self inductance respectively; R is the stator winding resistance; ω_r is the rotor electrical speed; v_d and v_q are the stator voltages expressed in the dqreference frame. The inverter frequency is related as follows

$$\omega_s = P\omega_r \quad (3)$$

where P is the number of pole pairs.

The electromagnetic torque equation is given by

$$T_e = \frac{3}{2}P\left(L_d - L_q\right)i_d i_q \quad (4)$$

For maximum torque per ampere $i_d = i_q$, therefore the currents i_d and i_q can be obtained from Equation (5) as follows:

$$i_d = \sqrt{\frac{T_e}{\frac{3}{2}P\left(L_d - L_q\right)}} \quad (5)$$

The equation of the motor dynamics is

$$T_e = T_L + B_r + J\left(\mathrm{d}w_r/\mathrm{d}t\right) \quad (6)$$

T_L stands for external load torque. B represents the damping coefficient and J is the moment of inertia of the rotor. The Equations (2), (4) and (6) constitute the whole control model of the SRM. A system configuration of a vector-controlled SRM drive system is shown in **Figure 7.**

The dynamic d q modeling is used for the study of motor during transient and steady state. It is done by converting the dqo variables to three phase currents by using inverse Parks transformation [16-18].

SRM is fed form a voltage source inverter (VSI) with current control. The control is performed by regulating the flow of current through the stator of the motor. Current controllers are used to generate gate signals for the inverter. Proper selection of the inverter devices and selection of the control technique will guarantee the efficiency of the drive.

In the vector control scheme, torque control can be carried out by suitable regulation of the stator current vector; this implies that accurate speed control depends on how well the current vector is regulated. In high-performance vector drives, a current-control loop, with a considerably high bandwidth, is necessary to ensure accurate current tracking, to shorten the transient period as much as possible and to force the voltage source inverter (VSI) to equivalently act as a current source amplifier within the current loop bandwidth. In this work, a hysteresis-band current controlled VSI is used. This controller will generate the reference currents with the inverter within a range which is fixed by the width of the band gap. In this controller the desired current of a given phase

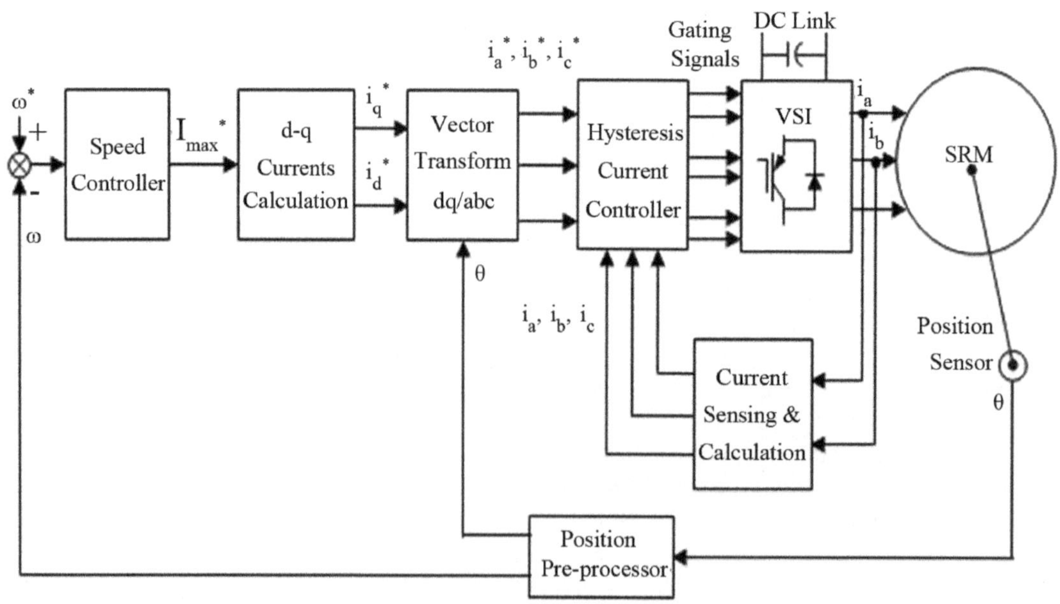

Figure 7. SRM vector-controlled drive system.

(i_a^*, i_b^* and i_c^*) is summed with the negative of the measured current (i_a, i_b and i_c). When the current error exceeds a predefined hysteresis band, the upper switch in the half-bridge is turned off and the lower switch is turned on. As the current error goes below the hysteresis band, the opposite switching takes place. The principle of hysteresis band current control is illustrated in **Figure 8** [4,16].

Speed controller calculates the difference between the reference speed (ω^*) and the actual speed (ω) producing an error, which is fed to the PI controller. PI controllers are used widely for motion control systems. Speed control of motors mainly consist of two loops the inner loop for current (band hysteresis current controller) and the outer loop for speed (speed controller) as shown in **Figure 7**. The order of the loops is due to their response, how fast they can be changed. This requires a current loop at least 10 times faster than the speed loop. An incremental encoder is used as a position sensor.

Control loops in the actual drive system, shown in **Figure 7**, are implemented in software on Texas Instruments (TMS320F24X) processor and executed with a cycle period of 70µs. The flow chart of this program is shown in **Figure 9**.

At switching on, the program initializes the hardware registers, I/O ports are then pre-set to their initial states, the inverter, ADC converters, position sensor (optical encoder) and software variables. Then it initializes the speed calculation and hysteresis regulators. The system now completes all initializations and starts the main program which requires a computational time of 70µs of period cycle.

The main program will first calculate actual speed of the motor (ω), read reference speed (ω^*) and actual currents (i_a, i_b and i_c) from ADCs. Errors are then saved and new errors are calculated. Speed regulator is realized by a discrete PI.

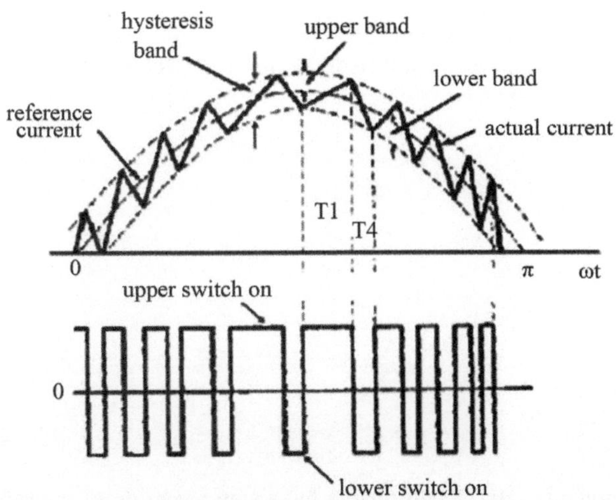

Figure 8. Hysteresis current controller.

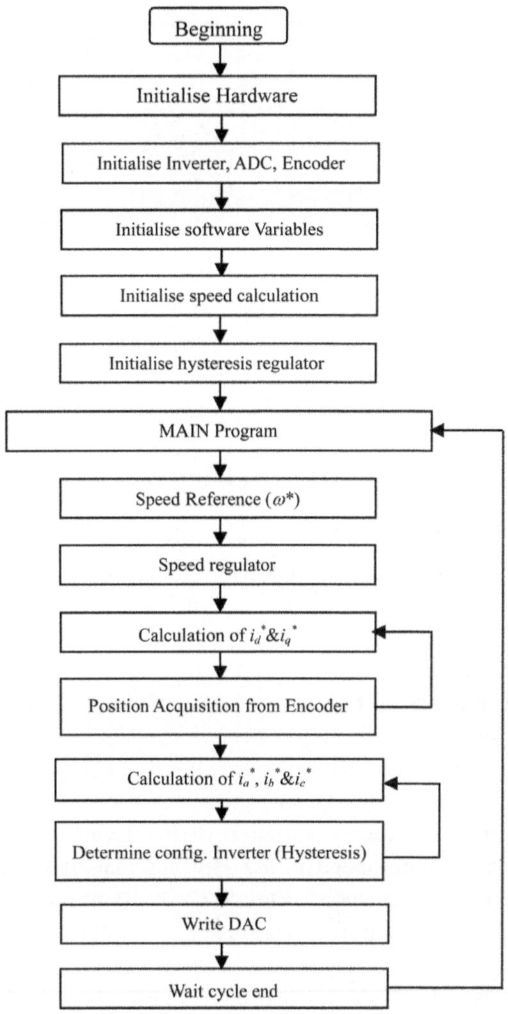

Figure 9. System flow diagram.

Calculation of i_q^* and i_d^* in the rotating reference frame is done. Then it will read position from the encoder. Based on rotor position, three-phase reference currents (i_a^*, i_b^* and i_c^*) in the stationary reference frame are calculated. This is followed by the execution of currents loop and resulting controlled signals are sent to the inverter. Write DAC and wait end cycle are finally executed.

4. Simulation in Simulink

The SRM drive simulation was built in several steps like dqo variables transformation to abc phase, calculation torque and speed, control circuit, inverter and SRM. The dqo variables transformation to abc phase is built using the reverse Parks transformation. For simulation purpose the voltages are the inputs and the current are output. Using all the drive system blocks, the complete system block has been developed as shown in **Figure 10**. The system built in Simulink for a SRM drive system has been tested with the Hysteresis current control method at

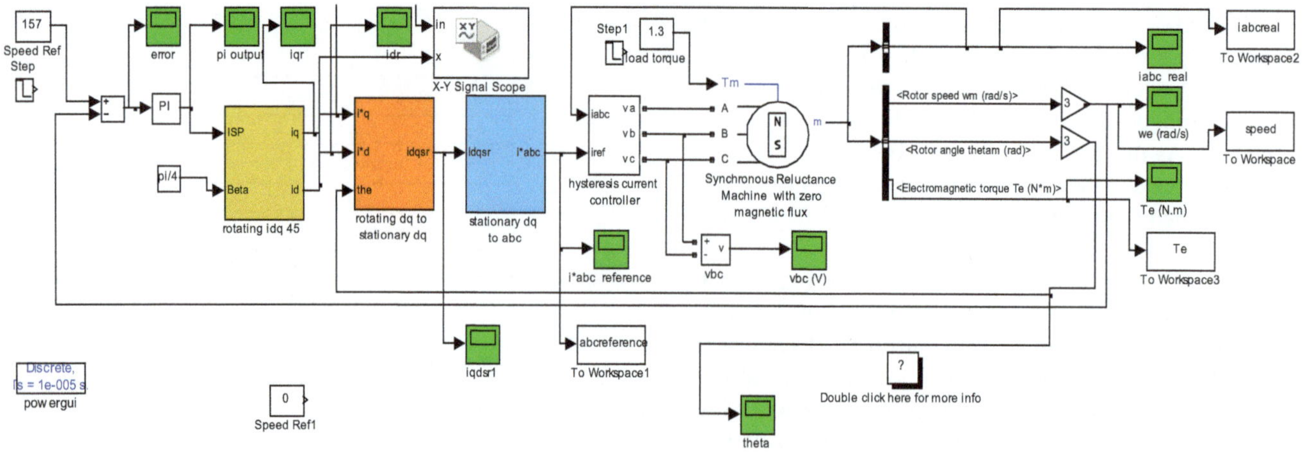

Figure 10. Synchronous reluctance motor drive system in Simulink.

the constant torque region of operation.

The motor parameters used for simulation are given in **Table 1**. **Figure11** shows the real three phase currents drawn by the motor as a result of the hysteresis current control. The currents are obtained using Park's reverse transformation. It is clear that the current is non-sinusoidal at the starting and becomes sinusoidal when the motor reaches the controller command speed at steady state.

Figure 12 shows a variation of the speed with time. The steady state speed is the same as that of the commanded reference speed. **Figure 13** shows the developed torque of the motor. The starting torque is the rated torque. The steady state torque is about 1.3 N·m.

Figure 14 shows the real three phase currents drawn by the motor as a result of the hysteresis current control, when the motor changes its speed from 1000 rpm to −1000 rpm with a load torque of 1.3 N·m. It is clear that the currents are inversed due to the speed variation from 1000 to −1000 rpm. The speed performance is shown in **Figure 15** for this case. The steady state speed is the same as that of the commanded reference speed.

Figure 16 shows the developed torque of the motor for the speed variation from 1000 to −1000 rpm. The starting torque is the rated torque (5 N·m). The steady state torque is about 1.3 N·m in positive and negative operation of the motor.

5. Experimental Results

A DSP based PC board integrated system (TMS320F24X DSP board), is used for vector control of SRM drive [14]. The schematic diagram of the hardware implementation is shown in **Figure 17**. Feedback signals to the controller board are the actual motor currents and the rotor position angle. The currents are measured by the Hall-effect transducers. The currents are then buffered and fed to the **A/D** ports of the controller board. The motor shaft posi-

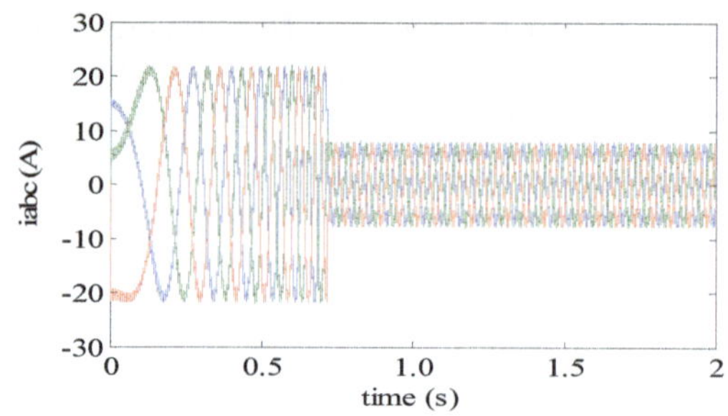

Figure 11. Actual phase currents with hysteresis control at 1000 rpm with 1.3 N·m.

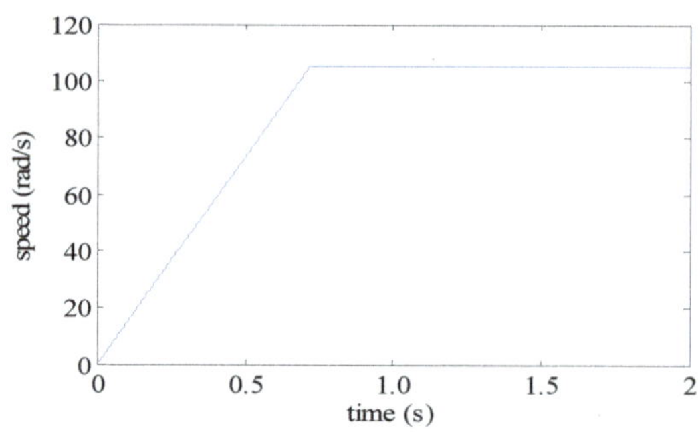

Figure 12. Dynamic performance for a step variation of the reference speed from 0 to 1000 rpm (ω = 105 rad/s) with a torque load of 1.3 N·m.

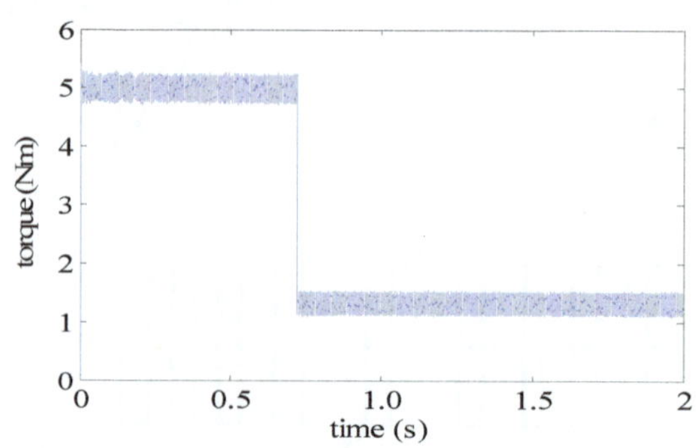

Figure 13. Developed torque for a step variation of the reference speed from 0 to 1000 rpm (1.3 N·m load torque).

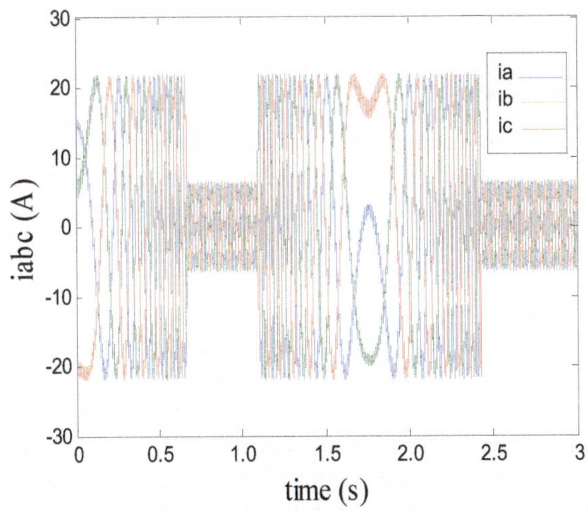

Figure 14. Inversion of actual phase currents due to a step variation of a speed from 1000 rpm to −1000 rpm with a torque load of 1.3 N·m.

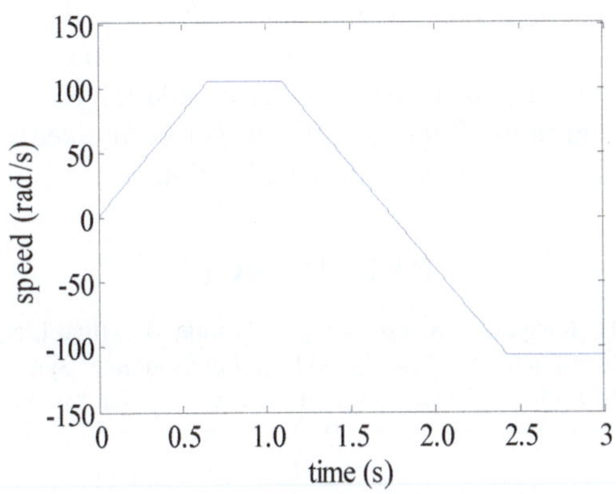

Figure 15. Dynamic performance for a step variation of the speed from 1000 rpm to −1000 rpm ($\omega = -105$ rad/s) with a torque load of 1 N·m.

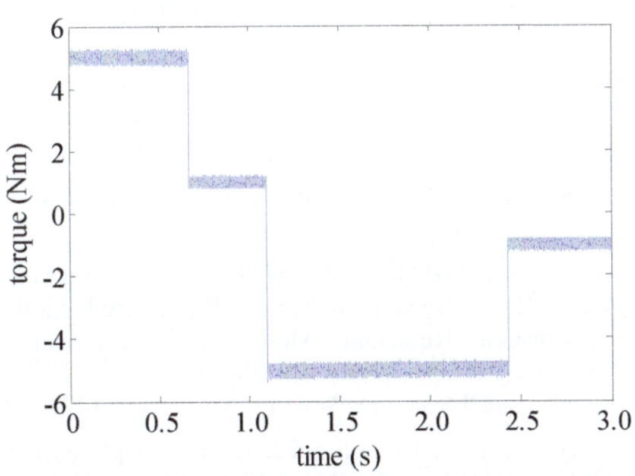

Figure 16. Developed torque with hysteresis control for a step variation of the speed from 1000 rpm to −1000 rpm (1 N·m load torque).

tion is measured by an optical incremental encoder installed at the motor shaft. The commutating signals for the drive pulses have also been generated by the hysteresis controller. The control algorithm has been implemented via the controller board using assembly language programming.

A series of experiments has been carried out to evaluate the performances of the proposed vector controlled SRM drive system. Different sample results are presented in the following figures. **Figure 18** demonstrates the actual phase current i_a wave form at speed 1500 rpm for a load of 2 N·m.

The experimental evaluation of speed with load as parameter of DSP based SRM drive is shown in **Figure 19.** It shows the step speed response of 1000 rpm of the proposed system for a load of 1.3 N·m.

In **Figure 20**, the behaviour of the current of phase A is shown during the inversion of speed from 1000 RPM to −1000 rpm with load torque of 2 N·m. **Figure 21** shows the response of the drive to a step variation of the reference speed from 1000 to −1000 rpm with a load

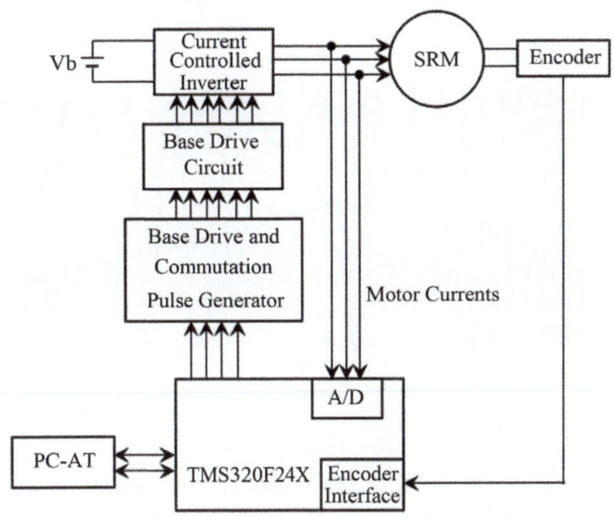

Figure 17. The hardware schematic of experimental system.

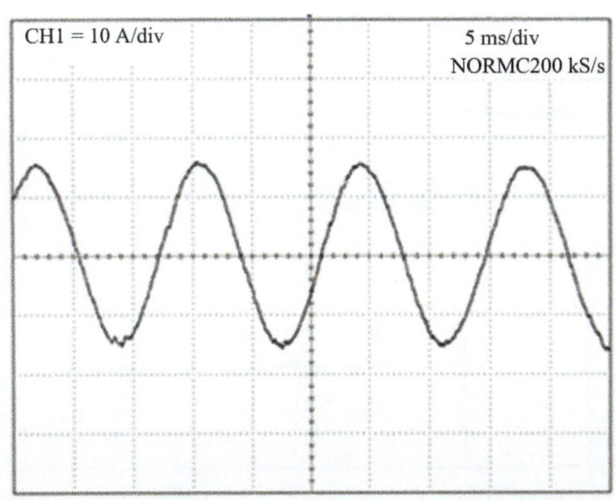

Figure 18. Actual phase current i_a wave form at 1500 rpm for a load of 2 N·m.

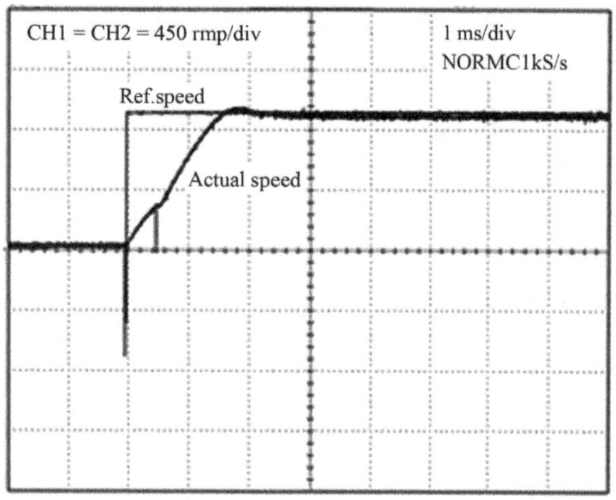

Figure 19. Experimental speed responses of SRM drive for step variation from 0 to 1000 rpm with load torque of 1.3 N·m.

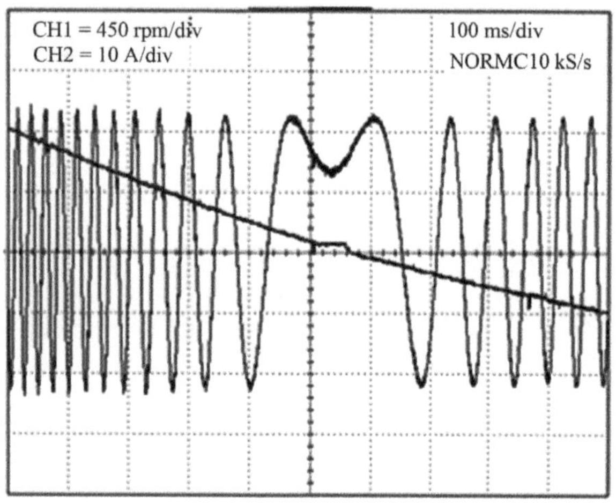

Figure 20. Inversion of actual speed and actual phase current i_a wave forms due to a step variation of the reference speed from 1000 rpm to −1000 rpm (load Torque of 2 N·m).

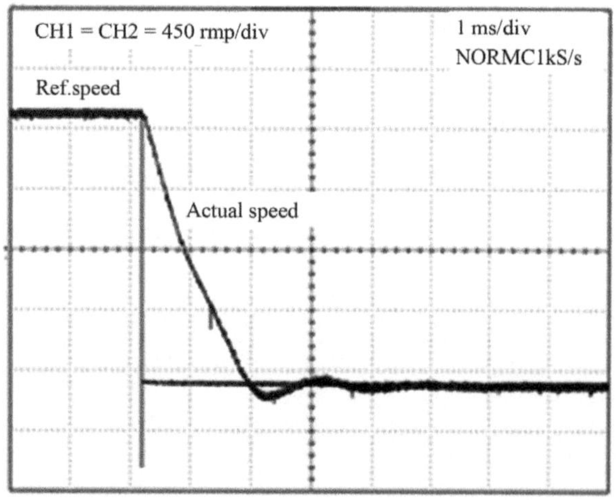

Figure 21. Dynamic performance for a step variation of the reference speed from 1000 rpm to −1000 rpm (load torque of 1 N·m).

torque of 1 N·m.

6. Conclusions

A rotor for a reluctance motor has been designed using 2D finite element method. The maximum torque has been found to be 4.2 N·m. The results obtained by FEA seem to be valid because, after the construction of the rotor, experimental results show that the maximum torque is about 4.2 N·m and the torque ripple with position is very closed to the results obtained by FEA.

A MATLAB/SIMULINK simulation model has been proposed for closed loop speed control with internal current loop using hysteresis controllers. The experimental results show that the proposed field oriented vector controlled SRM drive can handle the effects of step change in reference speed and parameter variations. The overall system performances are quite good in terms of dynamic, transient and steady-state responses.

Simulation and experimental results show that the proposed control scheme guarantees stable and robust response of the SRM drive, under a wide range of operating conditions. Subsequently, it can be utilized in high performance motion control applications.

REFERENCES

[1] H. Kiriyama, S. Kawano, Y. Honda, T. Higaki, S. Morimoto and Y. Takeda, "High Performance Synchronous Reluctance Motor with Multi-Flux Barrier for the Appliance Industry," *The* 1998 *IEEE Industry Applications Conference*, *Thirty-Third IAS Annual Meeting*, St. Louis, 12-15 October 1998, pp. 111-117.

[2] Malan, M. J. Kamper and P. N. T. Williams, "Reluctance Synchronous Machine Drive for Hybrid Electric Vehicle," *IEEE International Symposium on Industrial Electronics*, Pretoria, 7-10 July 1998, pp. 367-372.

[3] J. Malan and M. J. Kamper, "Performance of a Hybrid Electric Vehicle Using Reluctance Synchronous Machine Technology," *IEEE Transactions on Industry Applications*, Vol. 37, No. 5, 2001, pp. 1319-1324.

[4] G. K. Dubey, "Fundamentals of Electrical Drives," Alpha Science, Temple City, 2001.

[5] J.-M. Park, S.-J. Park, M.-M. Lee, J.-S. Chun and J.-H. Lee, "Rotor Design on Torque Ripple Reduction for a Synchronous Reluctance Motor with Concentrated Winding Using Response Surface Methodology," *IEEE Transactions on Magnetics*, Vol. 42, No. 10, pp. 3479-3481.

[6] S. B. Kwon, S. J. Park and J. H. Lee, "Optimum Design Criteria Based on the Rated Watt of a Synchronous Reluctance Motor Using a Coupled FEM and SUMT" *IEEE Transactions on Magnetics*, Vol. 41, No. 10, 2005, pp. 3970-3972.

[7] J. Rizk, M. H. Nagrial and A. Hellany, "Optimum Design Parameters for Synchronous Reluctance Motors," *Proceedings of the* 14*th International Middle East Power Systems Conference* (*MEPCON*'10), Egypt, 19-21 De-

cember 2010, Article ID: 290.

[8] N. Bianchi, S. Bolognani, D. Bon and M. D. Pre, "Torque Harmonic Compensation in a Synchronous Reluctance Motor," *IEEE Transactions on Energy Conversion*, Vol. 23, No. 2, 2008, pp. 466-473.

[9] F. Fernandez-Bernal, A. Garcia-Cerrada and R. Faure, "Efficient Control of Reluctance Synchronous Machines," *Industrial Electronics Society*, 1998. *IECON '98. Proceedings of the* 24*th Annual Conference of the IEEE*, Aachen, 31 August-4 September 1998, pp. 923-928.

[10] R. E. Betz, R. Lagerquis M. Jovanovic, T. J. E. Miller and R. H. Middleton, "Control of Synchronous Reluctance Machines," *IEEE Transactions on Industry Applications*, Vol. 29, No. 6, 1993, pp. 1110-1121.

[11] L. Xu and J. Yao, "ACompensated Vector Control Scheme of a Synchronous Reluctance Motor Including Saturation and Iron Loss," *IEEE Transactions on Industry Applications*, Vol. 28, No. 6, 1992, pp. 1330-1338.

[12] J. C. Kim, J. H. Lee, I. S. Jung and D. S. Hyun, "Vector Control Scheme of Synchronous Reluctance Motor Considering Iron Core Loss," *IEEE Transactions on Magnetics*, Vol. 34, No. 5, 1998, pp. 3522-3525.

[13] R. E. Betz, M. Jovanovic, R. Lagerquist and T. J. E. Miller, "Aspects of the Control of Synchronous Reluctance Machines including Saturation and Iron Losses," *Industry Applications Society Annual Meeting*, 1992. *Conference Record of the* 1992 *IEEE*, Houston, 4-9 October 1995, pp. 456-463.

[14] A.-K. Daud, B. Alsayid and A. Zaidan, "DSP Based Simulator for Field Oriented Control of the Surface Permanent Magnet Synchronous Motor Drive," *International Journal of Circuits, Systems and Signal Processing*, Vol. 6, No. 1, 2012, pp. 29-37.

[15] M. H. Rashid, "Power Electronics, Circuits, Devices and Applications," Pearson Prentice Hall, Upper Saddle River, 2004.

[16] B. K. Bose, "Modern Power Electronics and AC Drives," Prentice Hall, Upper Saddle River, 2002.

[17] C.-M. Ong, "Dynamic Simulation of Electric Machinery Using Matlab/Simulink," Prentice Hall, Upper Saddle River 1998.

[18] M. A. Fellani and D. E. Abai, "Matlab/Simulink-Based Transient Stability Analysis of a Sensorless Synchronous Reluctance Motor," *World Academy of Science, Engineering and Technology*, Vol. 68, 2010, pp. 1472-1476.

Experimental Study of Solar-Assisted Heating System

Jieting Wei[1], Yang Liu[2], Yu Gu[2]
[1]Changchun Institute of Technology,Changchun,China
[2]Northeast Dianli University,Jilin,China

ABSTRACT

Solar heating is a new energy saving technology. This paper studies the operating characteristics of solar-assisted heating system in Changchun area, the changes of the indoor temperature with the conditions of solar-assisted heating system .By analyzing these operating data we can see solar-assisted heating to meet the requirements of the indoor temperature and save energy.

Keywords: Solar; Assisted Heating; Energy Saving

1. Introduction

Solar heating is the most potential solar thermal utilization technology after solar heating the water, it should be popularized in the future. With the state has increased the use of renewable energy, and unveiled preferential policies, solar heating technology has an extensive application prospect [1]. As regards the present situation, there is a wide range of markets for solar energy in our country, but correlation study is still lacking, the lack of relevant data, so the experimental build a solar-assisted heating bench, finishing the test data, provided the basis for data for the practical application of solar-assisted heating engineering.

At present, the terminal device for most of the solar heating system are used radiant floor heating, however, terminal devices of most of the existing residential buildings are radiators, so this paper studies solar assisted radiator heating system.

2. The Experiment of Solar-assisted Heating System

In an energy-efficient building, selected a 58m^2 room as an experimental room, and selected another room as the comparison room that the size, towards and maintenance structure are the same with the experimental room. According to calculation the heating load of the experimental room is 3480W .The experimental system uses the Huayang vacuum tube solar water heater, collector area is 3.83, the tank capacity is 250 L. The system is lack of elevated water tank, not able to meet the conditions of the natural cycle (solar energy can be used as an elevated water tanks, the reason for it can not natural cycle is that the heat source and the heat dissipation center can not formed the height difference). Therefore, there is a 290w miniature pump in the system as a cycle power, the suction inlet of the pump is connected heat source, the direction of flow: heat source→ pump → radiator.

The experimental procedure is as follows: solar collector absorbs solar radiation, transforms it into heat energy stored in the storage tank for room heating, and composes of a set of sola- assisted heating system with the end of the heat dissipation device.

There are four long airfoil radiators, two intelligent temperature control strong convection radiators, two radiators through the valve can series parallel used, and can run separately, so that can satisfy the requirements of different indoor temperature. In summer, when the indoor temperature is too high, to pass the tap water into the intelligent temperature control strong convection radiators, after heat transfer treatment sending cold air to the room to eliminate waste heat, similar to the role of a simple split air-conditioning, to maintain the temperature of the interior cool. In winter, under normal circumstances, only use the long airfoil radiator heating to meet the requirements of indoor temperature, when outdoor temperature mutation or fluctuations in a very obvious situation can be appropriate to open the intelligent temperature control strong convection radiators, to ensure that the interior parameters are maintained within the desired temperature range. By two radiators combined operation (forced convection heat transfer and natural convection heat transfer) is sufficient to meet the heating requirements of the room.

Data collection included the supply water temperature of radiator t_1, the return water temperature of radiator t_2, indoor temperature of the heating roomt_3, indoor temperature of the non-heating room t_4, outdoor air temperature t_5, wind speed v, solar radiation intensity.

Heating system as illustrated below.

3. Experimental Data and Analysis

There are two cases of the experiment: one for fine weather, sunny, without opening the auxiliary heating device will be able to meet the basic needs of the interior; Another situation is the rainy weather, lacking of intensity of solar radiation, the need to open the auxiliary heating equipment to meet the basic needs of the interior. Now analysis the test data results of both cases.

Experimental results and analysis are as follows

3.1. Experimental Data and Analysis without Auxiliary Heating Equipment

Figure 2 shows contrast curves of the indoor temperature of the heating room, indoor temperature of the non-heating room and outdoor air temperature. **Table 1** lists the peak point on the curve. From above, it can be seen that the trend of the indoor and outdoor temperature is substantially same. The average temperature of the heating room is 22..23℃, and the average temperature of the non-heating room is 18.52℃. The average temperature of the heating room is higher than the non-heating room 1.1℃ ~5.7℃.

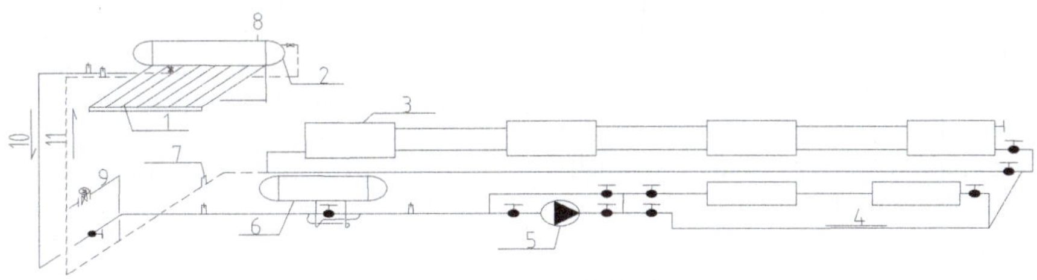

Figure 1. The system diagram of the solar-assisted heating.

Table 1. The equipments name of the solar-assisted heating system.

1	Solar collectors	7	Temperature Sensor Placement
2	Heat Storage Water Tank	8	Water Overflow Port
3	Long Airfoil Radiator	9	Connect Tap Water
4	Intelligent Temperature Control Strong Convection Radiator	10	Heating supply water
5	Circulating Water Pump	11	Heating return water
6	Electric Heater		

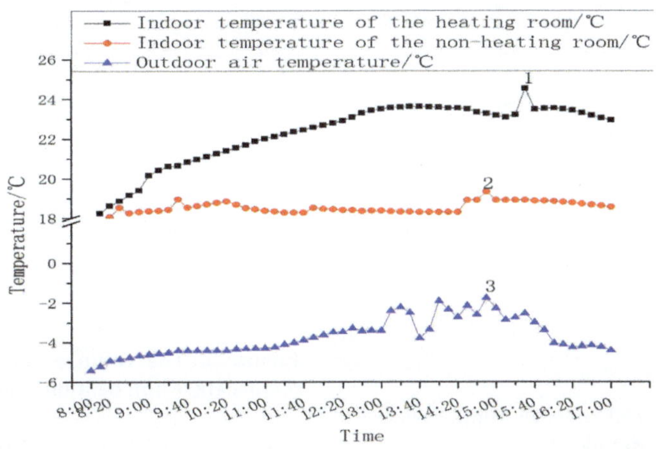

Figure 2. The comparison chart of indoor and outdoor temperature.

Table 2.

	X Axis (Time)	Y Axis (Temperature /℃)
1	15:20	24.62
2	14:50	19.34
3	14:50	-1.35

Figure 3 shows contrast curves of the radiator supply and return water temperature and heating load. **Table 2** shows the supply and return water temperature on the curve, the peak point of heating load. The loading heat Q is calculated after measured system flow rate.

$$Q = CM\Delta T\ /KW \tag{1}$$

According to the formula (1) the loading heat of radiator is calculated. We can educe that the trend of the radiator supply and return water temperature is broadly similar. The average temperature difference of the radiator supply and return water temperature is 7℃ ~10℃.

Figure 4 shows the contrast curves of the radiator supply and return water temperature and solar radiation intensity. **Table 1** lists the maximum value of solar radiation intensity and radiator supply and return water temperature. This shows that the trend of the radiator supply and return water temperature and solar radiation intensity is broadly consistent. The peak of the radiator supply and return water temperature delay compared to the intensity of solar radiation, which is caused due to the regenerative properties of the water tank.

Figure 5 shows comparison graph of the indoor and outdoor temperature, loading heat. **Table 4** lists the maximum worth of indoor temperature of the heating room, indoor temperature of the non-heating room, outdoor air temperature and the loading heat of the heating room. This shows that when the outdoor temperature is high, the room temperature and loading heat to obtain the maximum.

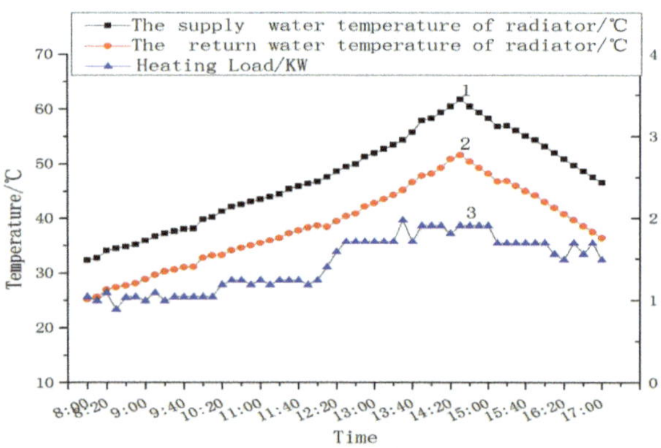

Figure 3. The comparison chart of radiator supply and return water temperature and heating load.

Table 3.

	X Axis (Time)	Y Axis (Temperature /℃)
1	14:30	61.679
2	14:30	51.679
3	15:00	1.9087 (Heating Load/KW)

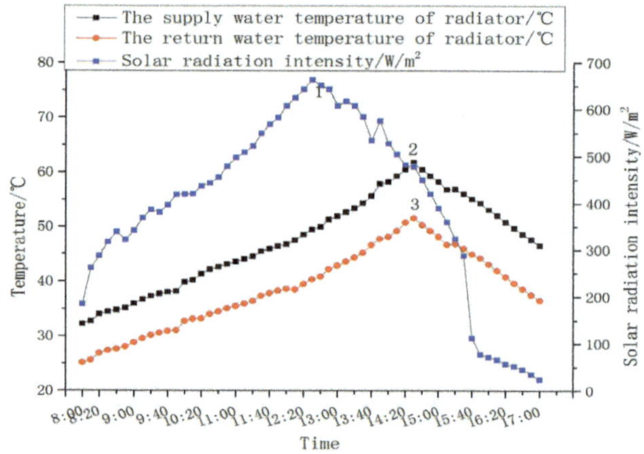

Figure 4. The comparison chart of radiator supply and return water temperature and solar radiation intensity.

Table 4.

	X Axis (Time)	Y Axis (Temperature /℃)
1	12:30	664(solar radiation intensity/W/m^2)
2	14:30	61.679
3	14:30	51.679

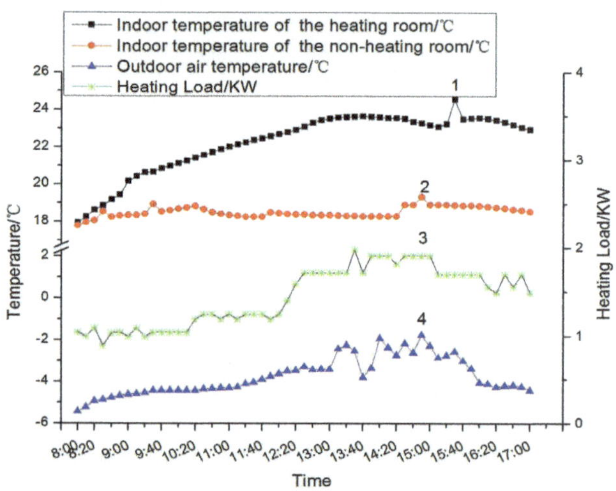

Figure 5. The comparison chart of indoor and outdoor temperature and heating load.

Table 5.

	X Axis (Time)	Y Axis (Temperature /℃)
1	15:30	24.539
2	14:50	19.335
3	15:00	1.9087 (Heating Load/KW)
4	14:50	-1.75

3.2. Experimental Data and Analysis with Auxiliary Heating Equipment

Figure 6 shows the contrast curve of the radiator supply and return water temperature and heating load when opening the electric heater. Opened the electric heater at 8:00, closed it after one hour. It reached a small peak of the radiator supply and return water temperature and heating load at 9:00.After that it operated independently rely on solar radiation .Opened the electric heater auxiliary heating again at 15:00 pm, closed it after one hour. It reached a small peak again at 16:00, but with the electric heater was turned off, all the value were gradually reduced. The value all showed in **Table 5**.

Figure 7 shows comparison chart of radiator supply and return water temperature and solar radiation intensity when opening the electric heater. **Table 6** shows the value of each point. Opened the electric heater at 8:00, closed it after one hour. The solar radiation was strongest and radiator supply and return water temperature reached the maximum value at 11:20. With the intensity of solar radiation, the supply and return water temperature gradually weakened, Turned on the electric heater again for one hour at 15:00. Thus, the radiator supply and return water temperature is controlled by the intensity of solar radiation.

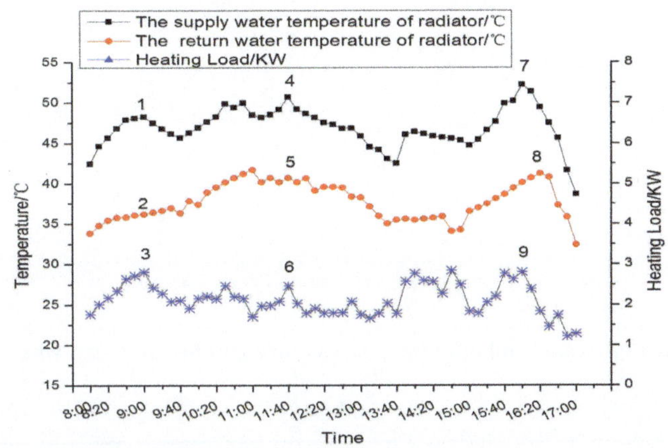

Figure 6. The comparison chart of radiator supply and return water temperature and heating load when opening the electric heater.

Table 6.

	9:00		11:40		16:00
	Y Axis (Temperature/℃)		Y Axis (Temperature/℃)		Y Axis(Temperature/℃)
1	48.265	4	50.695	7	52.203
2	36.226	5	40.734	8	40.156
3(Heating Load)	2.8164KW	6(Heating Load)	2.4694KW	9(Heating Load)	2.8182KW

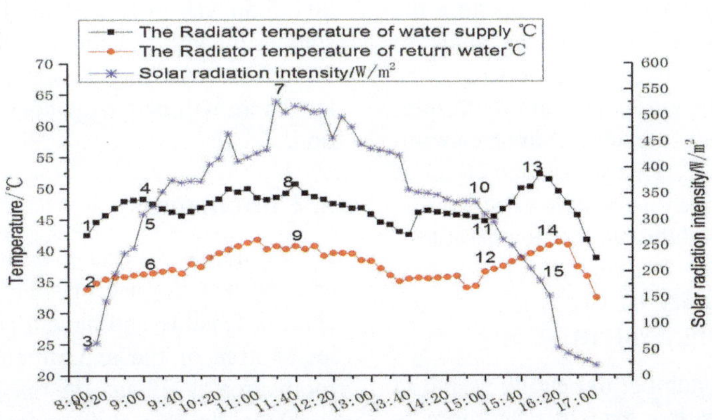

Figure 7. The comparison chart of radiator supply and return water temperature and solar radiation intensity when opening

the electric heater.

Table 7.

	8:00		9:00		11:20
	Y Axis(Temperature/℃)		Y Axis(Temperature/℃)		Y Axis(Temperature/℃)
1	42.484	4	48.265	7 (radiation intensity)	527 W/m^2
2	33.898	5	36.226	8	48.554
3 (radiation intensity)	53 W/m^2	6 (radiation intensity)	311 W/m^2	9	40.734
	15:00				16:00
	Y Axis(Temperature/℃)				Y Axis(Temperature/℃)
10 (radiation intensity)	310 W/m^2			13	52.203
11	44.725			14	40.156
12	36.585			15 (radiation intensity)	182 W/m^2

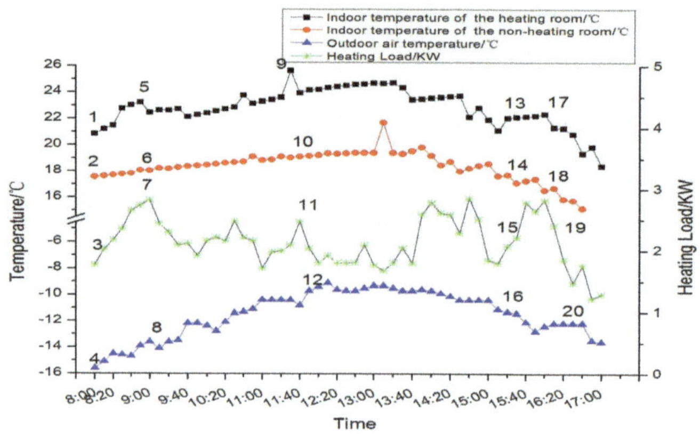

Figure 8. The comparison chart of indoor and outdoor temperature and heating load when opening the electric heater.

Table 8.

	8:00		9:00		11:30
	Y Axis(Temperature/℃)		Y Axis(Temperature/℃)		Y Axis(Temperature/℃)
1	20.843	5	22.453	9	25.648
2	17.546	6	18.015	10	19.054
3(Heating Load)	1.7687KW	7(Heating Load)	2.8164 KW	11(Heating Load)	2.4694 KW
4	-15.6	8	-13.6	12	-10.8
	15:00				16:00
	Y Axis(Temperature/℃)				Y Axis(Temperature/℃)
13	21.906			17	22.375
14	18.617			18	16.585
15(Heating Load)	1.8474 KW			19（Heating Load）	2.8182 KW
16	-10.4			20	-12.4

Figure 8 shows comparison chart of the indoor and outdoor temperature and heating load when opening the electric heater. **Table 7** respectively lists the value of each point. From the graph can be seen, opening the electric heater can increased significantly the room temperature and the heating load, the temperature can improve an average of 1℃ ~ 2℃. This shows, the effect of the solar- assisted heating system is common decision by the solar radiation intensity and the outdoor temperature

4. Energy Saving Quantity of the Solar-assisted Heating System

Take 40% solar fraction, the number of heating period in Changchun is 170 days, heating load of The room is 3480 W. Thus the heating load for one day is 3480 × 24/1000 = 83.52 Kwh = 300.67 MJ. Thus the total energy saving quantity is Q = 40% × 170 × 300.67 = 20445.56 MJ = 4891 Mcal. The heat of the definition of standard coal in China is 7000 kcal/kg (7 Mcal/kg). Therefore a heating season can save standard coal: 698 kg. There will be a huge contribution to energy conservation.

5. Conclusions

By the above experiments and the collation of data, we can see that the use of solar-assisted heating in Changchun is feasible and have a certain effect. The solar collector area of the experimental system is 3.83 m^2, the room area is 58 m^2 ,in the case of Changchun region sunshine, the rise of the room temperature can reach 4℃, and thus see, in energy-efficient buildings 1 m^2 solar collectors as an auxiliary heating can heat 15 m^2 room.

Through the experiment, conclusions are drawn as follows:

1) Solar-assisted heating can meet the temperature requirements of room, the average temperature of the heating room reaches 22℃ on average higher than non-heating room 4℃

2) The operating results of the solar-assisted heating system by the solar radiation, the outdoor temperature as so on. The system supply and return temperature difference maintain in 7℃.

3) In energy-efficient buildings, the proportion of solar collector area and room area is1:15.

4) Changchun area belongs to the area of solar energy resources are abundant, solar heating is both environmentally friendly and energy conservation, research should be focused.

REFERENCES

[1] Z. N. He and D. Z. Zhu, "Solar Heating Application Manual," Beijing: Chemical Industry Press, 2009.

Analysis on the Transfer Characteristics of Rogowski-coil Current Transformer and Its Influence on Protective Relaying

Yi Wang[1], Jiaman Li[2], Yulan Hu[1], Ranran An[1], Zexiang Cai[2], Ruiwen He[3]
[1]Electric Power Research Institute, Guangdong Power Grid Corporation, Guangzhou, China
[2]School of Electric Power, South China University of Technology, Guangzhou, China
[3]School of Automation, Guangdong University of Technology, Guangzhou, China

ABSTRACT

This paper systematically analyzes the transfer characteristics of the Rogowski-coil Current Transformer and its effect on protective relaying through theoretical analysis, experiments and simulations. The frequency characteristics and transient characteristics of Rogowski transducer and Rogowski-coil Current Transformer are deeply analyzed based on the physical structure of the transformer. It is revealed that broad bandwidth of the transformer can improve the performance of protective relaying, and the bandwidth is determined mainly by the parameters of the Rogowski transducer and signal processing circuits. It is also discovered that the measurement errors of transient current mainly depend on the abilities for the current transformer to reproduce an accurate replica of the decaying dc components, which is mainly decided by the decay time constant of the aperiodic component of transient current and the parameters of the integral unit. Finally, some measures are proposed for the performance improvement of Rogowski-coil Current Transformer to meet the requirements of protective relaying system in terms of structural design and testing standards.

Keywords: Current Transformer; Rogowski-coil; Transfer Characteristics; Protective Relaying

1. Introduction

The Rogowski-coil Current Transformer has attracted much attention and expectation as it can meet the requirements of power system protective relaying for its excellent performances, including a good linearity for measurement, broad bandwidth, excellent dynamic response and no magnetic saturation etc. [1, 2]

Scholars have done a lot of research on the Rogowski-coil current transformer about its sensors [3-6], power supply [7], electronic circuit [8], integral method [9] and merging unit [10], which have effectively pushed forward the Rogowski-coil current transformer to be practical. Reference [11] discussed the calculation method and measurement principle of distributed capacitance and its effects on dynamic characteristics of sensors in time domain and frequency domain. The integral time constant is the major factor affecting the response of transformer to the DC components of transient current [12]. In order to improve the performance of the transformers, PCB structure and T-integrator were adopted when measuring the transient current in power system in reference [13]. Simulation analysis and experiments on the frequency characteristics, steady-state performance and transient performance of PCB transformer have been done in reference [14]. These studies show that Rogowski-coil Current Transformer has a good steady-state and dynamic performance for measurement in terms of structural design, but they do not point out its problems on transfer characteristics from the aspect of application.

Rogowski-coil current transformer has a linear characteristic, which means the superposition principle can be applied. Thus the transfer characteristic of Rogowski-coil current transformer can be quantitatively analyzed, which is quite different from the electromagnetic current transformers. The outputs of the electromagnetic current transformers have uncertainty due to its nonlinear magnetic characteristics. The testing and evaluation standards [15, 16] of electronic current transformer have failed to reflect the structural and transfer characteristics of Rogowski-coil Current Transformer.

So far the application of Rogowski-coil Current Transformer is subjected to its stability and reliability. Besides this, the lack of research on transfer characteristics and the inadequate understanding of the existing problems, as well as the overgeneralizations of the relevant test standards, have seriously affected the maturity and extensive application of this new equipment. The protective relaying system has to adapt to the characteristics of electronic transformer in the practical application [17, 18] in this situation.

From the perspective of protective relaying system application, this paper deeply studies the frequency characteristics and transient response of the Rogowski transducer and the Rogowski-coil current transformer, combined with experimental testing and simulation results, analyzes the key factors affecting transfer characteristics of the transformer, and offers the proposals for Rogowski-coil Current Transformer to adapt to the needs of protective relaying.

2. Transfer Characteristics of Rogowski Transducer

The structure of Rogowski-coil Current Transformer can be divided into Rogowski transducer, analog signal processing unit, digital signal processing and transmission units, merging unit and power supply unit according to its function.

The equivalent circuit [12] of the Rogowski transducer is shown in **Figure 1** Rogowski transducer has a fairly small mutual inductance due to its air core. The Rogowski-coil Current Transformer dedicated to protective relaying works in the outside integral way, and the load resistance is very large. When the measured current has a low frequency, and coil internal resistance and interturn capacitance is tiny enough to be ignored, secondary side of the Rogowski transducer is close to be in open condition. At the moment, the output voltage is

$$u_1(t) \approx e(t) = -M \cdot \mathrm{d}i/\mathrm{d}t \,.$$

2.1. Frequency Characteristic

The transfer function for Rogowski transducer according to the equivalent circuit is

$$\mathrm{H}_1(s) = \frac{U_1(s)}{I(s)} = \frac{Ms}{L_0C_0s^2 + (\frac{L_0}{\mathrm{R}_L} + \mathrm{R}_0\mathrm{C}_0)s + (\frac{R_0}{\mathrm{R}_L} + 1)} \quad (1)$$

The undamped natural angular frequency

$$\omega_n = \frac{1}{\sqrt{L_0C_0}}\sqrt{\frac{R_L + R_0}{R_L}},$$

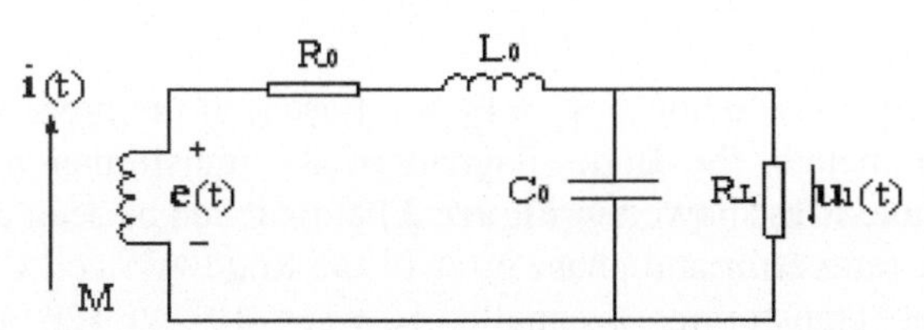

Figure 1. Equivalent circuit of Rogowski transducer.

damping ratio

$$\zeta = \frac{1}{2\sqrt{L_0C_0}}\left(\frac{L_0}{R_L} + R_0C_0\right)\sqrt{\frac{R_L}{R_L + R_0}},$$

and then the transfer function is

$$H_1(s) = \frac{M}{L_0C_0}\cdot\frac{s}{(s-s_1)\cdot(s-s_2)} \quad (2)$$

Where characteristic roots are $s_{1,2} = -\zeta\omega_n \pm \omega_n\sqrt{\zeta^2 - 1}$.

When primary current is the rated current I_N, the induction potential of Rogowski transducer is $E = \omega MI_N$, therefore, the manufacturers determine the number of turns of coil N depending on the desired value of the mutual inductance M. And the coil induction is is $L_0 = M \cdot N$. Then the main factors affecting the frequency characteristics of Rogowski transducer are the internal resistance of the coil and distributed capacitance.

The Bode diagram of the transfer function of Rogowski transducer is shown in **Figure 2**. Phase error increases as the frequency increases. When the internal resistance and the distributed capacitance vary individually with other parameters remaining unchanged, the Bode diagrams of the Rogowski transducer are shown in **Figures 2(a)** and **(b)**, from which we can know the impacts of the internal resistance and distributed capacitance on the bandwidth of Rogowski transducer should not be ignored.

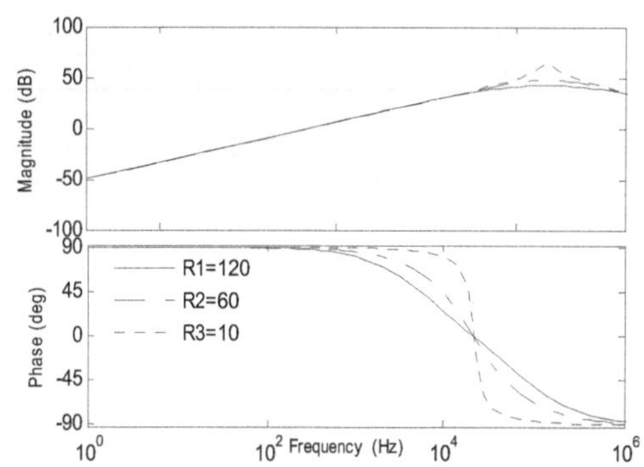

(a) characteristics when Internal resistance changes

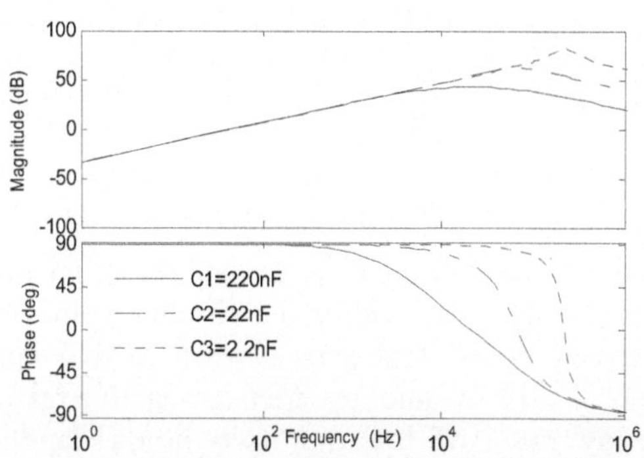

(b) characteristics when distributed capacitance changes

Figure 2. Bode diagram of Rogowski transducer.

To reduce the distributed capacitance can significantly increase the upper cutoff frequency. Changing the resistance will change the damping ratio, so that the frequency characteristics can be changed. When designing Rogowski transducer, select a suitable wire diameter and minimize the distributed capacitance under the promise of appropriately selecting the coil structure parameters, so as to improve the high-frequency characteristics of the Rogowski transducer.

2.2. Transient Characteristics

The Rogowski-coil Current Transformer has a good linearity for measurement, no saturation problem, so the transient response of Rogowski-coil Current Transformer can be analyzed using superposition principle, and the response of the transformer to each component of transient current can be studied respectively. In other words, the transient analysis of Rogowski-coil Current Transformer can focus on studying the effects of decaying aperiodic component on transformers based on frequency response analysis.

When the primary current is the aperiodic component decaying with the time constants τ, $i(t) = I_m e^{-t/\tau}$,

The corresponding Laplace transform is

$$I(s) = \frac{I_m}{s + 1/\tau},$$

then the sensor output is given by

$$\mathrm{U}_1(s) = \frac{M}{L_0 C_0} \times \frac{s}{(s - s_1)\times(s - s_2)} \times \frac{I_m}{s + 1/t} \tag{3}$$

Transform into the partial fraction

$$U_1(s) = \frac{MI_m}{L_0 C_0}\cdot\left(\frac{A_1}{s - s_1} + \frac{B_1}{s - s_2} + \frac{C_1}{s + 1/\tau}\right) \tag{4}$$

where $\mathrm{A}_1 = \frac{\mathrm{s}_1}{(\mathrm{s}_1 - \mathrm{s}_2)(\mathrm{s}_1 + 1/\tau)}$, $B_1 = \frac{s_2}{(s_2 - s_1)(s_2 + 1/\tau)}$,

$$C_1 = \frac{-\tau}{(1 + \tau s_1)(1 + \tau s_2)}.$$

By inverse Laplace transform, time domain expression of sensor output is

$$u_1(t) = \frac{MI_m}{L_0 C_0}(A_1 e^{s_1 t} + B_1 e^{s_2 t} + C_1 e^{-t/\tau}) \tag{5}$$

In the underdamping situation $\zeta < 1$, s_1 and s_2 are a pair of complex conjugate roots, and the decay rates of the first two items depend on the absolute values of the characteristic roots' real parts , when in overdamping situation $\zeta > 1$, s_1 and s_2 are two negative real roots, the decay rates of the first two items depend on $\left(\zeta \pm \sqrt{\zeta^2 - 1}\right)\omega_n$. As the distributed capacitance of the sensor is very small, then ω_n is generally above 10^5, the first two items decay very quickly regardless of the damping, with negligible impact on the output voltage. For $s_1 + s_2 = -2\zeta\omega_n$, $s_1 \cdot s_2 = \omega_n^2$, the output voltage corresponding to input decaying DC component is given by

$$u_1(t) = MI_m \frac{-\tau}{L_0 C_0 - \tau \cdot R_0 C_0 + \tau^2} e^{-t/\tau} \tag{6}$$

$L_0 C_0$ and $R_0 C_0$ are very small compared to τ, so the equation above is approximately

$$u_1(t) = MI_m \frac{1}{-\tau} e^{-t/\tau} \tag{7}$$

Simulation shows that after 0.5 ms, the output of the equation (7) is exactly the same with that of equation (5), which illustrates the simple differential model, can describe the response of sensor to the decaying aperiodic component. The error of the Rogowski-coil Current Transformer when transforming the a periodic component mainly depends on the following integral circuit whether can accurately restore the differential signal into decaying DC or not.

3. The Transfer Characteristics of Rogowski-coil Current Transformer

The integral unit in the prototype discussed in this paper is implemented by analog circuit. The A/D conversion and the following link in the signal acquisition system of transformers can only change the signal form or its carrier, although in which the inevitable errors also exit, they does not change the signal transfer characteristic. So this paper only discusses the model and characteristic of Rogowski transducer and analog signal processing circuits. In order to react Frequency characteristics more realistically, leave out the low-pass filter in the signal processing circuit.

3.1. Frequency Characteristic

The overall transfer function of Rogowski-coil Current Transformer is:

$$H(s) = H_1(s)H_2(s)H_3(s) \tag{8}$$

$H_1(s)$, $H_2(s)$, $H_3(s)$ is respectively the transfer function of Rogowski transducer, the integral amplification and phase compensation part .

$$H_2(s) = \frac{K_2}{1 + RCs}, \quad H_3(s) = \frac{1 - K_3 s}{1 + K_3 s}$$

where $RC = 0.066$, $K_2 = 19.56$. Based on the prototype parameters, the Bode diagram of the transformer as a whole is as showed in **Figure 3** below it can be seen that the ratio error and phase error of the Rogowski-coil Current Transformer is smaller between 10-200 Hz. The phase error is larger in high frequency range.

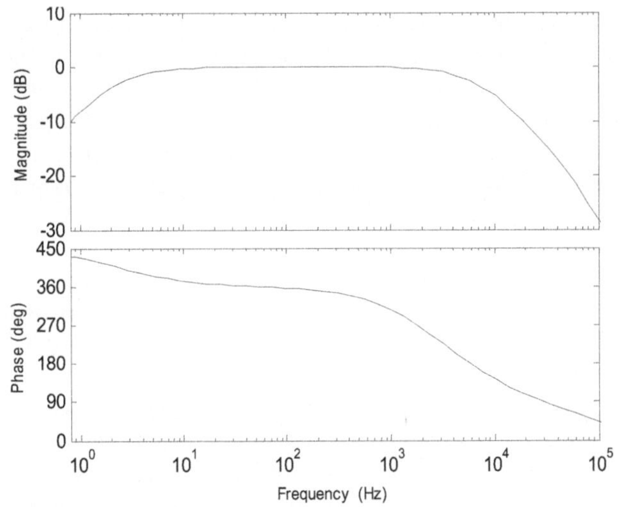

Figure 3. Bode diagram of Rogowski-coil Current Transformer.

3.2. Transient Characteristics

Similarly, set $I(s)=\dfrac{I_m}{s+1/\tau}$, then the transformer output is

$$U(s)=H(s)\frac{I_m}{s+1/\tau} \tag{9}$$

Rewrite it by

$$U(\mathrm{s})=\frac{A}{s-s_1}+\frac{B}{s-s_2}+\frac{C}{s+1/RC}+\frac{D}{s+1/K_3}+\frac{E}{s+1/\tau} \tag{10}$$

where

$$A=\frac{MI_mK_2}{L_0C_0RC}\cdot\frac{s_1}{(s_1-s_2)}\cdot\frac{1}{s_1+1/RC}\cdot\frac{1/K_3-s_1}{1/K_3+s_1}\cdot\frac{1}{s_1+1/\tau}$$

$$B=\frac{MI_mK_2}{L_0C_0RC}\cdot\frac{s_2}{(s_2-s_1)}\cdot\frac{1}{s_2+1/RC}\cdot\frac{1/K_3-s_2}{1/K_3+s_2}\cdot\frac{1}{s_2+1/\tau}$$

$$C=\frac{MI_mK_2}{L_0C_0RC}\cdot\frac{-1/RC}{(1/RC+s_1)\cdot(1/RC+s_2)}\cdot\frac{1/K_3+1/RC}{1/K_3-1/RC}\cdot\frac{1}{-1/RC+1/\tau}$$

$$D=\frac{MI_mK_2}{L_0C_0RC}\cdot\frac{-2/K_3^2}{(1/K_3+s_1)(1/K_3+s_2)}\cdot\frac{1}{-1/K_3+1/RC}\cdot\frac{1}{-1/K_3+1/\tau}$$

$$E=\frac{MI_mK_2}{L_0C_0RC}\cdot\frac{-1/\tau}{(1/\tau+s_1)\cdot(1/\tau+s_2)}\cdot\frac{1}{-1/\tau+1/RC}\cdot\frac{1/K_3+1/\tau}{1/K_3-1/\tau}$$

As s_1 and s_2 are very large, K_3 is very small, the three corresponding items in expression (10) decay rapidly and so can be ignored. After simplification according to the features of s_1, s_2 and K_3, the transformer output voltage

$$u(t)=\frac{MI_mK_2}{RC-\tau}\left(e^{-t/\tau}-\frac{\tau}{RC}e^{-t/RC}\right) \tag{11}$$

Thus it can be seen that the response of Rogowski-coil Current Transformer to the decaying a periodic component is mainly associated with decay time constants and integral time constant. Equation (11) equals to the output results of the sensor using differential model with the integrator.

As shown in **Figure 4**, the maximum instantaneous error increases as the decay time constant increases. Remain the dc decay time constant at 80ms, the relations between maximum instantaneous error current and integral time constant is as **Figure 5** shows, the error is smaller when the integral time constant RC is between 0.05-0.1. It can be found that the maximum instantaneous error at least above 30% in all situations. The experiment and simulation analysis both indicate that it is difficult for the simple RC integrator to satisfy the requirements of protective relaying on transient characteristics of transformer, namely the maximum peak instantaneous of Electronic transformer for transient protection cannot exceed 10%. It is necessary to improve the performance of Rogowski-coil Current Transformer transferring aperiodic components by improving the structure of the integrator.

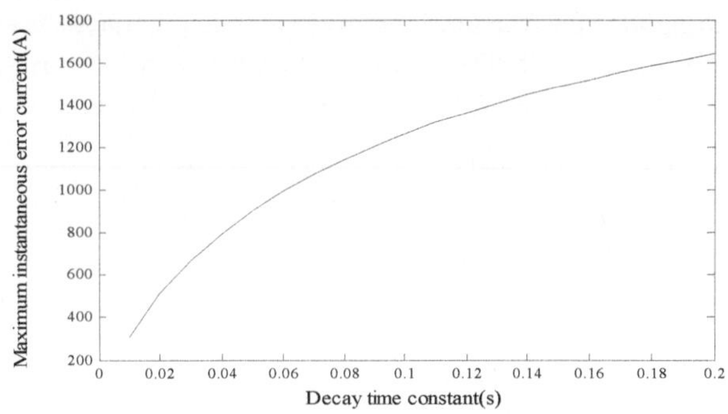

Figure 4. Relationship between the maximum instantaneous error current and the DC decay time constant.

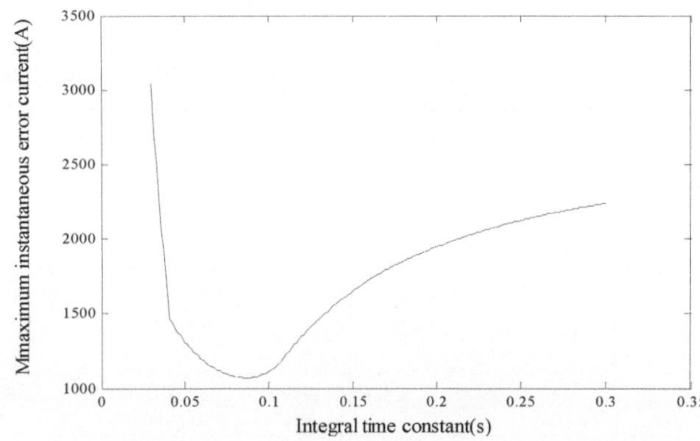

Figure 5. Relationship between the maximum instantaneous error current and the integral time constant.

4. Adaptability Analysis on Rogowski-coil Current Transformer to Protective Relaying

The unsaturation characteristics of Rogowski-coil Current Transformer have provided a solid foundation for its application in protective relaying system. But to replace electromagnetic current transformers not only requires the products have a very good maturity and stability, but also needs further improvement of Rogowski-coil Current Transformer in its performance, standards, and initiative in adapting to the demands of protective relaying.

1) The frequency characteristics of Rogowski-coil Current Transformer have a great effect on transient characteristics. So far the requirements of protective electronic current transformer on the accuracy of harmonics still follows the conventional standards of electromagnetic current transformers, which only consider the transfer accuracy of the harmonic within the 5th harmonic.

This paper believes that the requirements of Rogowski-coil current transformer on harmonic accuracy should not be limited to the harmonic times that protection principles and algorithms apply and should analyse the harmonic features of fault current or the magnetizing inrush current of transformer to determine the bandwidth of Rogowski-coil Current Transformer. To this end, the simulation for transformer no-load closing have been performed in RTDS, the magnetizing inrush current waveforms are shown in **Figure 6**. The harmonic analysis results show that the harmonic ratio of 11th harmonic is up to 2.14%. It indicates that the harmonics accuracy of transformers should at least take harmonics higher than 10th harmonics into account.

2) The lower-cut-off frequency of the band of Rogowski-coil Current Transformer is mainly limited by the integral unit parameters, and the upper cutoff frequency is mainly confined to the resonant frequency of the sensor, as shown in **Figure 3**.

Generally the manufacturers do not measure the distributed capacitance of the Rogowski coil; it is recommended that the resonance method should be used for measuring capacitance. And the specific bandwidth parameter of Rogowski-coil Current Transformer should be provided, which can provide reference for the selection of protective relaying applied.

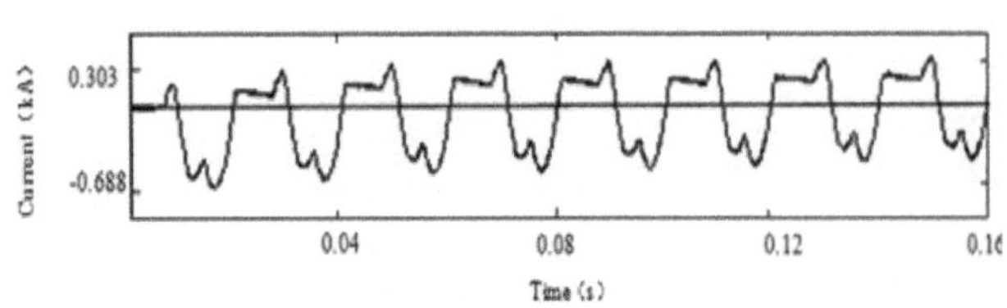

Figure 6. The excitation inrush current waveform from RTDS simulation.

3) The transient characteristics of Rogowski-coil Current Transformer can be quantitatively analyzed. If the frequency band is wide enough, the transient characteristics can be described by the ability to transfer decaying aperiodic component. Thus the design of the integrator for the differential signal of Rogowski transducer is particularly crucial.

This paper proposes that the existing test standards for Rogowski-coil Current Transformer can quantify the transfer characteristics of decaying a periodic component, and have no need to follow the transient standard for electromagnetic current transformer.

It is recommended that the test of transient characteristics can be performed by directly applying decaying dc. Considering the balance of the speed requirements of protective relaying and acceptable measurement errors, the transient errors should be limited to below a certain value (such as 5%) in 40ms after the failures occur.

4) Redundant configuration is widely used in the design and application of Rogowski-coil Current Transformer, such as double A/D design in sampling unit, double configurations for transformers and merging units, two independent synchronization sources, two power standby and so on. Redundant configuration is an effective measure if the hardware or software of devices fail, and the technical difficulty lies in achieving a high degree of coordination of double circuit.

The switch between the external synchronization and the internal synchronization mode in a merging unit device has been tested, the test results are as shown in **Figure 7** below. It can be found that when the synchronization mode of merging unit switches, a part of the data of the earlier cycle were resent and resulted in the distortion of output. Thus, transformer design should ensure effective stand-by and seamless switch; otherwise it may cause the decline of the protection reliability. It is necessary to analyze the most reasonable redundant configuration in the design of protective relaying system.

5. Conclusions

Rogowski-coil Current Transformer has the characteristics of a linear system, it will not saturate and have a

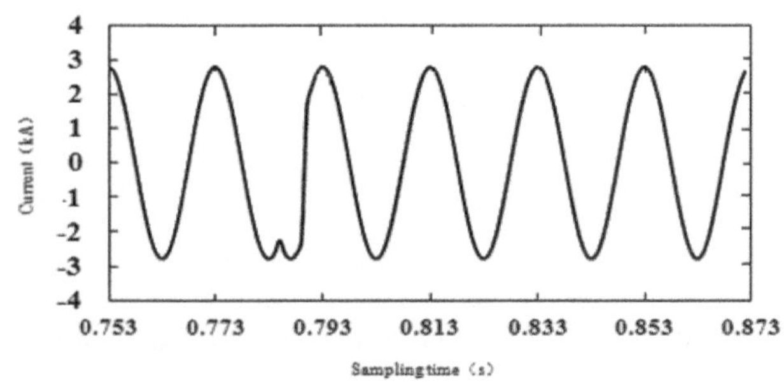

Figure 7. Synchronous mode switch of Merging unit.

good linearity. Internal resistance and distributed capacitance of the sensor has an impact on its high-frequency characteristics, the high-frequency characteristic of the overall transformer is greatly affected by electronic circuits, especially the phase compensation circuit, other than limited by the sensor.

The errors of Rogowski-coil Current Transformer transferring the transient currents largely depend on whether the integrator can accurately restore the differential signal of the sensor or not. This research suggests that first-order RC integrator cannot meet the protective relaying requirements on transient characteristics of transformer. Following research on appropriate way of integration will be performed facing the protective relaying application.

This paper put forward some proposals on structure design and testing standards of the Rogowski-coil Current Transformer to improve transfer performance of transformers, so as to better adapt to application requirements of protective relaying.

REFERENCES

[1] D. A. Ward and J. La T. Exon, "Using Rogowski Coils for Transient Current Measurements," *Engineering Science and Education Journal*, Vol. 2, No. 3, 1993, pp. 105-113.

[2] L. Kojovic, "Rogowski Coils Suit Protective Relaying and Measurement," *IEEE Computer Application in Power*, Vol. 10, No. 2, 1997, pp. 47-52.

[3] J. D. Ramboz, "Machinable Rogowski Coil Design and Calibration," *IEEE Transactions on Instrumentation and Measurement*, Vol. 45, No. 2, 1996, pp. 511-515.

[4] S. N. Luo, Z. B. Tian and X. C. Zhao, "Performance analysis of Rogowski-coil Current Transformer," *Proceedings of the CSEE*, Vol. 24, No. 3, 2004, pp. 108-113.

[5] Q. Chen, H. B. Li, M. M.Zhang, *et al.*, "Design and CharActeristics of Two Rogowski Coils Based on Printed Circuit Board," *IEEE Transaction on Instrumentation and Measurement,* Vol. 55, No. 3,2006, pp. 939-943.

[6] C. Y. Wang, Y. P. Chen, G. Zhang, *et al.*, "Analysis and Control on the Position Errors of PCB Air-core Coils," *Proceedings of the CSEE,* Vol. 55, No. 3, 2008, pp. 103-108.

[7] E. F. Donaldson, J. R. Gibson, G. R. Jones, *et al.*," Hybrid Optical Current Transformer with Optical and Power-line Energisation," *IEE Proceedings-Generation Transmission and Distribution*, Vol. 147, No. 5, 2000, pp. 304-309.

[8] G. Yang and X. R. Liu, "Improvements on Circuit Design for Highvoltage Side of Active Power Electronic Current Transducer," *Power System Technology*, Vol. 29, No. 14, 2005, pp. 81-84.

[9] K. W. Zhang,, N. Wang, X. Y. Duan, *et al.,* "A Digital Integrator for Electronic Current Transducer," *Proceedings of the CSEE,* Vol. 24, No. 12, 2004, pp. 104-107.

[10] Y. Wang, B. d. Cao and T. Jiang, "Fast Data Processing for Merging Unit of Electronic Transformer," *Power System Technology*, Vol. 33, No. 1, 2009, pp. 87-91.

[11] W.B. Li, C.X. Mao, J.M. Lu, *et al.*, "Study of the Influence of the Distributed Capacitance on Dynamic Property of Rogowski Coil," Transactions of China Electrotechnical Society, Vol. 19, No. 6, 2004, pp. 12-17.

[12] Y. Xu and B. Ran, "Analysis and Simulation of Air-core Coil Transducer's Response to Transient Current," *Transactions of China Electrotechnical Society*, Vol. 19, No. 11, 2004, pp. 61-65.

[13] G. Zhang, C.Y. Wang and Y. P. Chen, "Transient Behavior of PCB Air-core Coil Current Transformer," *Transactions of China Electrotechnical Society,* Vol. 25, No. 11, 2010, pp. 25-29.

[14] T. Wu, Y. Q. Zhou and W. Gong, "Performance Analysis on Electronic Current Transformer Based on PCB Planar-type of Air Core Coil," *Power System Technology*, Vol. 34, No. 6, 2010, pp. 210-214.

[15] IEC60044-8, "Instrument Transformers: Part 8 Electronic Current Transformers," 2002.

[16] GB/T20840.8-2007, "Instrument Transformers —Part:8: Electronic Current Transformers," 2007.

[17] J. Yu, X. F. Xiong and Y. Zhang, "Research and Simulation on New Reliability Measures for Digital Substation Protection System," *Power System Technology*, Vol. 33, No. 4, 2009, pp. 28-33.

[18] Z. Q. Li, Z.X. Zhou, Y. Huang, *et al.*, "Research on Applicability of Protective Relaying in Digital Substations". *Power System Technology*, Vol. 35, No. 5, 2011, pp. 210-215.

Comparison and Evaluation of Power Factor Correction Topologies for Industrial Applications

John Karatzaferis[1], Nick Papanikolaou[2], Emmanuel Tatakis[1], Michael Loupis[2], John Spanoudakis[3]
[1]Laboratory of Electromechanical Energy Conversion (LEMEC), Department of Electrical and Computer Engineering, University of Patras, Patras, Greece
[2]Department of Electrical Engineering, TEI of Lamia, Lamia, Greece
[3]MILTECH HELLAS S.A., Koropi, Greece

ABSTRACT

Power factor correction is a major issue for all industries, since a typical industrial load is causing current delays, as well as higher order current harmonics. Power factor correction is often mandatory from the power companies, usually by charging the reactive power that the company consumes. Many solutions for power factor correction have been presented in the bibliography; in this paper, the most significant power factor correction topologies will be reviewed and simulated with SABER RD software. Finally, a prototype design will be presented, based on a mass/cost analysis of the selected topologies and with an aim to manufacture 10 kW modules. The main outcome of this work is the feasibility for an SME to manufacture a competitive modular power factor correction product for industrial applications.

Keywords: Power Factor Correction; Industrial Load; Converter; Thermal Design; Mass/Volume Analysis

1. Introduction

A typical industrial unit contains a large number of non-unity power factor loads. Motor drives and ballast lighting are among the most significant industrial loads that cause current displacement angle, as well as higher order harmonic currents. Thus, power factor correction topologies are mandatory in order to reduce the impact of these loads to the grid. Most of the large industrial units have already installed reactive power compensators in order to reduce the reactive power consumed by the unit, and increase the power factor of the whole unit. However, the level of integration of such technologies in small and medium enterprises (SMEs) is not satisfactory. This is mainly due to the fact that purchasing such equipment is not expected to have a justifiable economic payback for enterprises of this size. Thus, a low budget solution needs to be implemented. In this paper, an analysis and evaluation of different three-phase power factor correction topologies for industrial use is carried out. The criteria used for the evaluation are mainly cost related; the topologies are examined with regard to their cost (€/kW), efficiency, and power density (kg/kW).

2. Topologies under Evaluation

The main industrial load is the induction machine, usually controlled by a motor drive, fed by a DC bus. The DC bus is formed by rectifying the grid voltage, normally using a diode bridge. For some special applications, a DC machine is used, which is usually controlled by a 6 pulse thyristor controlled rectifier. In either case, the load causes a delay of the input fundamental harmonic current, as well as higher order harmonics due to its non-linearity. The power factor of a load is given by:

$$pf = \frac{\cos\varphi}{\sqrt{1+THD^2}} \quad (1)$$

where φ is the displacement angle of the current fundamental harmonic and

$$THD = \frac{\sqrt{\sum_{k=2}^{n} I_{k,rms}^2}}{I_{b,rms}} \quad (2)$$

where $I_{k,rms}$ is the *rms* value of the *k*-th order current harmonic and $I_{b,rms}$ is the *rms* value of the fundamental current harmonic.

Ideally, the power factor should be equal to 1, thus, the

angle φ as well as the THD factor should be set to zero. This can be achieved by adding a power factor correction topology, either using a cascade connection scheme (**Figure 1**), or a parallel connection scheme (**Figure 2**). In the cascade connection scheme, a power factor correction rectifier is used is series with the motor drive, substituteing its input stage, which is normally a diode bridge. This leads to unity power factor at the input. On the other hand, using a parallel configuration, a power factor correction topology, (usually an active filter) is placed in parallel with the rectifier input, in order to supply a current, such that the total input current has zero distortion, and zero phase delay.

Here, the most significant topologies for use in a cascade or parallel connection scheme will be briefly presented and an evaluation will be performed, based on simulation results.

2.1. Topologies for Cascade Connection Scheme

2.1.1. Boost Topologies for PFC Operation

The main feature of this PFC family is the boost operation, meaning that the output voltage is higher than the peak line-to-line grid voltage. Depending on the control method and the complexity of the design there is a wide variety of converters.

1) Vector or Direct Power Controlled Boost PFC Rectifiers [1,2]

This type of PFC rectifiers (**Figure 3**) includes a vector or direct power controlled inverter, operating as a rectifier. By controlling the power flow at the d-q axes, unity power factor can be achieved. This family of converters offers high quality PFC operation and good power density and is a good solution for high power applications.

2) PFC Rectifiers with Parallel or Series Connected Dual Boost Topology [3-5]

This family of PFC converters (**Figure 4**) includes the rectifiers with parallel or series connected dual-boost topology. The basic configuration includes a bridge rectifier and three switches, delta or star connected, depending on the operation mode. For any given time interval, one of the three switches remains constantly on and the other two are controlled usually by the one shot control method. This results in two controlled boost rectifiers forming the line currents. Depending on the configuration of the switches, the boost converters may be parallel connected, or series connected (Vienna rectifier).

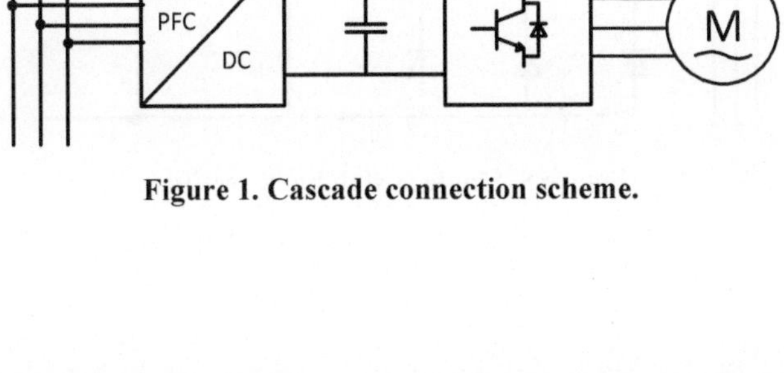

Figure 1. Cascade connection scheme.

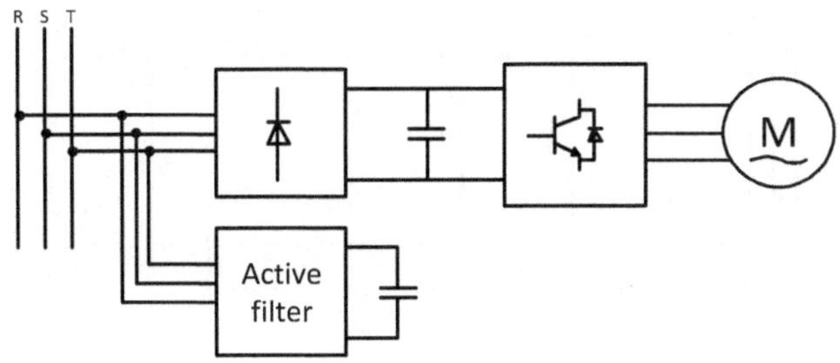

Figure 2. Parallel connection scheme.

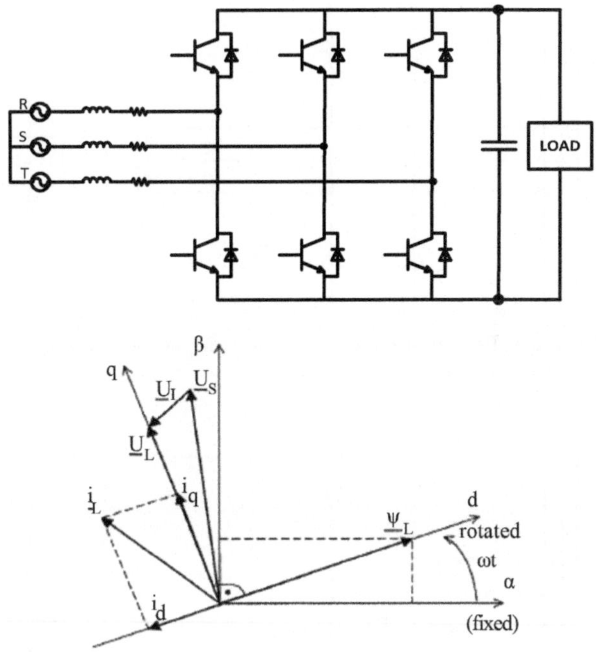

Figure 3. Vector or direct power controlled PFC rectifiers.

3) Half Controlled Boost Rectifiers [6]

This family of PFC rectifiers (**Figure 5**) contains fewer semiconductor switches, offering higher power density and simpler control. However, the control is not effective throughout the whole line period, thus higher order current harmonics appear.

4) Single Switch Boost PFC Rectifiers [7,8]

This topology includes only one PWM operated semiconductor switch (**Figure 6**), hence it results to high power density and simple control. The duty cycle of the switch is always such that the input inductors' currents are fully discharged during every switching cycle. When the switch is on, a virtual, neutral point is created and as a result, the inductors' currents are increased at a rate proportional to the instantaneous values of the respective line voltages. Consequently, at the end of each switching cycle the inductors peak currents can be considered proportional to the instantaneous line voltages, under the

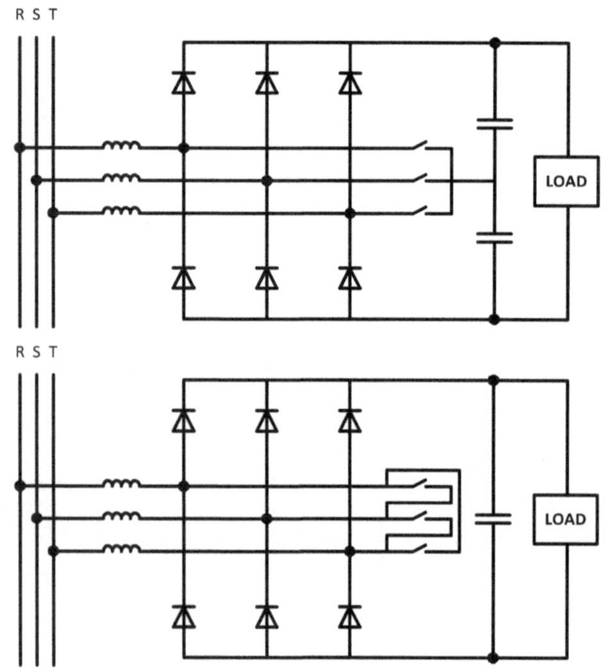

Figure 4. PFC rectifiers with series (upper case) or parallel (lower case) dual boost topology.

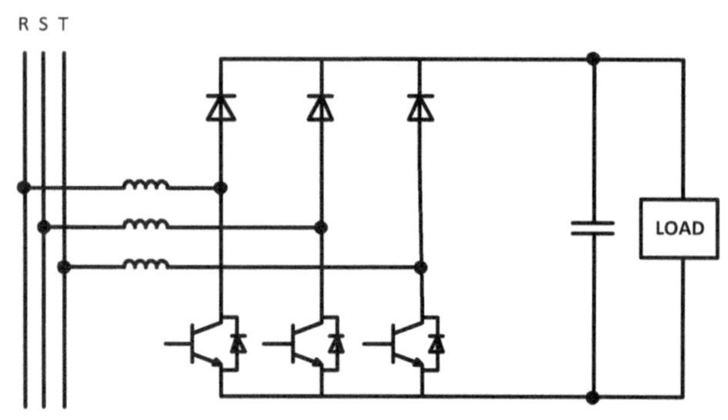

Figure 5. Half controlled boost PFC rectifier.

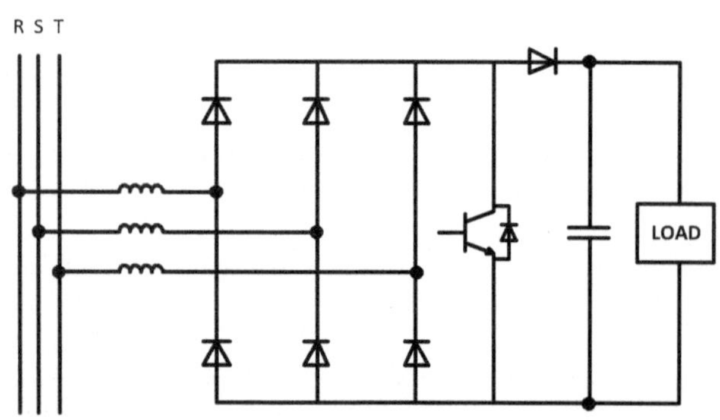

Figure 6. Single switch boost PFC rectifier.

assumption that the switching frequency is significantly higher than the line frequency and—after applying filtering—almost unity power factor can be achieved. It is also worth mentioning that no input voltage sensing is required for operation.

2.1.2. Buck PFC Topologies for Cascade Connection Schemes

When lower voltage is required for specific applications, (electrolysis, battery charging etc.), a buck PFC topology may be used.

1) Fully Controlled Buck Rectifier [9,10]

For this topology (**Figure 7**), the switches are placed at the input, to meet the case of a buck converter. During any given time interval, one switch should remain constantly on, and the rest should be PWM controlled, to achieve unity power factor. Usually the control method is the charge control.

2) Single Switch Buck Rectifier [11]

This topology contains only one semiconductor switch, offering higher efficiency and power density for buck PFC operation (**Figure 8**). However, due to the simple design, higher harmonic current elimination is not as effective as with other topologies.

3) Buck-Boost Topologies [12,13]

This family of converters offers buck-boost capability, meaning that the output voltage has no upper or lower limit (**Figure 9**). However, some topologies may suffer low efficiency, due to the large number of semiconductor switches required for buck-boost operation, and may be unsuitable for high power applications.

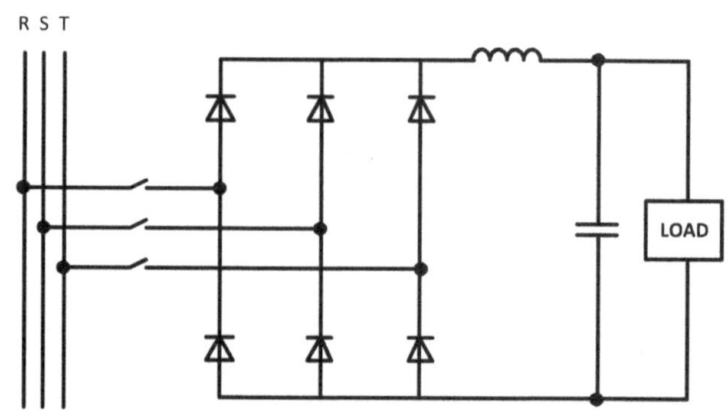

Figure 7. Fully controlled buck rectifier.

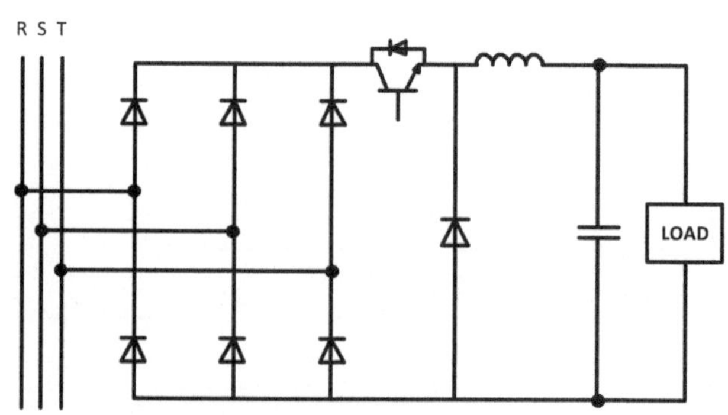

Figure 8. Single switch buck rectifier.

4) Series Active Filters [14]

This family of active filters is connected in series with the non-linear load, acting as a variable voltage source (**Figure 10**). The control method is usually a current control method, such as hysteresis control.

2.2. Topologies for Parallel Connection Schemes

2.2.1. Parallel Active Filter (PAF) [15-18]

The basic function of this topology is the injection of equal compensating currents (**Figure 11**), opposite in phase, to cancel harmonics and/or reactive components of the nonlinear load current at the point of connection. Many control methods have proposed in bibliography, such as direct power control, flux based control, P-I control, sliding mode control, etc.

2.2.2. Capacitor Mid-Point Four-Wire PAF [19,20]

This topology is a parallel active filter suitable for four-wire three phase loads (**Figure 12**). The dc-link consists of two capacitors connected in split. Thus the neutral point is connected directly to the midpoint of the dc-link capacitors.

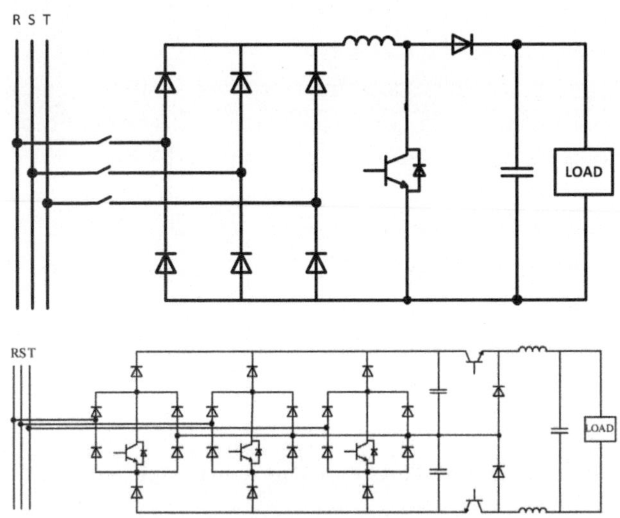

Figure 9. Buck-boost rectifiers.

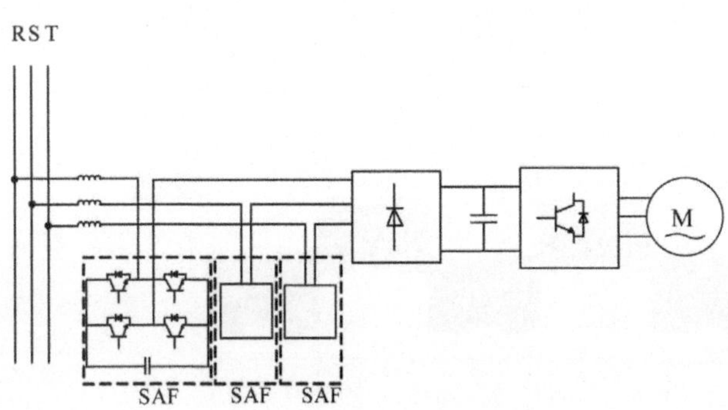

Figure 10. Series active filters.

2.2.3. Four-Pole Four-Wire PAF [21]

This is another variation of the parallel active filter for four-wire load, with an extra branch for the fourth wire (**Figure 13**).

3. Simulation of the above Topologies

All of the above mentioned topologies were simulated

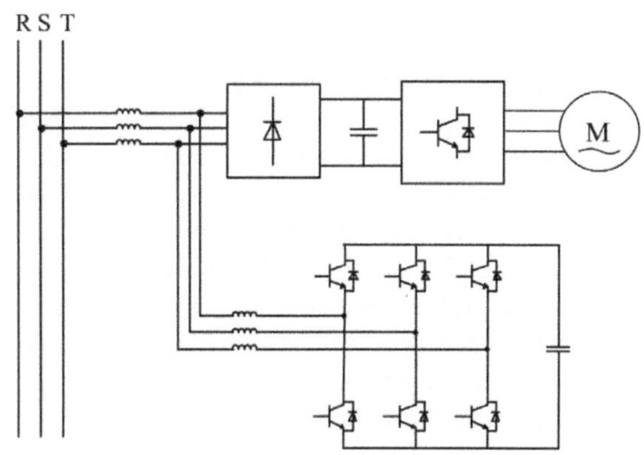

Figure 11. Parallel active filter.

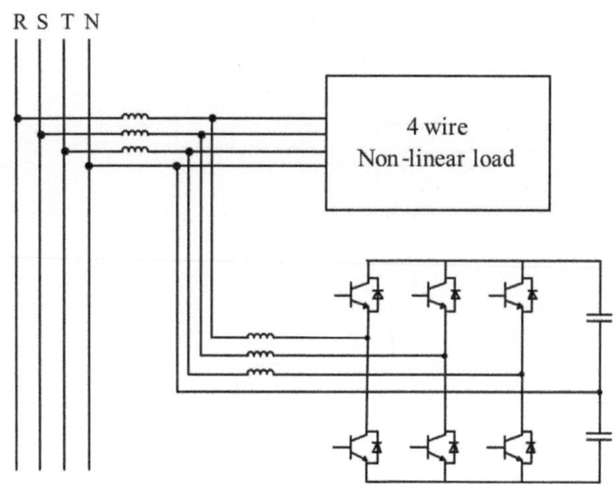

Figure 12. Capacitor Mid-Point active filter.

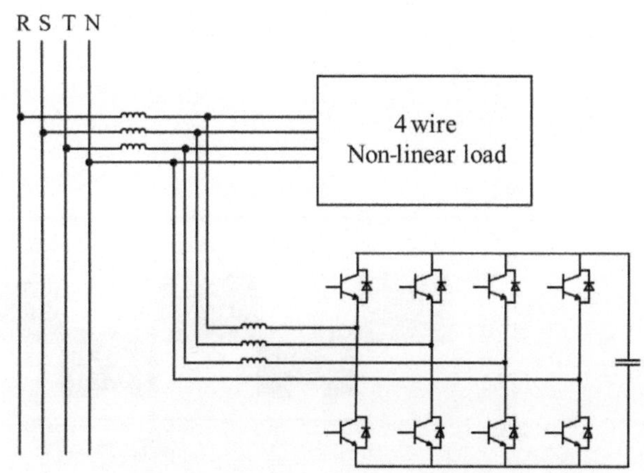

Figure 13. Four-pole four-wire parallel active filter.

using the SABER RD simulation tool. The SABER RD software is a powerful simulation suite that enables the simulation of electrical and mechanical systems. The use of detailed models for semiconductor devices and magnetic components allows the precise calculation of conductance and switching losses, as well as passive component losses (inductors, capacitors, etc.). Using the precise losses calculation, the efficiency as well as the power density (by calculating the coolant required) can be found. **Figures 14-16** present some typical simulation results concerning voltage and current waveforms, for the cases of a buck rectifier and a single switch boost rectifier and for a 10 kW load level. **Figure 14** demonstrates the simulation results for a semiconductor switch used on a buck rectifier. It can be clearly seen that the simulation tool can calculate conduction losses (while the switch is on) as well as switching losses (at the beginning and the end of conduction), thus the power dissipation estimation is adequately accurate. **Figures 15** and **16** demonstrate the good pfc operation of the rectifiers under study.

It should be noted that an extended simulation process for all the converters under study has been conducted and the efficiency calculation took place for the case of a 10 kW load. This power level has been decided in order to cover the majority of industrial loads with a few parallel operating modules (1 - 5 modules).

4. Mass Analysis Procedure

The main parameters that determine mass/volume figures are the necessary cooling system and the passive elements of each converter (filter capacitors and inductors). The present analysis is based on the semiconductor losses calculation through the simulation results given for each converter. It is worth mentioning that the cooling system design for each converter stands for the safe operation of its semiconductor devices. As it concerns filter capacitors and inductors, their mass/volume has been based on commercial products which are in line with the

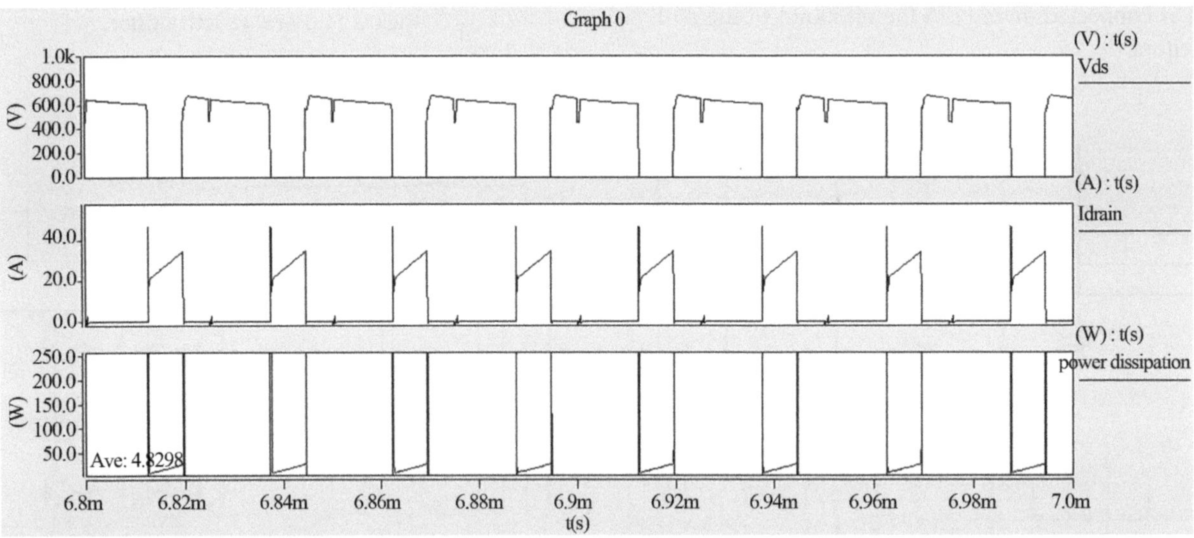

Figure 14. Drain-source voltage, drain current and power dissipation for a Mosfet used on a buck rectifier.

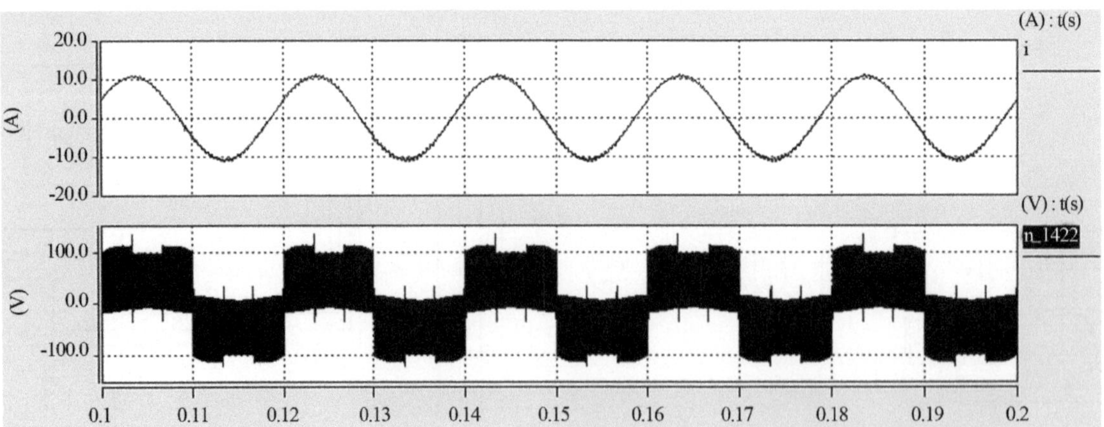

Figure 15. Input current of a buck rectifier unfiltered (lower case) and filtered (upper case).

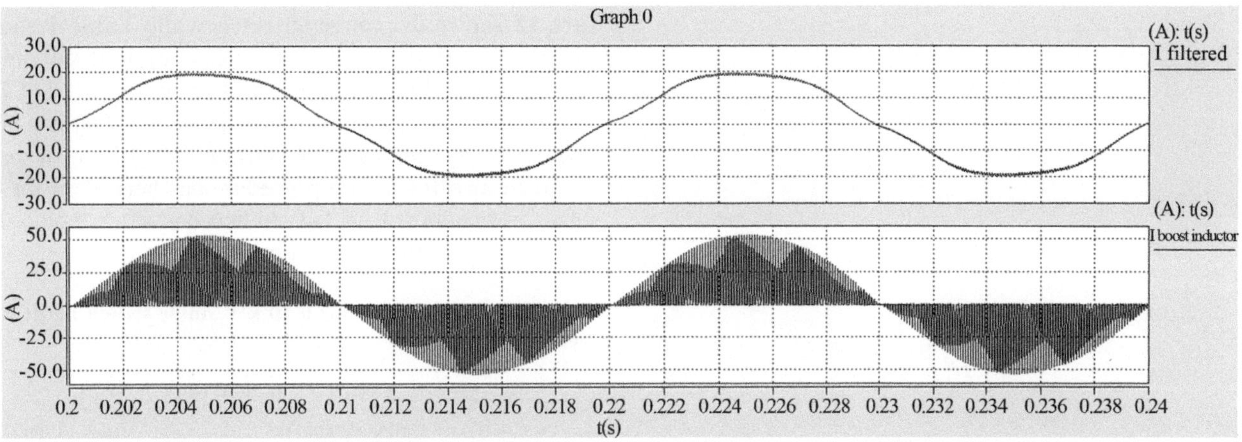

Figure 16. Input current of a single switch boost rectifier unfiltered (lower case) and filtered (upper case).

parameters' values that have used in the simulation process.

Regarding the initial target values, the power density aim has been set to 3,0 kW/kg while the final product price aim has been set to 300 €/kW. The power density critical value has been set by some recent customer specifications (MILTECH HELLAS SA); the final product price target has been set in cooperation with engineers of many well-known Greek industries, in the context of a commercial survey conducted by our team. The focus on the Greek industries is justified by the fact that they constitute the initial target group for this project.

4.1. Thermal Design; Basic Principles

The cooling system design for each converter begins with the definition of the following parameters:

Maximum Junction Temperature, T_j

Ambient Temperature, T_a

Power dissipation (heat, due to power losses), q

Then the following parameters are defined/calculated, according to the fundamental heat transfer calculations [22]:

Conduction Heat Transfer,

$$q_{cond} = \frac{k \cdot A \cdot \Delta T}{H} \tag{3}$$

q_{cond}, Heat flow due to conduction

k, Thermal Conductivity

A, Area normal to heat flow

ΔT, Temperature Difference

H, Height in flow direction

Conduction Thermal Resistance R_{cond},

$$R_{cond} = \frac{H}{k \cdot A} \tag{4}$$

Convection Heat Transfer,

$$q_{conv} = h \cdot A_h \cdot \Delta T \tag{5}$$

q_{conv}, Heat flow due to convection

h, Convection coefficient

A_h, Area of convection

ΔT, Temperature Difference

Convection Thermal Resistance, R_{conv},

$$R_{conv} = \frac{1}{h \cdot A_h} \tag{6}$$

Total system thermal resistance, R_{th},

$$R_{th} = R_{cond} + R_{conv} \tag{7}$$

4.2. Finned Heat Sink Thermal Resistance Calculation

The finned heat sink layout is presented in **Figure 17**, assuming forced cooling due to forced air or forced water circulation; nevertheless, for the case under study (10 kW modules for industrial use) forced air circulation is more convenient and so this will be considered. The finned heat sink thermal resistance can be calculated by Equations (3)-(7), assuming an aluminum heat sink (for lighter and cost effective constructions); R_{cond} can be calculated directly by Equation (3) since the geometry of the heat sink is standardized and the aluminum thermal conductivity is equal to $k = 205\frac{W}{m^{\circ}C}$. As it concerns R_{covn} Equation (6) can be rewritten as [22],

$$\left.\begin{aligned}
R_{conv,fin} &= \frac{1}{h \cdot \left(A_{base} + N_{fin} \cdot n_{fin} \cdot A_{fin}\right)} \\
A_{base} &= \left(N_{fin} - 1\right) \cdot b \cdot L \\
A_{fin} &= 2 \cdot H \cdot L \\
b &= \frac{W - N_{fin} \cdot t_{fin}}{N_{fin} - 1} \\
h &= \frac{N_{ub} \cdot K_{fluid}}{b}
\end{aligned}\right\} \tag{8}$$

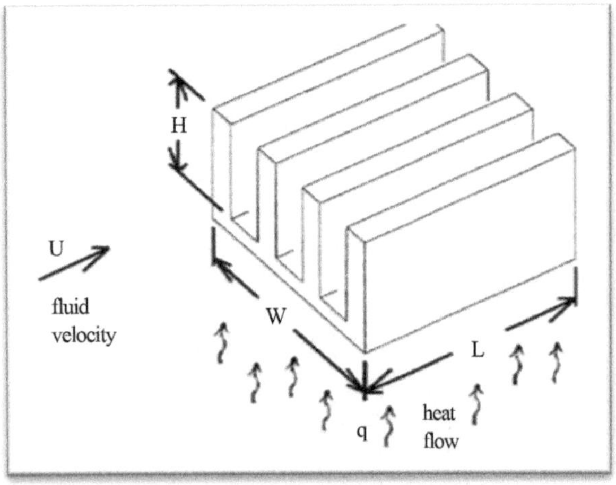

Figure 17. Finned heat sink thermal resistance calculation.

$R_{conv,fin}$, Finned heat sink convection thermal resistance
A_{base}, Exposed surface area
A_{fin}, Heat transfer area
N_{fin}, Number of fins
n_{fin}, Efficiency of Fins
u, H, L, W, as in **Figure 16**
b, Gap between fins
t_{fin}, Fin width
K_{fluid}, Thermal Conductivity of fluid (air for the specific application)
N_{ub}, Nusselt Number (Function of Reynolds and Prandtl Numbers, which depends also on the finned heat sink structure

Finally the fan selection (for forced air cooling) is based on the air pressure drop calculation, ΔP, as shown in **Figure 18**. Using the calculations of the thermal resistance and the pressure drop, the thermal design software tool used, comes up with the minimum geometric characteristics of the heat sink as well as the appropriate fan (among the ones that are registered in the relevant electronic library) that is needed in order to obtain a lower value than the power dissipation of the converter. The mass of the cooling system is calculated as a function of the geometric characteristics of the heat sink and its material, as well as the weight of the forced heat dissipation subsystem (if it is applied). A thermal design example for the converters under study is given in the following section.

4.3. Thermal Design Example

The previously presented thermal design procedure has been adopted by MILTECH HELLAS SA, for the development of its power converter products. In more details, a thermal design software tool has been developed, which calculates the necessary thermal management indexes (according to the customer settings) and comes up with the final cooling system setup. As an example, **Figure 18** shows the converter setup while **Table 1** shows the mass analysis details for the case of a 10 kW single switch boost PFC rectifier, according to MILTECH's prototype tool. The converter efficiency (97%) has been calculated by simulating it with the SABER software. It is noted that this design procedure has been used for the mass estimation of all the studied converter topologies, as presented in the next section.

Table 1. Mass analysis for a 10 kW single switch boost PFC rectifier.

Electrical Characteristics		
Output Power (W)		10000
Efficiency		97%
Maximum Junction Temperature (T_j)		100
Ambient Temperature (T_a)		55
Input Power (Calculated in W)		**10309,278**
Power Dissipated		**309,278**
Maximum Thermal Resistance (°C/W)		**0.146**
Inductor Data		
3 × (8 uH-100 A)	**1000**	**gr**
1 × (250 uH-45 A)	**1500**	**gr**
Total Weight:	**2500**	**gr**
Circuit PCB Data (Capacitors Included)		
Weight:	**200**	**gr**
Ventilation Characteristics		
Type FAN	Single inlet centrifugal fan	
Flow	0.05, 0.07, 0.09	m^3/s
Pressure	300.00, 200.00, 0.00	Pa
Height	246	mm
Width	98	mm
Length	226	mm
Weight	**2200**	**Kgr**
Heat Sink Characteristics		
Material	Aluminium 6061	
Themal Conductivity	250	W/m°C
Density	2.71	gr/cm^3
W	98	mm
H	55	mm
L	150	mm
T_{fin}	2	mm
Number of Fins	15	mm
Weight	**796**	**gr**
Total Thermal Resistance	**0.097**	(°C/W)
Total Weight (10 kW)	**5696**	gr
Total Weight (30 kW)	**17,088**	gr
Power Density	**1756**	kW/kgr

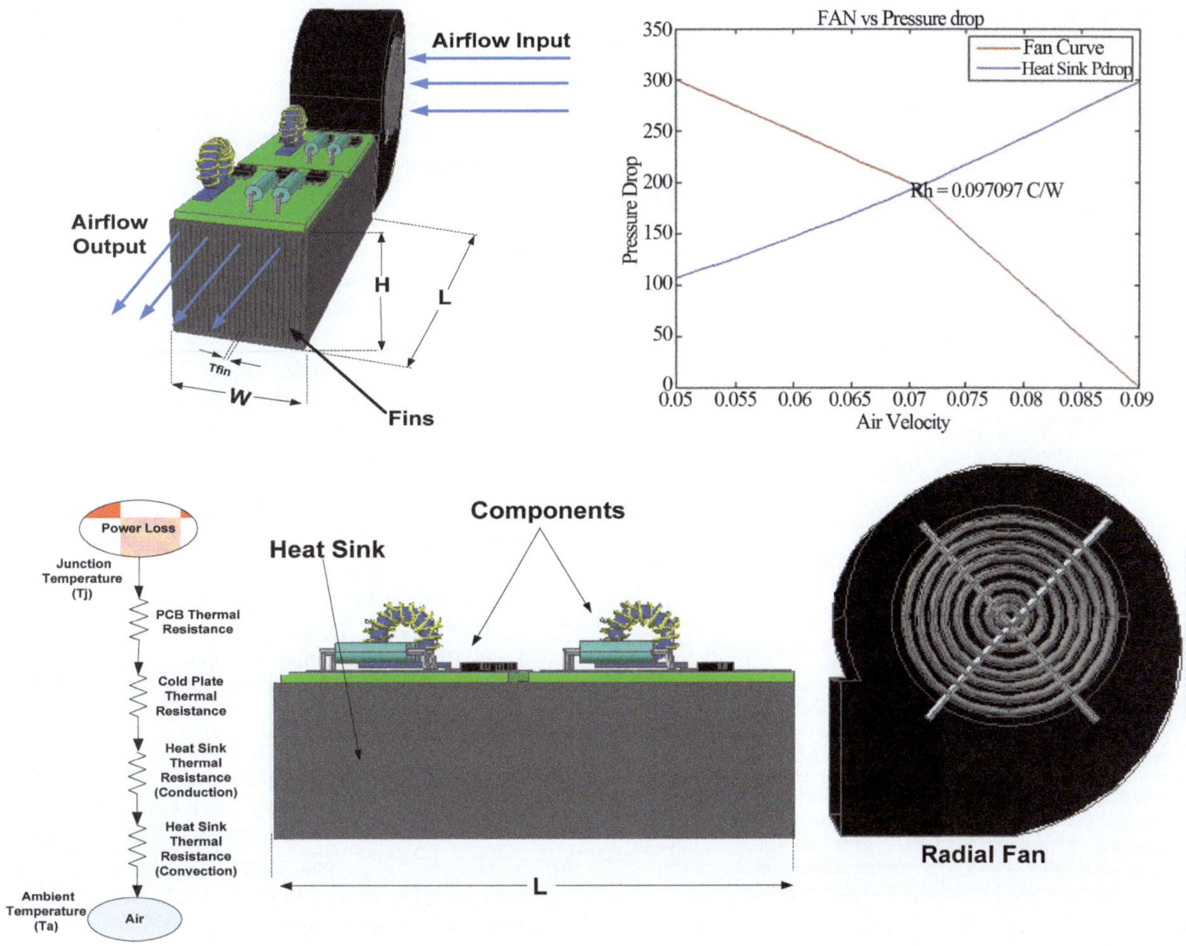

Figure 18. Thermal design example of a 10 kW single switch boost PFC converter.

5. Comparison of the Selected Topologies

A brief demonstration of the simulation as well as the mass/cost analysis results can be seen in **Table 2**, for all of the topologies under study. The simulations and the efficiency calculations were performed using the SABER software, as already discussed in Section 3, for a 10 kW load power and the criteria used for the evaluation of the topologies were cost related. Furthermore, final product price calculation has been done considering three years payback time for the manufacturing company. This is a very critical point due to the fact that the manufacturing company is an SME and so its resources are limited.

6. Conclusions

A detailed study of power factor correction topologies for industrial applications has been presented in this paper. The most appropriate converter topologies have been analyzed, simulated with SABER RD software and designed as industrial prototypes. The outcome results of the mass/cost analysis show that 3.0 kW/kg power density is a realistic target even without using expensive forced water cooling methods. It is noted that the power density aim of 3.0 kW/kg has been initiated by the electric vehicle and the aircraft industry and it is now a trend that is followed by the majority of the competitive converter manufacturers. However, the most prominent converter topology seems to be the Parallel Active Filter, since it demonstrates the maximum power density, while its expected product price is in the same range with the PFC rectifiers. It is noted that this result stands only for the case of industrial loads (supplied by the three phase low voltage network) and for 10 kW modules, mainly due to the fact that PAF topologies handle only the reactive power part of the industrial load. Nevertheless, it can be deduced that for lower power levels (e.g. PC-TV power supplies, electronic ballast) PFC rectifiers may become more competitive in terms of power density and cost. Additionally, this topology gives the flexibility to manufacture a modular product that can be used in existing industrial loads (parallel connection) and build up simply by adding multiple modules (the only limitation is an interconnection signal among the parallel modules in order to share the load reactive power). Finally, with re-

Table 2. Comparison of the above mentioned topologies.

	Topology	Efficiency	Power density (kW/kg)	Expected Product Price (€/kW)
1	Vector or direct power controlled PFC rectifiers	97%	1.8 - 2.3	300 - 400
2	PFC rectifiers with parallel or series connected dual boost topology	97%	1.8 - 2.3	200 - 250
3	Half controlled Boost rectifiers	95%	1.2 - 1.5	150 - 200
4	Single switch boost PFC rectifiers	97%	1.8 - 2.3	150 - 200
5	Fully controlled buck rectifier	92%	1.1 - 1.5	200 - 250
6	Single switch buck rectifier	93%	1.3 - 1.6	200 - 250
7	Buck boost topologies	91%	1.1 - 1.3	350 - 400
8	Series active filter (*)	93%	1.9 - 2.3	200 - 250
9	Parallel active filter (*)	98%	3.5 - 4.5	200 - 250
10	Capacitor Mid-point four-wire PAF (*)	97%	3.5 - 4	200 - 250
11	Four-pole four-wire PAF (*)	96%	3.5 - 4	250 - 300

(*)These converters are handling only reactive power; the 10 kW active power in these cases refers to the nominal active power of the industrial load that is compensated. It is assumed that the industrial load is supplied through a three phase diode rectifier, which is the common case for industrial inverters (VVVF converters for speed control) as discussed in Section 2.

gard to the expected final product price it has been proven to be competitive for the Greek industries—which are the initial target group of this project—according to a relevant commercial survey conducted by MILTECH.

Finally, as it concerns the future direction in the field of power factor correction topologies for industrial use, the most prominent target is the introduction of silicon carbide switches (both diodes and transistors); this may lead to substantial power losses reduction and conesquently to power density values higher than 5 kW/kg.

7. Acknowledgements

This work was supported by the "Reinforcement program of small and medium enterprises for research and technological development activities" under research project 23SMEs2009 and funded by the European social fund and by national funds.

REFERENCES

[1] J. L. Duarte, "Reference Frames Fit for Controlling PWM Rectifiers," *IEEE Transactions on Industrial Electronics*, Vol. 46, No. 3, 1999, pp. 628-630.

[2] M. Malinowski, *et al.*, "Virtual Flux Based Direct Power Control of Three-Phase PWM Rectifiers," *IEEE Transactions on Industry Applications*, Vol. 37, No. 4, 2001, pp. 1019-1027.

[3] Q. Chongming and K. M. Smedley, "A General Three-Phase PFC Controller for Rectifiers with a Series-Connected Dual-Boost Topology," *IEEE Transactions on Industry Applications*, Vol. 38, No. 1, 2002, pp. 137-148.

[4] Q. Chongming and K. M. Smedley, "A General Three-Phase PFC Controller for Rectifiers with a Parallel-Connected Dual Boost Topology," *IEEE Transactions on Power Electronics*, Vol. 17, No. 6, 2002, pp. 925-934.

[5] J. C. Salmon, "Reliable 3-Phase PWM Boost Rectifiers Employing a Stacked Dual Boost Converter Subtopology," *IEEE Transactions on Industry Applications*, Vol. 32, No. 3, 1996, pp. 542-551.

[6] J. Kikuchi, M. D. Manjrekar and T. A. Lipo, "Performance Improvement of Half Controlled Three-Phase PWM Boost Rectifier," *The 30th Annual IEEE of Power Electronics Specialists Conference (PESC 99)*, Vol. 1, Charleston, August 1999, pp. 319-324.

[7] A. R. Prasad, P. D. Ziogas and S. Manias, "An Active Power Factor Correction Technique for Three-Phase Diode Rectifiers," *IEEE Transactions on Power Electronics*, Vol. 6, No. 1, 1991, pp. 83-92.

[8] N. Takeuchi, "A Novel PFC Circuit for Three-Phase Utilizing a Single Switching Device," *IEEE 30th International Telecommunications Energy Conference (INTELEC 2008)*, San Diego, 14-18 September 2008, pp. 1-5.

[9] Y. Nishida, "Three-Phase PWM-Current-Source Type PFC Rectifier (Theory and Practical Evaluation of 12kW Real Product)," *Power Conversion Conference (PCC 2002)*, Vol. 3, Osaka, 2002, pp. 1217-1222.

[10] K. Wang, D. Boroyevich and F. C. Lee, "Charge Control of Three-Phase Buck PWM Rectifiers," *The 30th Annual IEEE of Applied Power Electronics Conference and Exposition (APEC 2000)*, Vol. 2, New Orleans, 2000, pp. 824-831.

[11] J.-H. Song, *et al.*, "A Pulse Frequency Modulation Control Method for Single-Switch Three-Phase Buck Rectifiers," *The 20th International Telecommunications Energy Conference*, 1998, San Francisco, pp. 231-236.

[12] D. Bortis, *et al.*, "25kW 3-Phase Unity Power Factor Buck Boost Rectifier with Wide Input and Output Range for Pulse Load Applications," *The 16th IEEE International Pulsed Power Conference*, Albuquerque, 17-22 June 2007, pp. 1505-1508.

[13] T. Nussbaumer and J. W. Kolar, "Comparison of 3-Phase Wide Output Voltage Range PWM Rectifiers," *IEEE Transactions on Industrial Electronics*, Vol. 54, No. 6, 2007, pp. 3422-3425.

[14] Z. G. Pan, F. Z. Peng and S. L. Wang, "Power Factor Correction Using a Series Active Filter," *IEEE Transactions on Power Electronics*, Vol. 20, No. 1, 2005, pp. 148-153.

[15] S. Saetieo, R. Devaraj and D. A. Torrey, "The Design and Implementation of a Three-Phase Active Power Filter Based on Sliding Mode Control," *IEEE Transactions on Industrial Applications*, Vol. 31, No. 5, 1995, pp. 993-

1000.

[16] S. Bhattacharya, A. Veltman, D. M. Divan and R. D. Lorenz, "Flux-Based Active Filter Controller," *IEEE Transactions on Industrial Applications*, Vol. 32, No. 3, 1996, pp. 491-502.

[17] S. Väliviita and S. J. Ovaska, "Delayless Method to Generate Current Reference for Active Filters," *IEEE Transactions on Industrial Electronics*, Vol. 45, No. 4, 1998, pp. 559-567.

[18] A. Chandra, B. Singh, B. N. Singh and K. Al-Haddad, "An Improved Control Algorithm of Shunt Active Filter for Voltage Regulation, Harmonic Elimination, Power-Factor Correction, and Balancing of Nonlinear Loads," *IEEE Transactions on Power Electronics*, Vol. 15, No. 3, 2000, pp. 495-507.

[19] B. Singh, K. Al-Haddad and A. Chandra, "A Review of Active Filters for Power Quality Improvement," *IEEE Transactions on Industrial Electronics*, Vol. 46, No. 5, 1999, pp. 960-971.

[20] R. Griño, "Digital Repetitive Control of a Three-Phase Four-Wire Shunt Active Filter," *IEEE Transactions on Industrial Electronics*, Vol. 54, No. 3, 2007, pp. 1495-1503.

[21] P. Salmeron and S. P. Litran, "A Control Strategy for Hybrid Power Filter to Compensate Four-Wires Three-Phase Systems," *IEEE Transactions on Power Electronics*, Vol. 25, No. 7, 2010, pp. 1923-1931.

[22] F. Incropera and D. DeWitt, "Fundamentals of Heat and Mass Transfer," 6th Edition, J. Wiley & Sons, New York, 2007.

15

Wind Power Forecasting using an Artificial Neural Network for ASPCS

Kazuma Hanada[1], Takataro Hamajima[1], Makoto Tsuda[2], Daisuke Miyagi[2], Takakazu Shintomi[3], Tomoaki Takao[4], Yasuhiro Makida[5], Masataka Kajiwara[6]

[1]Department of Electrical and Electronics Systems, Hachinohe Institute of Technology, Aomori, Japan
[2]Department of Electrical Engineering, Tohoku University, Miyagi, Japan
[3]Advanced Research Institute for the Sciences and Humanities, Nihon University, Tokyo, Japan
[4]Department of Engineering and Applied Sciences, Sophia University, Tokyo, Japan
[5]High Energy Accelerator Research Organization, Ibaraki, Japan
[6]Iwatani Corporation, Tokyo, Japan

ABSTRACT

In order to use effectively renewable energy sources, we propose a new system, called Advanced Superconducting Power Conditioning System (ASPCS) that is composed of Superconducting Magnetic Energy Storage (SMES), Fuel Cell-Electrolyzer (FC-EL), hydrogen storage and DC/DC and DC/AC converters in connection with a liquid hydrogen station for fuel cell vehicles. The ASPCS compensates the fluctuating electric power of renewable energy sources such as wind and photovoltaic power generations by means of the SMES having characteristics of quick response and large Input-Output power, and hydrogen energy with FC-EL having characteristics of moderate response and large storage capacity. The moderate fluctuated power of the renewable energy is compensated by a trend forecasting method with the Artificial Neural Network. In case of excess of the power generation by the renewable energy to demand it is converted to hydrogen with EL. In contrast, shortage of the electric power is made up with FC. The faster fluctuation power that cannot be compensated by the forecasting method is effectively compensated by SMES. In the ASPCS, the SMES coil with an MgB_2 conductor is operated at 20 K by using liquid hydrogen supplied from a liquid hydrogen tank of the fuel cell vehicle station. The necessary storage capacity of SMES is estimated as 50 MJ to 100 MJ depending on the forecasting time for compensating fluctuation power of the rated wind power generation of 5.0 MW. As a safety case, a thermo-siphon cooling system is used to cool indirectly the MgB_2 SMES coil by thermal conduction. In this paper, a trend forecasting result of output power of a wind power generation and the estimated storage capacity of SMES are reported.

Keywords: Renewable Energy; Artificial Neural Network; Forecasting; SMES; Liquid Hydrogen; Fuel Cell and Electrolyzer

1. Introduction

The applications of renewable energy are needed in the context of climate change and energy resource. But, when renewable energy generations, such as wind and photovoltaic power generations, affected by weather condition are introduced in large quantities, an existing power system may become unstable. Therefore, we propose a new system shown in **Figure 1**, called Advanced Superconducting Power Conditioning System (ASPCS) that is composed of Superconducting Magnetic Energy Storage (SMES), Fuel Cell-Electrolyzer (FC-EL), hydrogen storage and DC/DC and DC/AC converters in connection with a liquid hydrogen station for fuel cell vehicles [1-3].

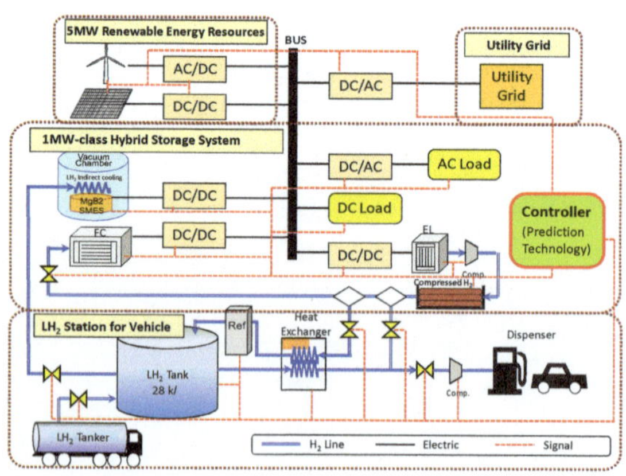

Figure 1. Advanced Superconducting Power Conditioning System.

The ASPCS compensates the fast fluctuating electric power of the renewable energy sources by means of the SMES having characteristics of quick response and large Input-Output power, and hydrogen energy with FC-EL having characteristics of moderate response and large storage capacity. The moderate fluctuated power of the renewable energy is compensated by a trend forecasting with the Artificial Neural Network (ANN) [4, 5]. In case of excess of the power generation by the renewable energy to demand it is converted to hydrogen with EL. In contrast, in case of shortage the electric power is made up with FC. The faster fluctuation that cannot be compensated by the forecasting method is effectively compensated by SMES. In the ASPCS, the SMES coil with an MgB_2 conductor is operated at 20 K by using liquid hydrogen supplied from a liquid hydrogen tank of the fuel cell vehicle station. The necessary storage capacity of SMES is estimated as 50 MJ to 100 MJ depending on the forecasting time for compensating fluctuation of the rated wind power generation of 5.0 MW. As a safety, a thermo-siphon cooling system is used to cool indirectly the MgB_2 SMES coil by thermal conduction.

In this paper, a forecasting result of output power of a wind power generation and the necessary storage capacity of SMES are reported.

2. Artificial Neural Network

ANN is a model of biological neural network, and is composed of a lot of artificial neurons that collect output signals through a transfer function of Equation (1) when they are received input signals.

$$f(x) = \tanh(x) = \frac{e^{x} - e^{-x}}{e^{x} + e^{-x}} \quad (1)$$

The ANN shown in **Figure 2** is a layered ANN. The layered ANN is formed three layers; an input layer that receives input signal, an output layer that produces outputs signals and a hidden layer that makes processes of signals.

Weights are set in connection between neurons. The outputs of each neuron inputs into other neurons after they are multiplied by the weights. Since the weights are usually initialized with random values, they are adjusted to get desirable outputs. This is a learning process. A back propagation method that is one of the learning methods of ANN is used in this paper. The ANN can find relations between input data and the output data easily by the learning process.

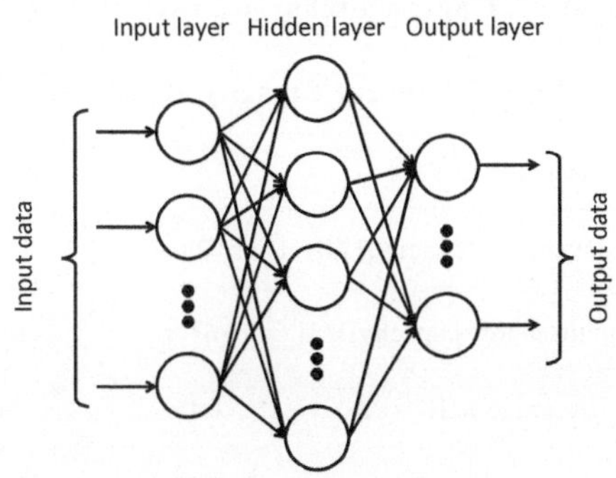

Figure 2. Layered ANN.

3. K-means Method

A k-means method is one of clustering methods to perform data classification [6, 7]. And the method classifies given data into k clusters. The step of k-means method is as follows:

(i) Determine number k of the clusters.

(ii) Assign clusters for the data at random.

(iii) Calculate the center of each cluster with the assigned data.

(iv) Calculate distances between the data and the cluster-centered, and assign the data to the nearest cluster again.

(v) When the allotment of all data into the clusters does not change by the process mentioned above, the calculation is over. Otherwise, the above process is repeated after the cluster centers are recalculated from newly assigned clusters.

4. Simulation

4.1. Output Power Forecasting

In this study, a wind power generation of rated output 5 MW having time series data of the output power shown in **Figure 3** is used. The future trend output is forecasted for 50 seconds needed for control of FC-EL by the ANN model shown in **Figure 4**.

Let the output power time series of the wind power be $\{x_1, x_2, \ldots, x_n\}$. The ANN learns relation between x_t, $x_{t-10}, \ldots, x_{t-60}$ that are the output power at present time t to 60 seconds before and x_{t+50} that is the output power at 50

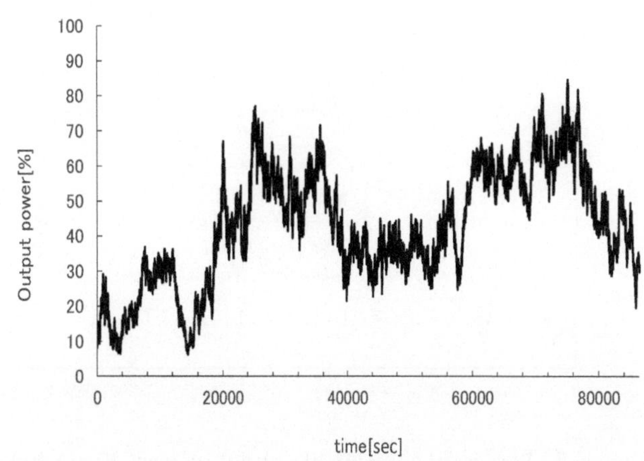

Figure 3. Output power of a wind power generation.

seconds from the present time, and forecasts output power 50 seconds in future. The forecasting result by the ANN is shown in **Figures 5**, **6** and **Table 1**. **Table 1** includes the result of moving average that is calculated by $x_t = (x_{t-10}+ x_{t-20}+ x_{t-30}+ x_{t-40}+ x_{t-50})/6$ to compare with the result by the ANN. The ANN can forecast the output power of the wind power generation at 50 seconds in future, and it is shown that the result of the ANN is better than that of the moving average.

4.2. The Necessary Storage Capacity of SMES

When FC-EL is operated based on the forecast mentioned above, charging or discharging power by SMES compensates the difference between actual output power and the forecasted output power as shown in **Figure 7**. The required storage capacity of SMES shown in **Table 2** is determined by alternately charging or discharging power. In addition, distribution of the charging or discharging power by SMES is shown in **Figures 8** and **9**. From the above result, the storage capacity of SMES using the ANN is smaller than that using moving average.

5. Conclusions

In this paper, the output power of the wind power generation is forecasted for ASPCS that is composed wind

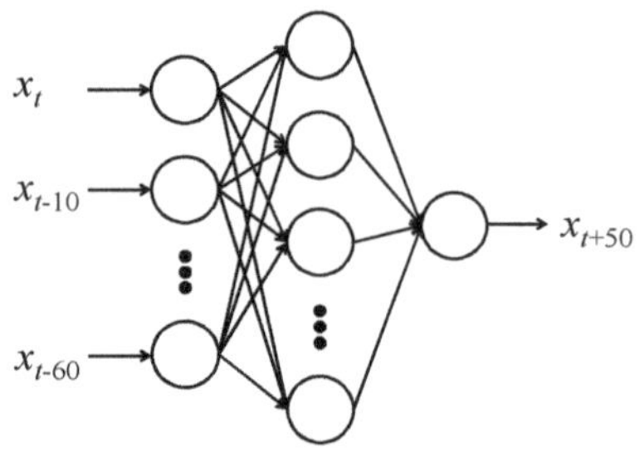

Figure 4. The ANN forecasting output power of a wind power generation.

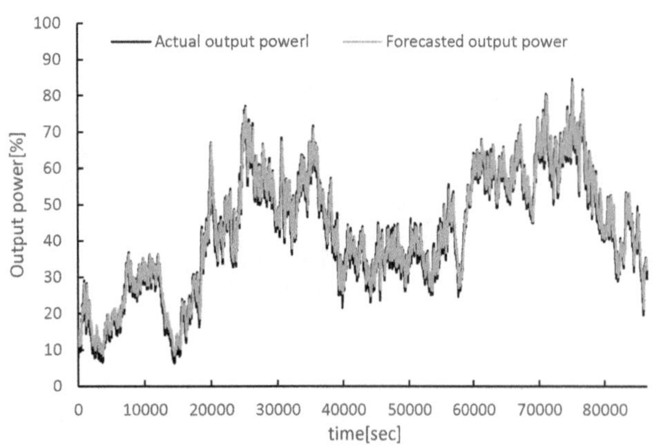

Figure 5. Forecasting result of output power of a wind power generation by ANN.

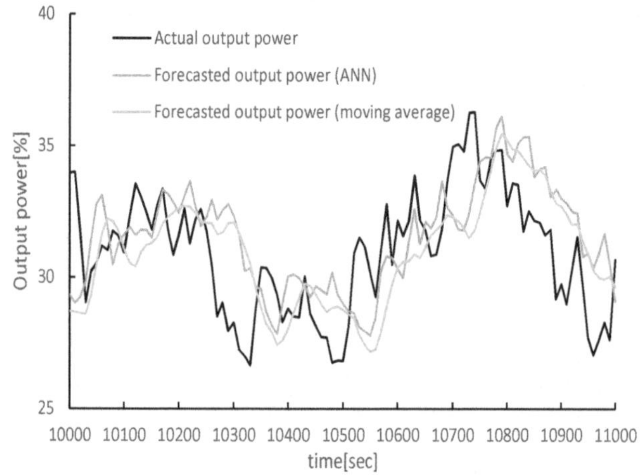

Figure 6. Forecasting result of output power of a wind power generation by ANN (between 10000 and 11000 seconds).

Table 1. Forecasting result of output power of a wind power generation.

Forecasting method	ANN	Moving average
Average error [MW]	0.028	–0.001
Distribution [(MW)^2]	0.025	0.028
Maximum absolute error [MW]	0.685	0.725

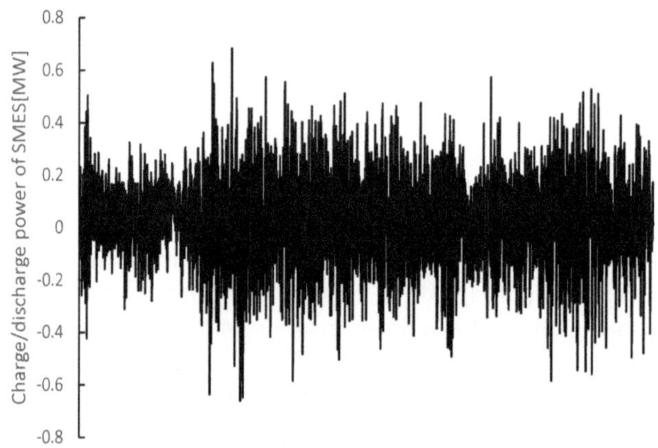

Figure 7. Charge/discharge power of SMES.

Table 2. Storage capacity of SMES.

Forecasting method	ANN	Moving average
Maximum charge capacity[MJ]	72.0	73.5
Maximum discharge capacity[MJ]	-61.8	-83.4
Average[MJ]	1.45	-0.05
Absolute average[MJ]	6.47	7.83

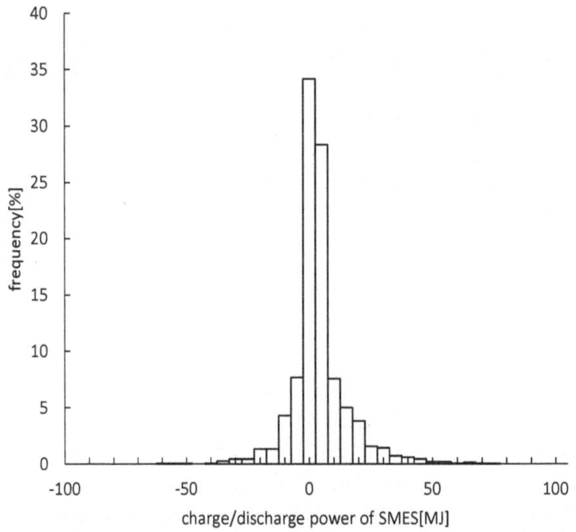

Figure 8. Distribution of charging/discharging energy of SMES (ANN).

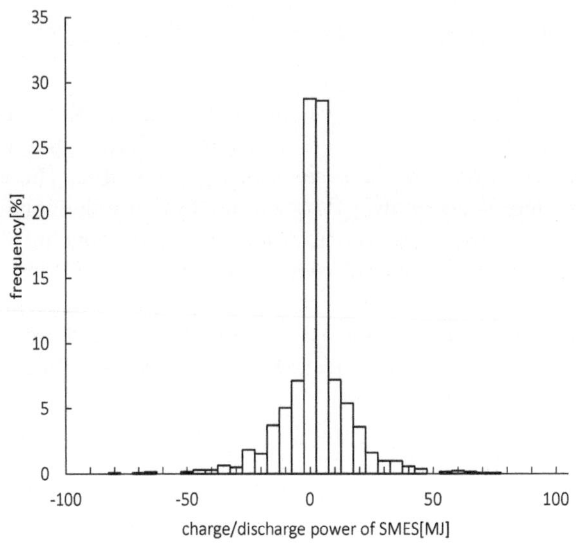

Figure 9. Distribution of charging/discharging energy of SMES (moving average).

power generation, SMES and FC-EL and supplies smoothed power from this system. As a result, the ANN has better accuracy than the moving average method and is able to forecast the output power, and the required storage capacity of SMES is smaller. In the future study, we optimize structure of the ANN and improve the selection method of learning data.

6. Acknowledgements

This work is supported by the Advanced Low Carbon Reduction Technology R&D of Japan Science and Technology Agency.

REFERENCES

[1] T. Hamajima, *et al.*, "Application of SMES and Fuel Cell System Combined with Liquid Hydrogen Vehicle Station to Renewable Energy Control," *IEEE Transactions on Applied Superconductivity,* Vol. 22, No. 3, 2011, p. 5701704.

[2] T. Shintomi, *et al.*, "Design Study of SMES System Cooled by Thermo-siphon with Liquid Hydrogen for Effective Use of Renewable Energy," *IEEE Transaction Supercond.*, Vol. 22, No. 3, 2011, p. 5701604.

[3] T. Nakayama, *et al.*, "Optimization of SMES Compensation Capacity for Stochastic Power Using a Kalman Filter," *TEION KOGAKU*, Vol. 45, No. 3, 2010, pp. 99-106.

[4] D. E. Rumelhart, *et al.*, "Parallel Distributed Processing," MIT Press, 1986.

[5] T. J. Sejnowski, *et al.*, "Parallel Networks that Learn to Pronounce English Text," *Complex System,* Vol. 1,1987, pp. 145-168.

[6] J. B. MacQueen, "Some Methods of Classification and Analysis of Multivariate Observations," Proceedings of 5th Berkeley Symposium on Math, Stat. and Prob., 1967, pp. 281-297.

[7] E. W. Forgy, "Cluster Analysis of Multivariate data: efficiency vs. Interpretability of Classifications," Biometrics, Vol. 21, 1965, pp. 768-769.

16

Accommodating High PV Penetration on the Distribution System of Kinmen Island

Yuan-Kang Wu[1], Shao-Hong Tsai[2], Ming-Yan Zou[1]
[1]Department of Electrical Engineering, National Chung-Cheng University
[2]Department of Electrical Engineering, Hwa Hsia Institute of Technology

ABSTRACT

A large-scale PV installation has a profound effect on the power quality and operation of the grid. Therefore, as installations of PV systems increase, changes to both technologies and interconnection standards are needed to address these high PV penetration systems. This paper investigates the main technical impacts of large-scale PV generations on power grids and associated interconnection standards. A case study by using the PSS/E software in the Kinmen power system was implemented to study the steady-state and transient characteristics for the high penetration PV systems.

Keywords: PV; Penetration; Interconnection Standard; PSS/E; Kinmen

1. Introduction

There is an increasing introduction of renewable energy generation into power grids, like wind power and photovoltaic (PV) generations, which causes control and operation problems to the utilities since the networks have not been pre-designed to take into account these distributed or intermittent generations. The electricity from PV is difficult to predict and mainly dependent on weather conditions. Furthermore, they are usually connected to grid through power electronic converter, and this also presents a significant difference from conventional generators. As the penetration of the solar PV plant increases, its impact on the stability and operation of the grid needs to be well understood. Because of the intermittency of the output from solar plants, these impacts might be significant, especially when the solar irradiance drops from 100% to 20% in a minute, causing numerous problems in a high penetration scenario. Additionally, since traditional feeders are commonly designed for radial unidirectional power flows, it is expected that some of the most significant impacts occur for large PV penetration levels.

The main technical impacts of large scale PV generations on power grid are: voltage, current profiles, power quality, protection, electric losses, power factor, power balancing, reliability, protection and operability of the system. Solar PV generation impacts can be of steady state or dynamic in nature. These impacts vary in severity as a function of the degree of penetration and location of solar PV distributed generation. For instance, PV generation is generally at its optimal around moon, and generally this is a period of light load condition; therefore, power export through the distribution feeder and the substation back to the system becomes possible. The associated reverse power flow will tend to raise the voltage on the distribution feeder. Additionally, introducing PV at the load side would reduce the load demand and in turn leads to reduced losses and improved voltage profiles on the feeder. This is an important benefit from a capacity planning perspective since it allows utilities to defer capital investments.

Typical system integration problems in large PV systems include [1]: no dispatch capability of PV solar farms without storage, ultra-fast ramping requirements (400 – 1000 MW/min), reactive power management of feeders are not designed with high PV production, power quality, especially voltage fluctuations, flicker and harmonics may be out of IEEE-519 and other standards. Several research works [2] indicate that the impact is lower in the case of distributed PVs compared to concentrated PVs with the same penetration level, and a larger grid with slow-response generators experiences larger variations in system.

In the USA, the department of energy (DOE) solar energy conducted a competitive Funding Opportunity Announcement, and six competitively selected projects have been completed [3]. For example, several projects used actual distribution feeders as the basis for model development and testing high PV penetration; according to the research results, PV inverter capabilities may be useful in mitigating system impacts. Intelligent inverters are expected to provide significant benefit to future electric distribution system operations. As PV penetration rises, new functionality from the inverter would need to be incorporated. Advanced PV inverters should have the following features: reactive power control, low voltage ride through, dynamic control, and works like a traditional synchronous generator.

Some of the projects demonstrated how the high penetration of PV will affect operations of an electric distribution grid. Other important research items in these DOE projects include the evaluation of the effects of energy storage in combination with distributed PV generation, the benefits from high resolution data collection, the impact of PV system integration on power system reliability indices, the impact of cloud movement on power quality, the impact of feeder configuration and the PV location on the maximum PV penetration.

This paper investigates several potential problems associated with high penetration levels of grid-tied PV. Additionally, interconnection codes or standards for PV generation and major requirements will be studied. Finally, a case study by using the PSS/E software in the Kinmen power system was implemented to study the steady-state and transient characteristics for the high penetration PV systems.

2. Potential Problems Associated with High Penetration Levels of Grid-tied PV

The International Energy Agency (IEA) has produced a series of reports on Task V of the PV power systems (PVPS). The demonstration projects issued by the IEA have revealed that voltage rise is one of the primary concerned problems [4]. The study in [5] also indicates that PV penetration is limited to approximately 33% by voltage rise issues in the UK. A study from Japan [6] suggested that voltage rise is the most serious problem and would limit penetration levels between 5% and 20%. Any injection of power into a distribution system would cause a voltage rise at the point of connection, and this must be considered by network operators.

Voltage dips due to cloud transients might be an issue at high PV penetration. Cloud cover and morning fog require fast ramping and power balancing. Several research works [7] examined cloud transient effects if the PV were deployed as a central-station plant, and the results indicate that the maximum tolerable system level penetration level of PV would be limited by the transient following capabilities (ramp rates) of the conventional generators. Additionally, slow voltage regulation and fast voltage regulation in high penetration PV scenarios should also be addressed. To mitigate the effects of PV output variability, energy storage systems are another option, in which the PV inverter control system is integrated into the system. The storage system will generate or absorb energy as the PV power fluctuates according to the available solar resource.

In terms of power quality, the results of several research works reveal that the PV contribution to voltage distortion is far less than the contributions made by many customer loads [8]. Therefore, harmonics would not be a problem as long as the inverter has a well design.

3. Interconnection Codes and Standards

Grid connection guidelines are a major subject with regard to PV generation. The connection rules and technical requirements that differ from region to region make it more complicated. IEC has been developing many standards related to individual DG technologies including PV generators. These guidelines would serve as input to develop a more general connecting guideline for all types of distributed generation systems and their interactions with the grid. In 2000, the IEEE published *IEEE Standard 929 IEEE Recommended Practice for Utility Interface of Photovoltaic (PV) Systems* up to 10 kW [9]. This standard was updated from IEEE Standard 929-1988 by incorporating developments in the PV industry and by coordinating with *UL1741 1999*; meanwhile, the IEEE Standards Coordinating Committee 21 on Fuels, Photovoltaics, Dispersed Generation, and Energy Storage formed a working group to develop the *IEEE Standard for Interconnecting Distributed Resources with Electric Power Systems*, or IEEE Standard 1547 [10]. It provides a uniform standard for interconnection of distributed resources with electric power systems, and also provides requirements relevant to performance, operation, testing, safety and maintenance of the interconnection. In Europe, every country has its own technical requirements. However, major requirements in these interconnection codes for distributed generators or PV generation include general requirements and specifications, safety and protection requirements, and power quality requirements.

For general requirements and specifications, most standards use a power size to limit the scope. IEEE 929 uses 10 kW and limits the technology to PV systems, while IEEE 1547 uses 10 MVA and attempts to cover all technologies. In addition, different PV interconnection standards specify different requirements on voltage variation, system frequency, synchronization, and voltage unbalance. As for safety and protection requirements, PV generators must detect and respond to abnormal conditions occurring on the power distribution systems. For instance, in almost all technical requirements and standards, unexpected islanding operation is not wanted. PV units must be disconnected as soon as possible when the main grid is not energized. All interconnected PV plants must have a readily accessible, lockable, visible-break isolation switch shall be located between the DG unit and the grid. This disconnection switch must accessible to the distribution company. In addition, after grid fault, the reconnection of the DG should be automatic once the grid voltage and frequency have returned to and maintained normal ranges.

PV generators should provide electric power to the grid with acceptable power quality. Generally, the requirements of interconnection standards include the limits on harmonic, DC current injection, flicker, and power factor. PV units using power electronic converters may inject harmonic currents in the network. The level of harmonics produced by the PV should not cause any disturbances on the distribution system. Many interconnection codes require specific values for different order harmonics and the total harmonic distortion. Flicker can be the result of fast variations of power output of generators or rapid changes in load current like arc furnaces or induction motor starting leading to significant voltage changes on the feeder. In order to reduce or avoid flicker, many countries require the maximum installed power is several times smaller than the level of the short-circuit capacity at the power common coupling (PCC) point. The PV unit shall not create objectionable or observable flicker for other customers on the power system. DC injection is a problem because the increased DC voltage has the potential to increase saturation of magnetic components, such as cores of distribution transformers. This saturation, in turn, causes increased power system distortion. Following IEEE 1547, DC currents injected by DG must be smaller than 0.5% of its rated current at the connection point.

4. Dynamic Model of PV Arrays

4.1. Modeling of PV Arrays

The physics based relationship between the output current and voltage of the PV cell is given as [11]:

$$I = I_{ph} - I_O\left(e^{\frac{q(V+R_sI)}{nKT}} - 1\right) - \frac{V+R_sI}{R_{sh}} \quad (1)$$

where, I is output current from the cell, V is output voltage of the cell, I_{ph} is the generated current from photovoltaic action, Io is the reverse saturation current, q is the charge of an electron, k is Boltzmann's constant, T is the ambient temperature and n is the ideality factor.

4.2. Dynamic PV Model in PSS/E

The PV model was built using the user defined model integration feature of PSS/E. The PSS/E Solar PV Unit dynamic stability model was developed to simulate performance of a photovoltaic (PV) plant connected to the grid via a power converter. The model is based on the generic type 4 wind model, with the added ability to simulate output changes due to solar irradiation. The PV Generic Model comprises the following modules: power converter/generator module, electrical control module, linear model of a panel's output curve, and linear solar irradiance profile.

5. Case Study

The Kinmen power system (**Figure 1**) is chosen as the test system to integrate PV plants in PSS/E. Before adding any PV plants to the system, the load flow and dynamic study have been conducted in this work. Then, a few case studies were performed where PV plants are integrated at different buses (1101, 1104 and 1105) and the effect of the changes in solar irradiation and loss of PV plants for various penetration levels was observed.

5.1. Effect of Sudden Change of Solar Irradiance on System Stability

PV power is characterized as an intermittent source of energy as it is largely dependent on the environmental conditions. The solar irradiance can have 70~80% drop in a minute. The drop out of a large PV power in such a short time could have significant impact on the system.

To investigate the impact on the system due to sharp changes in the level of radiation, the solar irradiance has been reduced from 1000 W/m^2 to 0 W/m^2 using different ramping down rate. The objective is to find out the critical rate beyond which the system might become unstable or require some load shedding action. **Figure 2** shows the rate at which output power was changed due to the drop in solar irradiance at different rate, where PV penetration is 30%. While the change in power for PV is shown in **Figure 2**, the change in the system frequency is shown in **Figure 3**. In this work, various scenarios with different ramping down rates on radiation and PV penetration have been simulated and the results are shown in **Table 1**. It was found that in 30% PV penetration, the frequency of the PV connected bus for the case of solar

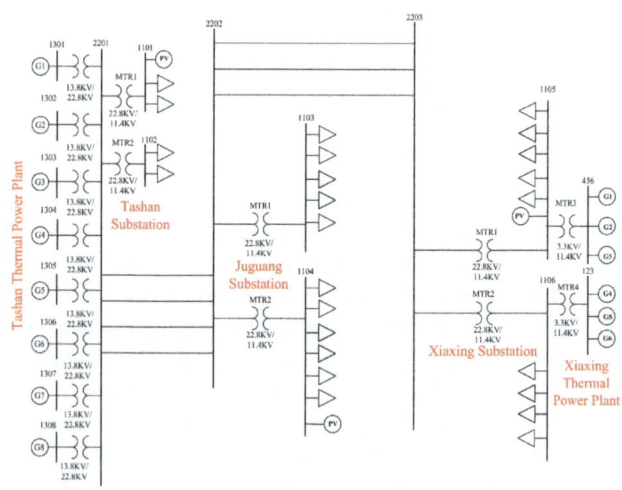

Figure 1. One-line diagram of the Kinmen power system.

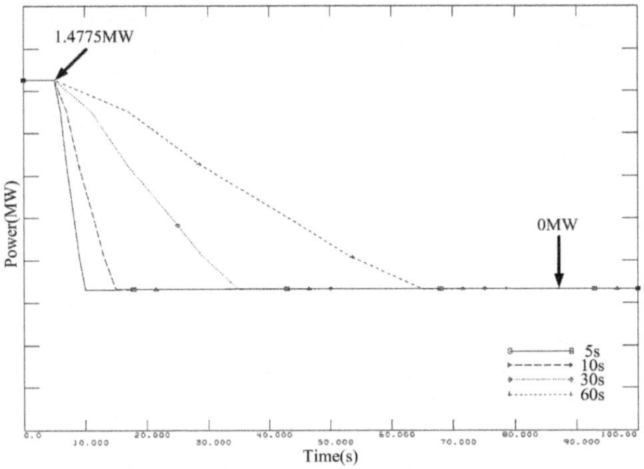

Figure 2. PV output power with 30% PV penetration under different ramping down rate.

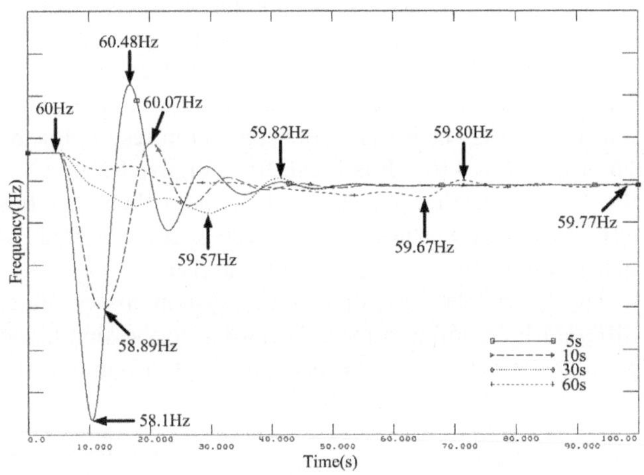

Figure 3. Frequency response with 30% PV penetration in the Kinmen system.

Table 1. Frequency response under various scenarios.

Penetration	5 s	10 s	30 s	60 s
5%	59.68 Hz	59.81 Hz	59.93 Hz	59.95 Hz
10%	59.35 Hz	59.62 Hz	59.85 Hz	59.89 Hz
20%	58.73 Hz	59.26 Hz	59.71 Hz	59.79 Hz
30%	58.10 Hz	58.89 Hz	59.57 Hz	59.68 Hz

irradiation change (from 1000 W/m^2 to 0 W/m^2) in 5s drops to 58.1 Hz which is above the limit for a low-frequency load-shedding relay in the Kinmen power system to operate. If the PV penetration is continuously increased, we can find the critical PV penetration at which the frequency response indicated that the load-frequency load-shedding relay started to operate. The critical PV penetration in this study case under the solar irradiance drop of 1000 W/m^2 in 5 s is 42.5%.

5.2. Effect of Tripping all of PV Generation

In this case, the expected worst case scenarios are demonstrated by assuming PV generator tripping. The PV penetration under 5%, 10%, 20% and 30% is simulated respectively. It is assumed that all of the PV generation are tripped out at time t = 2 s. **Figure 4** and **Figure 5**, show the system frequency and the voltage at bus 2201 respectively when all of the PV generators are tripped. From **Figures 4** and **5**, it can be observed that penetration level of PV power has significant impact on the system frequency and voltage. With the increase of penetration level, the severity of impact on system stability also increases. A trip of the PV generation creates momentary oscillations in the frequency. In **Figure 4**, the magnitude of the frequency perturbation is influenced by the penetration level. The higher the penetration, the higher is the magnitude of oscillation. For instance, for 30% penetration level, the system frequency goes down to 57.63 Hz.

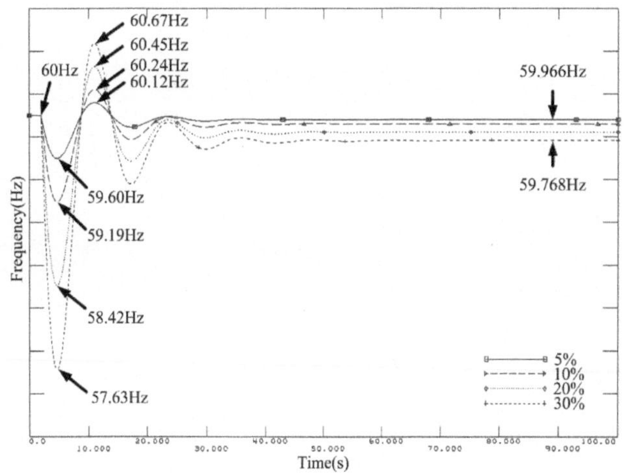

Figure 4. Frequency response in the Kinmen system if PV generation trips offline.

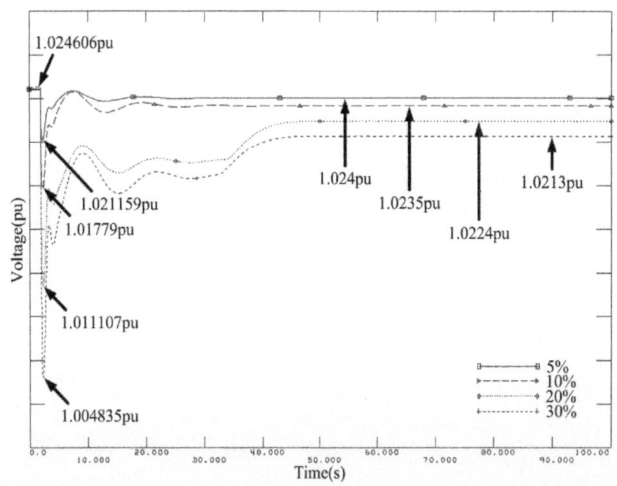

Figure 5. Transient voltage at bus 2201 in the Kinmen system if PV generation trips offline.

If the PV penetration is continuously increased, we can find the critical PV penetration at which the frequency response indicated that the load-frequency load-shedding relay started to operate. The critical PV penetration in this study case (all of PV trip offline) is 34%.

5.3. Effect of a Three-phase Short Circuit Fault

Another worst case scenario, a three phase short circuit fault, is applied at bus 2201 where the PV plant is connected. The fault was applied through very small impedance and lasted for 0.2 s. The solar irradiance is assumed to remain constant at 1000 W/m^2 and therefore, the output power from the PV plant was at its maximum.

Due to the fault, the voltage at bus 2203 goes down to 0.058 pu until the fault is cleared. As shown in **Figure 6**, when the fault was cleared, the voltage curve shows some transient with a maximum to a value of 1.11 p.u. volt. However, the oscillation damped down shortly and goes back to a normal value. **Figure 7** shows the system

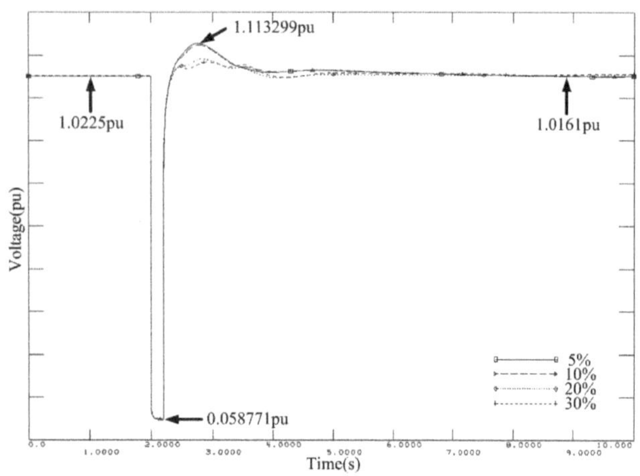

Figure 6. Voltage response in the Kinmen system if a three-phase short circuit fault appears.

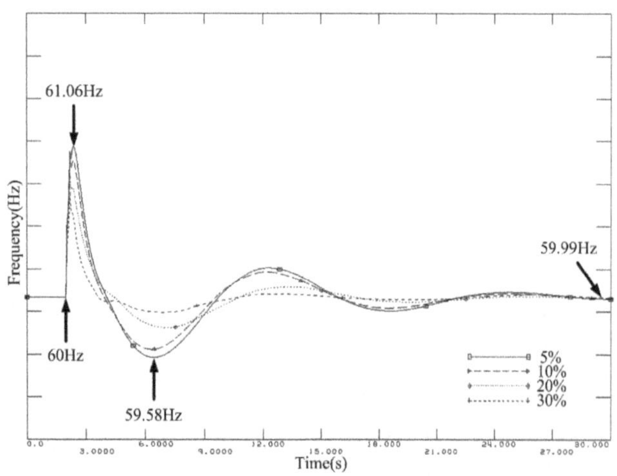

Figure 7. Frequency response in the Kinmen system if a three-phase short circuit fault appears.

frequency response. The frequency shows oscillation and varies between 61.06 Hz and 59.58 Hz under 5% penetration, which does not trigger a series of loads shedding relays. If the PV penetration increases, the oscillation amplitude reduces.

5.4. Effect of PV on Steady-state Voltage and System Loss

Load current flowing through line impedances and distribution transformers causes voltage drops along the feeder. Introducing distributed PV reduces the current being drawn from the substation and improves the voltage drop. The performance of grid at the point of connection for PV generators is an important factor for increasing the PV penetration. In this work, two types of PV generation configurations, i.e., concentrated and distributed, are considered in the simulation studies. For the concentrated configuration, all of the PV generators are installed on bus 1101. For the distributed configuration, PV plants are integrated at different buses (1101, 1104 and 1105). **Figure 8** shows the simulation results for the voltage on bus 1101. It is found from **Figure 8** that when the PV penetration increases, then the voltage on bus 1101 increases. Furthermore, the type of concentrated PV generation would cause larger voltage rise.

This work also investigates the system losses under different PV configurations. According to the simulation results, as 20% and 30% penetration is discussed, the

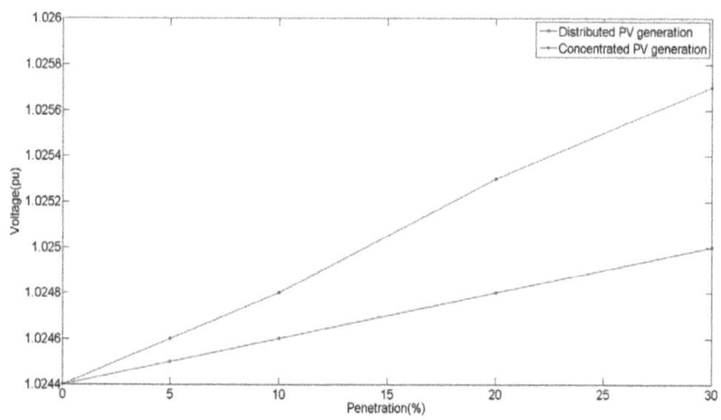

Figure 8. Voltage characteristic under different PV penetration and two types of PV generation configurations.

Table 2. System loss under different PV penetration and configurations.

Penetration (%)	Distributed Type (MW)	Concentrated Type (MW)
5	0.04	0.04
10	0.04	0.04
20	0.03	0.04
30	0.03	0.04

system power loss with distributed configuration type is lower than that with concentrated one. Additionally, the system power loss is reduced when the PV penetration is increased under the distributed configuration. As shown in **Table 2**.

6. Conclusions

As the integration of large and commercial scale PV on the distribution level, such systems can have considerable effect on the operation and protection of the system. This work investigates the main technical impacts of large scale PV generations on power grids and associated interconnection standards. Additionally, a case study by using the PSS/E software in the Kinmen power system was implemented to study the steady-state and transient characteristics for the high penetration PV systems. From the simulation results, several system disturbances, such as change of solar irradiance, tripping of PV generation, and three-phase short circuit fault would affect the maximum PV penetration as a grid. Moreover, PV generation penetration and configuration types would affect system voltage and power losses.

7. Acknowledgements

The authors gratefully acknowledge the financial support by Bureau of Energy, Ministry of Economic Affairs, R.O.C. under the Project "Development of Key Control Technologies for Distributed Energy System", and the National Science Council, Department of Executive, R.O.C. under the grant no. NSC-101-2221-E-146-009.

REFERENCES

[1] J. H. R. Enslin and P. J. M. Heskes, "Harmonic Interaction between a Large Number of Distributed Power Inverters and the Distribution Network," *IEEE Transactions on Power Electronics*, Vol. 19, No. 6, 2004, pp. 1586-1593.

[2] S. A. Pourmousavi, A. S. Cifala and M. H. Nehrir, "Impact of High Penetration of PV Generation on Frequency and Voltage in a Distribution Feeder," *North American Power Symposium*, 2012.

[3] H. Thomas, K. Lynn and A. Razon, "Current Results of the US DOE High Penetration Solar Deployment Project," *38th IEEE Photovoltaic Specialists Conference (PVSC)*, 2012, pp. 731-736.

[4] A. Povlsen, "International Energy Agency Report," IEA PVPS T5-10: 2002, February; 2002. Available online at www.iea.org.

[5] M. Thomson and D. Infield, "Impact of Widespread Photovoltaics Generation on Distribution Systems," *IET Journal of Renewable Power Generation,* 007;1 (March (1)):33–40.

[6] H. Kobayashi and M. Takasaki, "Demonstration Study of Autonomous Demand Area Power System," *IEEE PES Transmission and Distribution Conference and Exhibition*, 2006, pp. 548-555.

[7] R. F. Yan and T. K. Saha, "Investigation of Voltage Stability for Residential Customers Due to High Photovoltaic Penetrations," *IEEE Transactions on Power Systems*, Vol. 27, No. 2, 2012, pp. 651-662.

[8] I. T. Papaioannou, M. C. Alexiadis, C. S. Demoulias, D. P. Labridis and P. S. Dokopoulos, "Modeling and Field Measurements of Photovoltaic Units Connected to LV Grid. Study of Penetration Scenarios," *IEEE Transactions on Power Delivery*, Vol. 26, No. 2, 2011, pp. 979-987.

[9] IEEE, Standard 929-2000 IEEE Recommended Practice for Utility Interface of Photovoltaic (PV) Systems, 2000.

[10] IEEE, Standard 1547, Standard for Interconnecting Distributed Resources with Electric Power Systems, June 2003.

[11] M. G. Villalva, J. R. Gazoli and E. R. Filho," Comprehensive Approach to Modeling andSimulation of Photovoltaic Arrays," *IEEE Transactions on Power Electronics*, Vol. 24, No. 5, 2009, pp. 1198-1208.

The Energy Supply Chain Net

Albana Ilo
Siemens AG Austria, Vienna, Austria

ABSTRACT

The integration and the effective use of all available energy resourpces are possible only under a global view of the power systems. A holistic approach of the power system control, which includes all voltage levels from high to low voltage and the corresponding commercial model, is represented in the following.

Keywords: Smart Grid; Transmission; Distribution; High Voltage; Medium Voltage; Low Voltage; Conservation Voltage Reduction; Operator Role; TSO; DSO

1. Introduction

Power systems have operated successfully for so long now in the presence of large generators, transmission and distribution grid. Recently the strong environmental movement and the technological advances are just some of the factors propelling the installation of renewable energy resources (RES). In those conditions, the efficient and safe operation of the power systems is not warranted any more. Consequently, the strategic use and integration of RES are the hottest and the most challenging topics nowadays. Power system characteristics, combined with the manifold behavior of the volatile renewable energy, make the matter quite complex. There are many research projects running over own demo regions with very good results, but that can not be rolled out because of the tight interdependences that do exist in the power systems. As a result, on one hand, there are many research projects focused on low voltage in which the medium voltage grid is also taken into account [1] and, on the other hand, there are other research projects focused on the transmission and the medium voltage of the distribution grid [2,3].

By examining this complexity, it comes out that the integration and the effective use of all available resources are possible only under a global view of the power systems [4]. A holistic approach of the power system control called "Energy Supply Chain Net", which includes all voltage levels from high to low voltage and the corresponding commercial model, is represented in the following.

2. Definition

An "Energy Supply Chain Net" is a set of automated power grids, intended for links, which fit into one another to establish a flexible and reliable electrical connection. Each individual link or a link-bundle operates independently and have contractual arrangements with other relevant boundary links, link-bundles, and suppliers which inject directly to their own grid. Under specific conditions each automated power grid or a couple of them can split, thus creating a "Microgrid" and vice versa.

3. Power Grid Control Model

Most of the transmission utilities, which operates the High and very High Voltage Grids (HVG) are interconnected into a gigantic power grid. They operate independently fulfilling the existing contractual arrangements. In emergency cases, each of them can split and operate isolated. The HVG interconnections create a horizontal connection axis, while the connection with the corresponding distribution grids creates a vertical one. These connections are nowadays not flexible at all.

With regard to network operation, the electric power grid is divided into two parts: transmission (HVG) and distribution, known as Medium Voltage Grid (MVG) and Low Voltage Grid (LVG). Within the power system, only the HVG is fully monitored, as well as automatically regulated and controlled, with the primary, secondary and sometimes tertiary control implemented for both major quantities of power systems, frequency and voltage. **Figure 1** shows the vertical overview of the power grid in the past and according to "The Energy Supply Chain Net" model. Power grid in the past is shown with its transmission and distribution part. The primary and secondary control loop of the HVG is represented with blue lines, while the respective control area is represented with a blue surface. To facilitate the voltage control in the transmission grid, in some cases the HV/MV transformers are included in the TSO control area. Unlike transmission network, currently the distribution networks are almost neither automated nor monitored. The monitoring area of the distribution network operator is shown in **Figure 1(a)** by a green surface. Both, power and communication flow normally from transmission to distribution. The power flow in the intersection points is shown with red and blue arrows to mark active and reactive power respectively.

The increasing penetration of RESs in the distribution network requires a number of steps to move toward automation. Under the heading "The Energy Supply Chain Net" a new holistic regulatory approach is shown in **Figure 1(b)**. The electric power grid is divided into HVG, MVG and LVG. In the new representation, each of these elements has the same control scheme. That means that MVG and LVG are also designed to have primary and secondary control for both frequency and voltage. HV/MV and MV/LV transformers are included respecttively in the control area of MV and LV grid, because they have a big impact in keeping the voltage within the limits in those areas. As shown in **Figure 1(b)**, the primary objects of regulation in MVG are the distributed generators, while in LVG the customers facilities. The power system is conceived as an "Energy Supply Chain Net" with each grid (HVG/MVG/LVG) being considered as a link on its own.

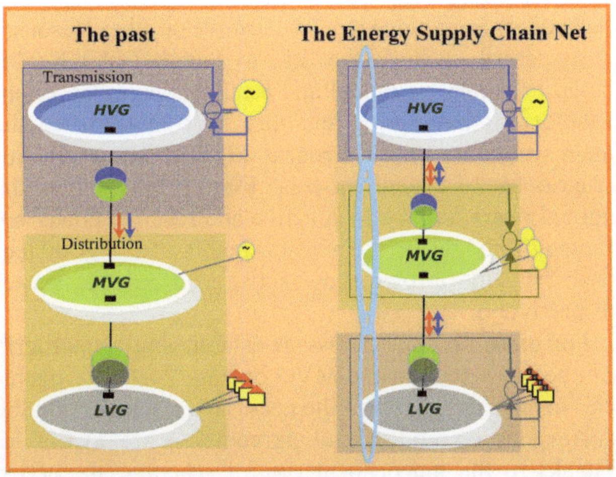

(a) (b)

Figure 1. Power grid overview in vertical axis: (a) in the past; (b) in "The Energy Supply Chain Net" model.

The heart of this model involves the following core principle:

HVG, MVG and LVG are managing themselves, and react flexible with each other like the links in a chain net. The MVG represents the central, strategic link-bundle.

That means that each individual link or a link-bundle member operates independently, however they have contractual arrangements with other relevant boundary links or link-bundles and suppliers which inject directly into its own grid. The MVG takes over a central, strategic role because of its position in the middle (it has intersecttion points with both HVG and LVG) and only when the decentralized generators integrated in the MVG reach a critical mass with a reasonable diversity.

The base of "The Energy Supply Chain Net" model is the decentralization of different grid parts. Each grid part, be it HVG, MVG, or LVG, represents a black box for the others. The links or the so-called black boxes interact in real time in a flexible way with each other through the active and reactive power flow at corresponding intersection points. **Figure 2** shows a general overview of the "Energy Supply Chain Net" model. In the horizontal axis, HVG areas are already linked with each other via interconnections and build up a somehow flexible connection with each other. Actually, the flow in the interconnection is neither automated nor coordinated from the interconnected system level. Instead, the flow schedules are primarily established for economic reasons. In "The Energy Supply Chain Net" model, the decentralized control is used to preserve the frequency and voltage quality, as well as the optimized operation within a certain administrative area. The control design is subject to constraints on the generators power, interconnection line flows, and on the MV/HV and MV/LV transformers for security, reliability, and economic purposes.

3.1. Voltage Control

Power systems are designed and operated so that the service voltages fall within the allowed limits, which are defined in different international or national standards.

Figure 3 shows the automatic voltage control based on "The Energy Supply Chain Net" model. The operators of each grid (HVG, MVG, and LVG) are responsible for keeping the voltage on the corresponding grids within the allowed limits by using their own reactive power resources, transformer steps and respecting the constraints on the neighboring intersection points.

"The Energy Supply Chain Net" model has been field proofed and constitutes the foundation of the industrial research project called ZUQDE (Zentrale Spannungs-U& Blindleistungs-Q Regelung Dezentraler Erzeuger-Central Voltage and Reactive Power Control of Decentralized Generators) [5], as shown in **Figure 4**. The voltage in the MVG is automatically controlled and at the same

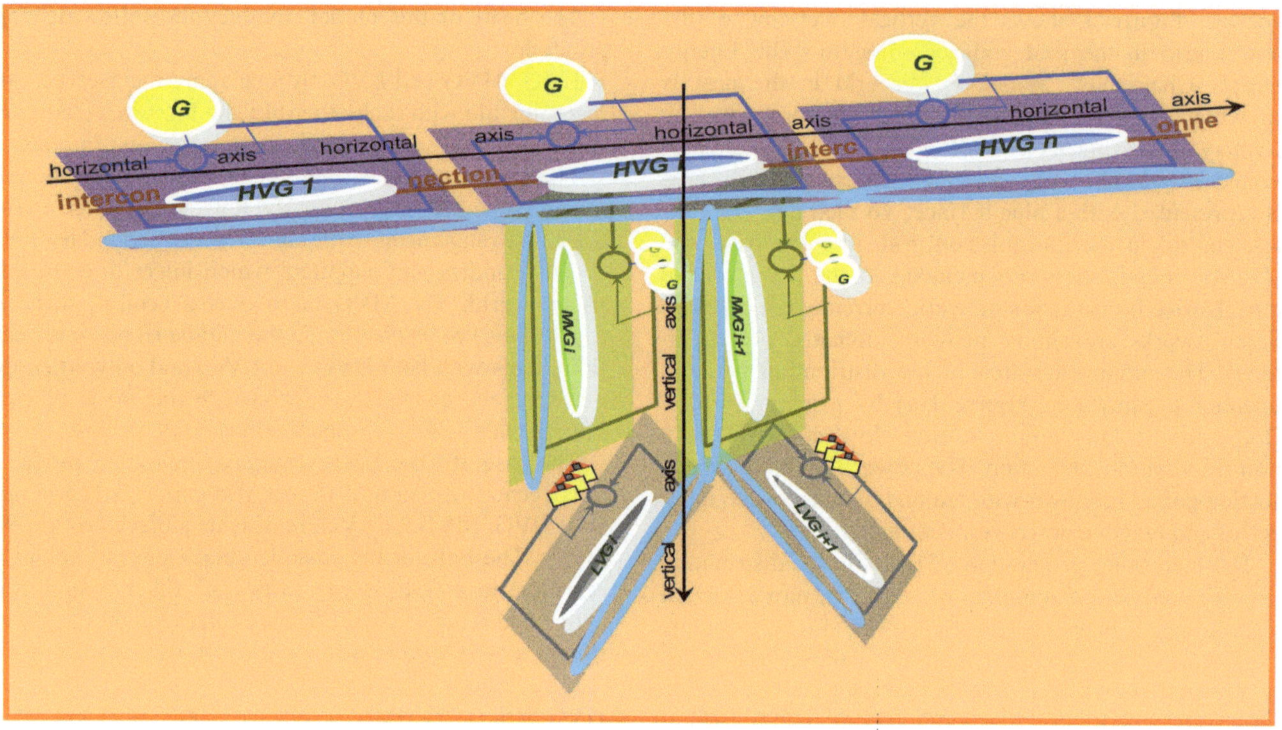

Figure 2. General overview of "The Energy Supply Chain Net" model.

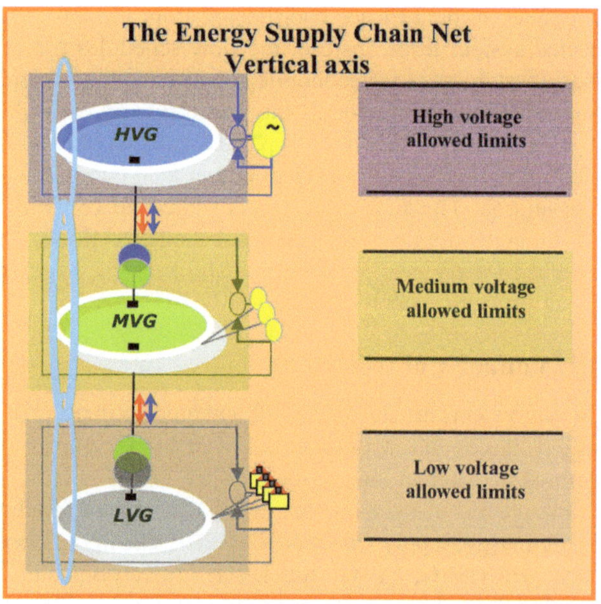

Figure 3. Automatic voltage control.

time the operation is being dynamically optimized in real-time having clear interaction with the HVG. The object of the primary regulation is the reactive power of the decentralized generation and the voltage on the low side of the transformers changed by taps. The secondary control is realized by the Volt var control (VVC) application, which is based on the distribution system state estimator (DSSE). The reactive flow in the neighboring intersection point HV/MV is taken into account in the form of a static cos constraint. Since January 2012, the MVG has been operating in a closed loop in the region of Lungau in Salzburg, Austria. Amongst other things, the exploitation results show that the conservation voltage reduction can be applied smoothly.

3.2. Microgrid Establishment

Microgrids are regionally limited power systems designed to operate semi-independently, usually operating by being connected to the macrogrid or being separated (islanding) from it due to cost effective or other reasons.

As described above, the "Energy Supply Chain Net" is based on a generic model and easily supports the establishment of microgrids and their re-connection to the main system, *i.e.*, the macrogrid. **Figure 5** shows the microgrid establishment process. **Figure 5(a)** shows the normal operation in which active and reactive power are flowing through the intersection points of the grids part (red and blue arrows show active and reactive power respectively).

For example, the process of establishing a microgrid can begin as soon as the MVG operator has the exigentcies to split up and when the necessary resources and the corresponding contractual agreements are available. Thanks to the independent control schemas, the active and reactive power in the intersection points can be controlled up to 0 values. That means that the HVG and MVG are still synchronized, as in **Figure 5(b)**, but no

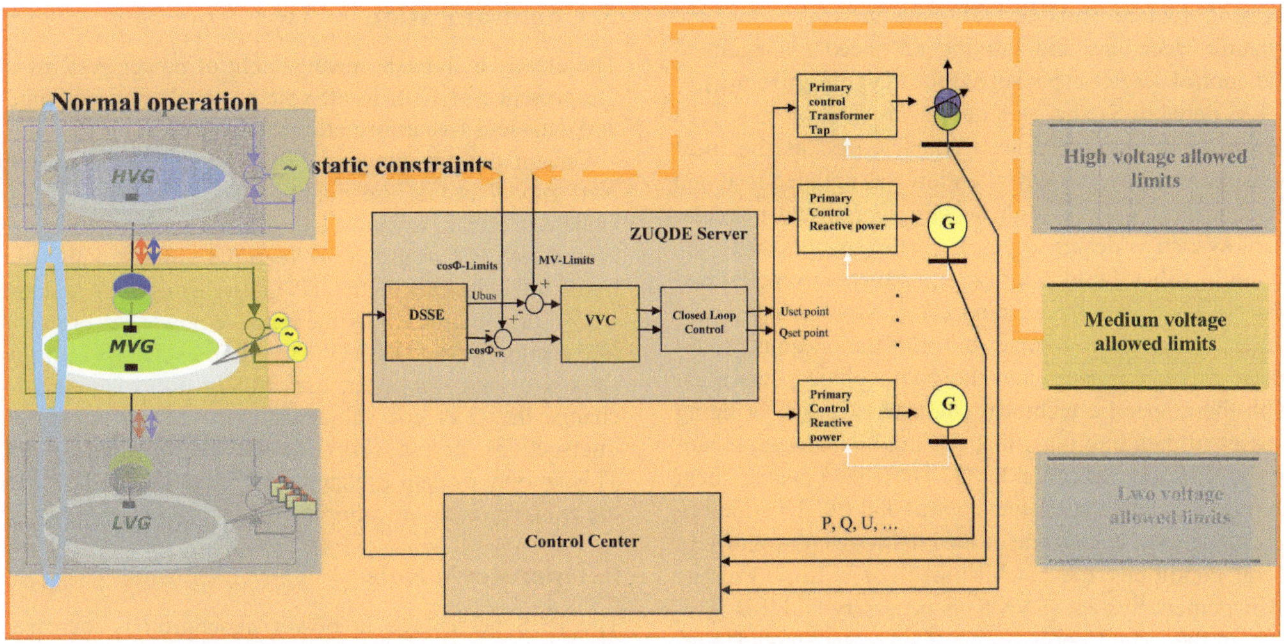

Figure 4. ZUQDE—project scheme imbedded in the "Energy Supply Chain Net" model.

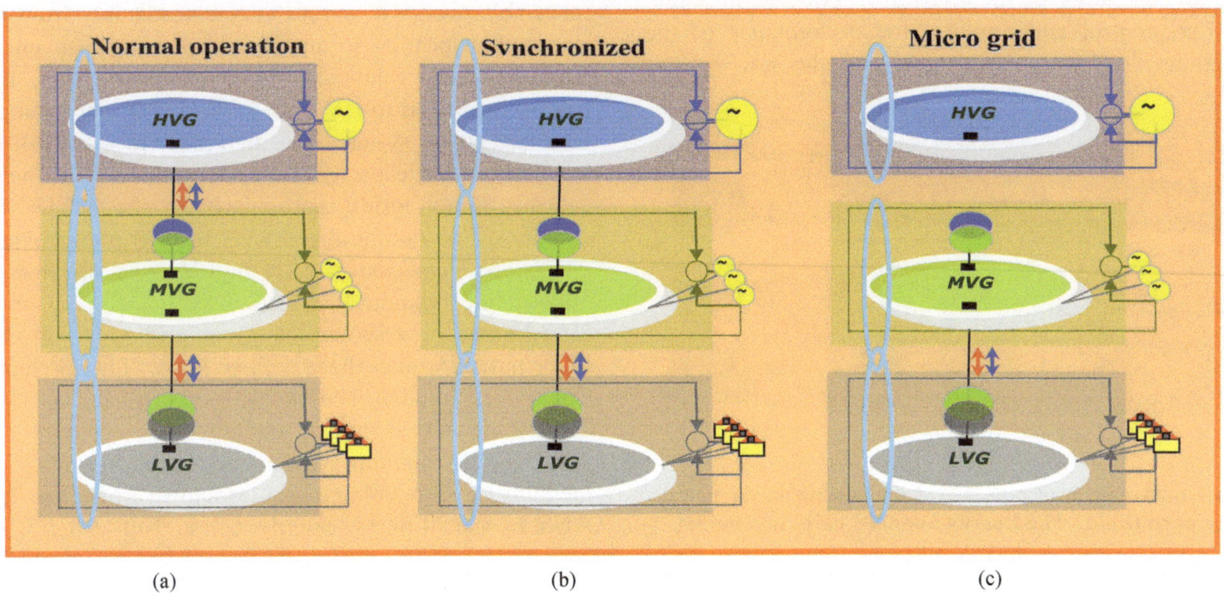

Figure 5. Establishment of Microgrids: (a) Normal oper- ation; (b) Synchronized; (c) Microgrid.

power is flowing between them. After that, the relevant CB can be switched off and the microgrid operation is established as shown in **Figure 5(c)**. The microgrid reconnection to the main system can be realized as usual through the opposite way, first the synchronization occurs and than the interchange power starts to flow through. The same method can be used for every link or link-bundle of the Chain Net.

4. Power Grid Management Systems

A Management System is a system of computer-added tools that provides management and control services. They are used by operators of electric utility to monitor, control and optimize the performance of the power grid. **Figure 6** shows an overview of power grid management systems in the past and according to "The Energy Supply Chain Net" model.

Energy Management Systems (EMS) are the first management system in the power systems history and provide advanced management and control services specialized for HVG. Meanwhile Distribution Management Systems (DMS) are developed recently due to the increasing requirements on real-time network view and dynamic decisions and provide advanced management and control services specialized for MVG, **Figure 6(a)**.

LV network is the part of the distribution networks which has traditionally been characterized as the most unglamorous one. No any attention was devoted to it and its elements, and as result no any appropriate management system is designed yet. Different manufactures are trying to solve the increasing requirements to model and observe LVG with the existing DMS, which are designed almost for medium voltage grid. But the on going smart meter roll out process and the photovoltaic penetration with there specific technical characteristics are bringing masses of data into play, that can not be managed meaningful with the traditional DMS. Those data management and their use for the on line control and the dynamic optimization of LVGs is one of the challenges nowadays. In those conditions the establishment of a Low Voltage Management System (LVMS) is necessary and it is foreseen in the "The Energy Supply Chain Net" model, **Figure 6(b)**. A LVMS is a system of computer-added tools used by operators of electric utility grids to measure, monitor, control and optimize the performance of the injections, loads, storing devices and the low voltage grid.

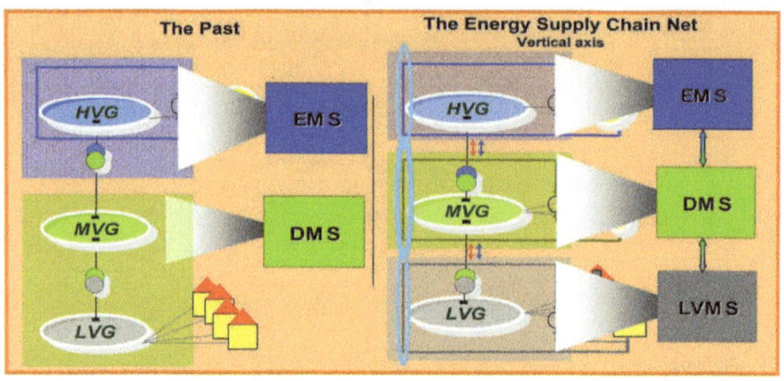

(a) (b)

Figure 6. Overview of power grid management systems: (a) in the past; (b) in "The Energy Supply Chain Net" model.

5. Communication

The effective and safe management of power systems in the presence of RES in all voltage levels requires time- and data-sensitive information exchange.

One of the advantages of the "Energy Supply Chain Net" model is that the number of the needed data exchange to ensure a secure and effective grid operation is minimized. For example, to keep the voltage within the limits in real-time in the HVG (transmission) and the MVG (distribution), only the reactive power, *Q*, or cos in intersection points HV/MV should be exchanged. In this case, a normal European distribution utility should exchange the *Q* or cos value with the TSO in about 500 intersections points (HVG/MVG intersection points). This amount of data exchange can also be managed with the existing communication methods.

6. Operator's Role

The RES percentage in power generation is increased recently according to the environment and political targets. The type and the size of the power plants based on renewable energy resources vary considerably from one geographic location to another. Depending also on the environmental, technical, and economical issues, they can be connected to HVG, MVG or LVG. For example, there are power systems in which more than 50% of the total production capacity is dispersed throughout the local distribution grids, particularly in the MVG. The amount of the decentralized RES such as photovoltaics, which are connected directly to LVG via invertors, is increasing continuously as well.

In **Figure 7** are shown the power grid operation areas in the past as well as those defined in the "Energy Supply Chain Net" model. The transmission system operator, TSO, is actually managing the high and very high grid, while the distribution system operator is managing the medium and low voltage grid, as shown in **Figure 7(a)**. Also in the "Energy Supply Chain Net" model, the

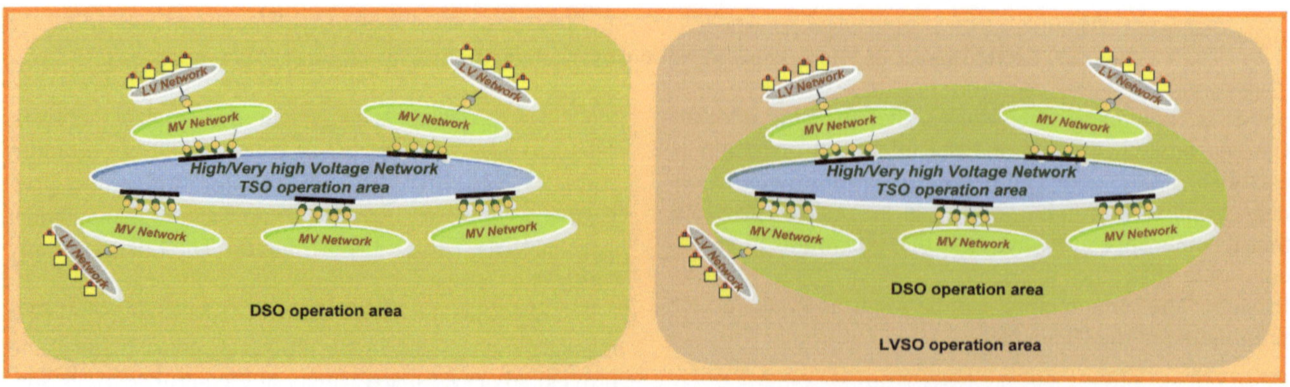

(a) (b)

Figure 7. Power grid operation areas: (a) in the past; (b) in "The Energy Supply Chain Net" model.

TSO will keep the same control area as it is now and will be further responsible for managing the HVG (including the very high voltage grid too). Meanwhile, because of the strategic position described above and the new targets aimed at facilitating effective and well-functioning retail markets, the DSO will take over other responsibilities and restrict the control area only to the MVG. The managing of the LVG in this model is taken over from a new actor. The need to introduce this actor has already been stressed in various research projects such as the Beywatch [6], Address [7], Fenix [8], etc., which have called for the necessity of this actor, sometimes named Supervisor or Aggregator, to be in charge of the interface between the electricity company and the prosumer portfolio. Based on "The Energy Supply Chain Net" model, in which each of the grid parts HVG, MVG, or LVG represents its own link or link-bundle, this actor is named Low Voltage System Operator, LVSO, as shown in **Figure 7(b)**. All system operators in the "Energy Supply Chain Net" model have the same mission to operate, maintain, and develop an efficient electricity system, and to facilitate effective and well-functioning retail markets.

The role of the operators is to act like an orchestra of three renowned violinists—in our case being the TSO, DSO, and LVSO, who play in full harmony and grandeur the world's most wonderful composition—in our case operate the electric power system, as shown in **Figure 8**. In a perturbation case, the DSO can play an important role because of his strategic position, which gives him the possibility to help most effectively in solving the problem as smoothly as possible.

7. Market Accommodation

As discussed above, one of the missions of each actor is to facilitate effective and well-functioning retail markets. The "Energy Supply Chain Net" model accommodates the presence of different market actors, the TSOs, DSOs, LVSO, or big/small suppliers and prosumers, and their respective responsibilities.

Figure 9 shows the accommodation of the different market actors. In the past, **Figure 9(a)**, the TSO, beside the technical targets, has also the mission to facilitate effective and well-functioning retail markets, while the DSO has actually almost only technical targets.

All energy producers should trade their production into the market. The trading market is already established for all large energy producers. While costumer facilities and

Figure 8. Operators' role.

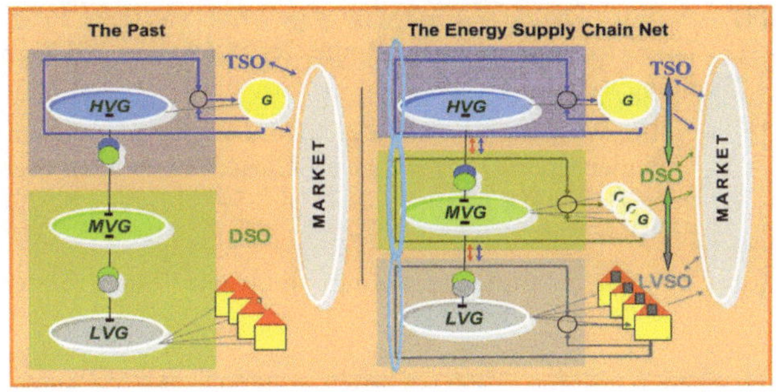

(a) (b)

Figure 9. Accommodation of the different market actors: (a) In the past; (b) In "The Energy Supply Chain Net" model.

small distributed energy producers can go into the market through the well known model of Commercial Virtual Power Plant (CVPP) which is established in the EU research project FENIX. Based on the definition of the "Energy Supply Chain Net" model, not only the TSOs but also the other operators, DSO and LVSO, have the mission to facilitate effective and well-functioning retail markets for all actors who are connected to their own grids, as shown in **Figure 9(b)**. In addition, they have to coordinate the operation between each other based on contractual arrangements at a technical and economical level. Thus the TSO should establish contractual arrangements at both levels, technical and economical, with the DSO. Similarly DSO should establish contractual arrangements with DSO and LVSO at the both levels too. As a result, DSO is keeping his strategic position also in the commercial point of view, because he is the only one having contractual agreement with two operators, TSO and LVSO.

8. Fazit

The "Energy Supply Chain Net" model therefore ensures a dynamic system control in RES presence and in a competitive electric power industry environment.

REFERENCES

[1] A. Bertani, A. Borghetti, C. Bossi, L. De Biase, O. Lamquet, S. Massuco, A. Morini, C.A. Nucci, M. Paolone, E. Quaia and F. Silvestro, "Management of Low Voltage Grids with High Penetration of Distributed Generation: Concepts, Implementations and Experiments," C6-304, CIGRE 2006.

[2] P. Lund, "The Danish Cell Project—Part 1: Background and General Approach," *IEEE Power Engineering Society General Meeting*, Tampa, 24-28 June 2007, pp. 1-6.

[3] N. Martensen, H. Kley, S. Cherian, O. Pacific and P. Lund, "The Cell Controller Pilot Project: Testing a Smart Distribution Grid in Denmark," *Proceedings of Grid Interpretability* 2009: *The Road to an Interoperable Grid*, 2009, pp. 216-222.

[4] A. Ilo, "The New Electricity Age," *6th Spectrum Power User Group Conference*, Rome, 26-28 October 2011, pp. 1-16.

[5] A. Ilo, W. Schaffer, T. Rieder and I. Dzafic, "Dynamische Optimierung der Verteilnetze—Closed Loop Betriebergebnisse," *VDE Kongress*, Stuttgart, 5-6 November 2012, pp. 184-189.

[6] "Design and Implementation of the BeyWatch Superwisor," FP7-ICT-22388, BeyWatch.

[7] ADD-WP1_ADDRESS_scenarios-v1.0. www.addressfp7.org

[8] J. Corera, "Virtual Power Plant Concept in Electrical Networks," *2nd International Conference on Integration of Renewable and Distributed Energy Resources*, Napa, 4-8 December 2006, pp. 1-16.

18

Research on Microgrid and its Application in China

Xing Wu, Xiaogang Yin, Quan Wei, Yifan Jia, Jing Wang
Research and Development Center, Xi'an High Voltage Apparatus Research Institute Co., Ltd, Xi'an, China

ABSTRACT

This paper presents the state of the art research and application of microgrid in China, and then introduces the major concerns for the development of microgrid. The research related to microgrid in China arose around 2004, focused on the connection of distributed energy resources (DERs) to grid and its influence on distribution network, and then followed the microgrid concept of Consortium for Electric Reliability Technology Solutions (CERTS). However, the microgrid in China has taken a more active path with the development of DER and the encouragement of the development of smart grid and resource and energy needing. Now, nearly every province in China has typical researches and applications of microgrid. Besides, there are already two major patterns in the development of microgrid, the institute-dominating research and the enterprise-dominating application. Research institutions like Tianjin University, Hefei University of Technology, Xi'an Jiaotong University, Institute of Electrical Engineering (Chinese Academy of Sciences), etc., are leading the research on microgrid. The companies like State Grid Corporation of China (SGCC), China Southern Power Grid (CSPG), etc., have accomplished several demonstration projects of microgrid, and give a further boost to its application. At the end of the paper, according to the present situation of the development of microgrid in China, the facing problems and the way to deal with it are put forward.

Keywords: Microgrid; Distributed Energy Resources; Research; Demonstration

1. Introduction

Since the entrance of 21st century, developing clean energy and ensuring the safety of energy have become the keywords in the energy industry. It is a common goal to build cleaner and more reliable power systems. In order to overcome the negative impact of distributed energy resources (DERs) on power system, and maximize the potential of DERs, the microgrid concept was proposed by Lassster *et al* at the University of Wisconsin–Madison (UWM). The microgrid can be operated as a single system with DERs and loads [1-3]. The Consortium for Electric Reliability Technology Solutions (CERTS) then presented the white paper on integration of DERs. The CERTS microgrid concept assumes an aggregation of loads and DERs operating as a single system providing both power and heat [4]. The proof of the concept was accomplished at UWM [5], the microgrid test bed was then constructed at Dolan Technology Centre [6].

Typical CERTS microgrid is illustrated in **Figure 1** [4]. It consists of micro sources, power and voltage controller, point of common coupling (PCC), circuit breaker, separation device and so on. The CERTS microgrid concept is the most popular microgrid concept around the world. The CERTS microgrid concept is an important reference for the researches and the applications in China.

2. Background of Microgrid in China

In China, the microgrid develops with the boom of DER technology, especially the new energy (renewable energy) technology. The Chinese government attaches great importance to the development of distributed energy generation based on new energy. It is clearly stated that China will promote the low-cost large-scale development and utilization of renewable energy in the National Outline for Medium and Long Term Science & Technology

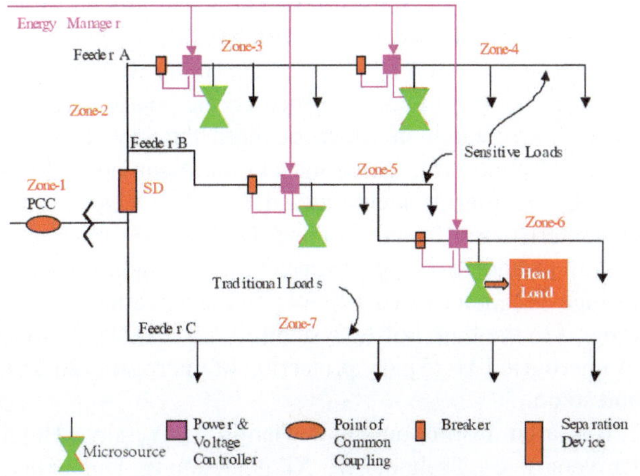

Figure 1. CERTS microgrid architecture.

Development (2006-2020). The National Outline for 12th Five-Year Science and Technology Development also put forwards a vision for the development of the new energy technology. The latest released Special Planning of 12th Five-Year Plan on Smart Grid Major Science & Technology Industrialization Projects emphasizes the importance to break through the technology of large-scale intermittent new energy and its storage. Chinese government issued Renewable Energy Law of the People's Republic of China, and medium & long-term Renewable Energy Development Plan to further accelerate the development of DER. With the support of the national policies mentioned above, the amount of investment in renewable energy in China has reached $51billion in 2011, the total installed capacity of wind power has exceeded 62 GW, and newly installed capacity of solar power has increased to 2.2 GW. The development of DER technology based on renewable energy in China has been an inevitable trend.

Meanwhile, the research and application of micro-grid in China also continues to heat up. The microgrid in China started by following others, however, it takes a more active way with the support of national policies on DER. Within less than one decade, there are already two developing patterns of microgrid in China, the institute-dominating research and enterprise-dominating application. With financial support of National Basic Research Program (973 Program), National High Technology Development Program (863 Program), National Natural Science Foundation of China (NSFC), etc., the research of microgrid in China includes various aspects of the microgrid technology. At the same time, the state-owned enterprises and the private enterprises increase their investment in the application of DER and microgrid, and nearly every province in China has already had typical DER and microgrid applications.

3. Research on Microgrid

The research of microgrid in China began around 2004. It is carried out in accordance with the CERTS microgrid concept, focusing on DERs interconnecting with the bulk power system and their influence on the power network. Now, the research in China covered the key technical issues of microgrid, including: (1) the planning and design of microgrid (including DER)[7-12]; (2)operating characteristics of micro sources [13-15]; (3) microgrid operation and its energy management (including energy storage technology)[16-19]; (4) interconnection of microgrid to the bulk power system[20-22]; (5)island mode of microgrid [23-25]; (6) protection of microgrid [26-28], and so on.

Research institutions like Tianjin University, Hefei University of Technology, Xi'an Jiaotong University, Institute of Electrical Engineering (Chinese Academy of Sciences), etc., are leading the research on microgrid.

Scholars in Tianjin University started their research from the perspective of planning and design of power system. With the financial support of 973 Program and 863 Program, they carried out research on the planning and design of the microgrid and DERs, the operation, control and protection of the microgrid and the distribution network with DERs. They have built a small microgrid experimental system and an integrated microgrid system [29]. The small microgrid experimental system is illustrated in **Figure 2**. This is single-phase 230 V, 50 Hz system. It is composed of energy storage system, photovoltaic (PV) simulator, wind power simulator and related control systems. The microgrid central controller (MCC) is used for the energy management. The system can be operated at two steady-state operation modes (interconnection and island), and two transient-state operation modes (transition between interconnection mode and island mode).

Hefei University of Technology developed a microgrid system with multi-energy generators through collaborative research with the University of New Brunswick, Canada [30, 31]. They carried out research on the optimization of the design, control and scheduling strategies of the microgrid. The microgrid structure is shown **Figure 3**. The system consists of a 10 kW single-phase and a 30 kVA three-phase PV generators, two 30 kW three-phase wind generation simulators, 5 kW fuel cell, 300 Ah battery bank, 1800F ultra-capacitor bank and two sets of 15kW conventional generators used to simulate the small hydro and small fossil generators, various loads (resistors, capacitor, inductors, AC and DC motors and other electronic loads). This microgrid has two-layer hierarchical control system, the local controllers and the central controller. An energy management system that conforms to the IEC61970 standards was developed.

The research on microgrid in Xi'an Jiaotong University started with the application of power electronics technology. With the financial support of 863 program and

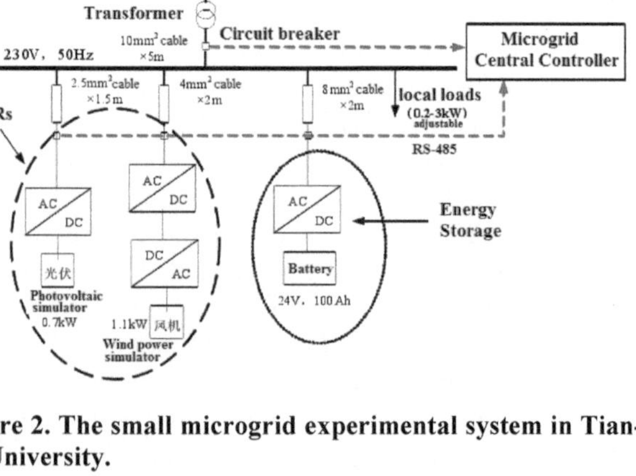

Figure 2. The small microgrid experimental system in Tianjin University.

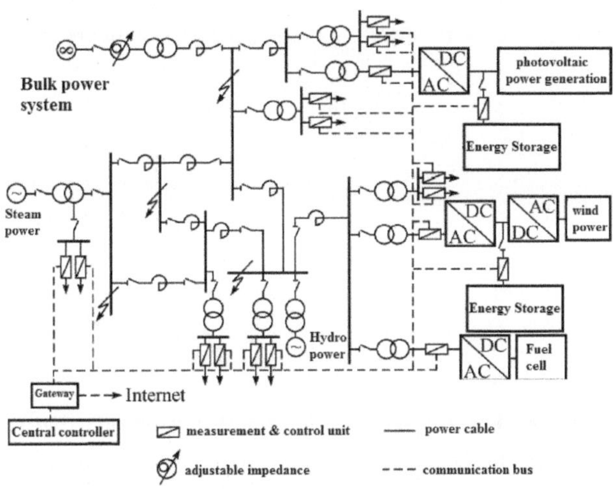

Figure 3. The microgrid system in Hefei University of Technology.

NSFC, they have done extensive and intensive research on the circuit topology and control strategy of power electronics devices in the microgrid, and have developed a fast simulation platform for microgrid based on PSCAD. During this process, they proposed a seamless control strategy to evaluate the platform [32], developed a dynamic reactive power compensation device and its control method for microgrid [33], and proposed a method to use virtual impedance to improve the stability of wireless connection between multiple inverters in microgrid [34]. They also studied the topology of microgrid. The microgrid that they proposed consists of wind, solar, gasoline generators, and energy storage, including battery and super capacitor.

Institute of Electrical Engineering, Chinese Academy of Sciences (IEECAS) is a pioneer of the research on microgrid in China. With the financial support of 863 Program and NSFC, they have done intensive research on the basic theory, key technologies, and experimental platform of microgrid. They also developed a series of control and protection technologies and devices, including a common interface for interconnection between DER and microgrid [35], a hybrid energy storage controller of super capacitor and battery [36], a small-signal model of super capacitor energy storage to improve voltage stability [37], microgrid operation control strategies [17] and so on. Moreover, they built a microgrid test platform, completed the design and commissioning of the grid-connected inverter and other accessories unit, and accomplished combine heat and power (CHP) system in their laboratory. Meanwhile, they built several test platforms for analysis and testing of super capacitor, simulation of dynamic voltage, and AC-DC-AC motor drive system, etc. In addition, the institute has built a number of renewable-energy-generation technology lab, including variable speed constant frequency wind turbine converter laboratory stand-alone PV, wind and solar control laboratory, large-scale wind turbine control system laboratory, large-scale grid-connected PV inverter laboratory, power electronics laboratory, and solar thermal power generation test bed.

4. Application of Microgrid

Smart grid is the key word in the power industry of China. The implementation of smart grid requires multi-level, robust and reliable electricity network. Microgrid is an important part of the network. Typical Chinese enterprises, like SGCC and CSPG, have done a lot of work to apply the technology of microgrid, and have built several demonstration projects of microgrid with world-leading level.

SGCC is the largest electric utilities company that transmits and distributes power in China and in the world. It proposed the slogan to build a strong smart grid in 2009, and wished to complete the grid in 2020. With the inspiration of this slogan, SGCC has completed many demonstration projects of microgrid and DER, including the pilot project for distributed PV power generation and microgrid operation control in Henan province, the pilot projects for control of microgrid in Hangzhou city and Xi'an city, and the microgrid projects in the rural area of north China.

Zhongxin Tianjin Eco-city microgrid system is the first microgrid system in the constructions for smart grid of SGCC. It realized the communication between microgrid systems and distribution automation system, and data exchange between microgrid and intelligent buildings for the first time. The microgrid system includes a 380V/30kW PV system, 5 kW wind power, 25 kW × 2 h energy storage, and monitoring and protection devices. It can achieve autonomous control protection and management, and can be operated at interconnection mode and island mode.

With the support of 863 Program, SGCC led the research on development of key technologies in microgrid with DERs, and developed an integrated platform for control/protection of microgrid. They built an off-grid microgrid demonstration project in Nanlu Island and a grid-connected microgrid demonstration project in Luxi Island. The microgrid in Nanlu Island consists of 10 × 100-kW wind turbines, 545 kW PV power generation system, 1600 kW diesel power generation system, 20 kW ocean current power generation systems, and energy storage system, with integration of electrical vehicle charging station, smart meters and other advanced smart-grid technologies. Luxi Island microgrid is composed of 2 × 780 kW wind turbines, 300 kW PV power generation systems and energy storage system. It can achieve flexible transition between grid-connected mode and off-grid mode.

To resolve the key issues in the large-scale application of DERs, Zhangbei wind/solar/battery demonstration project is built by SGCC. It is a key project of National Science and Technology Support Program and the National "Golden Sun" Program, and is the first pilot project of SGCC for the strong smart grid. The total investment of this project exceeds ¥12 billion. The entire project contains 500 MW wind power, 100 MW PV and 110 MW batteries, and a 220 kV intelligent substation. The project is an aggregation of all the advanced technologies in power design, wind power, PV, energy storage, control and protection. The first stage of this project had been completed by the end of 2011. The investment of this stage is ¥3.3 billion, including 100 MW wind power, 40 MW PV and 20MW battery. The framework of the system is shown in **Figure 4**. The wind turbines in this project include the doubly-fed wind turbines, direct-drive wind turbines and the vertical-axis wind turbines. The PV cells include the monocrystalline silicon cell, the polycrystalline silicon cell, amorphous thin-film solar cell and dye-sensitized solar cell. The solar trackers include the fixed tracker, horizontal single-axis tracker, tilted single axis tracker and dual-axis tracker. The batteries contain lithium iron phosphate batteries, vanadium redox batteries and sodium-sulphur batteries. The project will become the world's largest demonstration project with the highest level of intelligent operation of wind power, PV, energy storage.

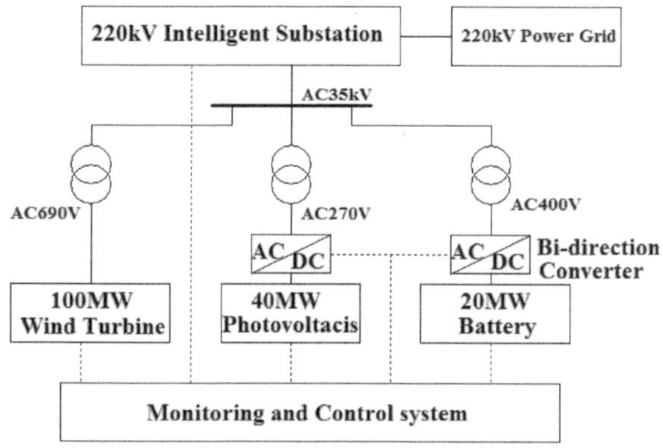

Figure 4. The framework of Zhangbei wind/solar/battery demonstration project.

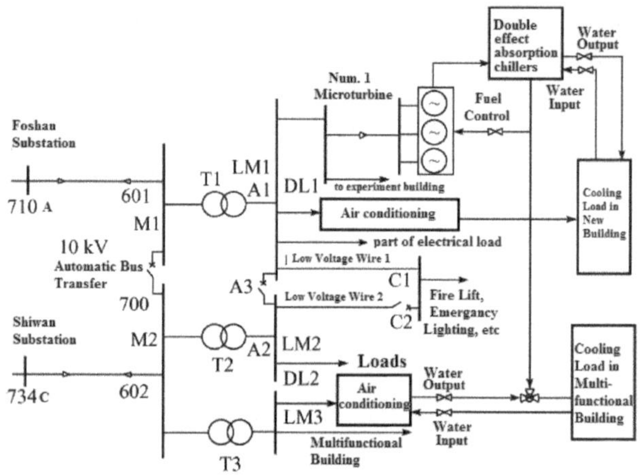

Figure 5. CSPG microgrid system in Foshan city.

CSPG pursues the goal to build an intelligent, efficient, reliable and green smart grid. They keep improving the stability and reliability of power system, the utilization efficiency of resources. They have made great efforts to investigate the DER and microgrid technology.

CSPG built a typical microgrid system with combined cooling, heating and power (CCHP) in Foshan city, Guangdong province, providing electricity and cooling for three office buildings. This project belongs to the 863 Program "Study on integration and demonstration of distributed CCHP technology with MW- class gas turbine" and "Study on key technology and demonstration of interconnection of microgrid with MW-class distributed CCHP system". It focuses on the optimization of design, operation of microgrid, and the impact on the distribution network. The demonstration project started operating in February, 2010. The structure is shown in **Figure 5** [38]. The experiment building, part of the loads in dispatch building, and the miroturbines were connected with the low voltage bus LM1, the rest of the loads in dispatch building were connected to LM2. Bus LM3 will provide electricity for the loads in multifunctional building. The CCHP system consists of the microturbine set (three 200kW microturbines) and LiBr double-effect absorption chiller. This project studies protection, control and supervision of power quality in microgrid system with CCHP, provides the foundation for the application of DER.

CSPG paid special attention to the energy storage technology. Baoqing, the MW-class energy-storage power station was built in Shenzhen, and came into service in September, 2011. The power station can output 110 kV electricity for industry, the total energy-storage capacity exceeds 12 MWh. Compare with Zhangbei demonstration project focusing on the combination of new energy and energy storage, Baoqing power station pays much attention to load shifting, regulation of voltage/frequency, and microgrid islanding.

5. Microgrid at Xi'an High Voltage Apparatus Research Institute Co., Ltd

Xi'an High Voltage Apparatus Research Institute Co., Ltd (XIHARI) has the National Engineering Laboratory for High Voltage Electric, National Energy Research and Development (Testing) Centre for Power Transmission and Distribution Equipment, and National Quality Supervision & Inspection Test Centre for high voltage apparatus, insulator and surge Arrestor, power capacitor, and intelligent power equipment. It is responsible for the research and development of power equipment, power test and establishment of national standards. XIHARI plays a very important role in smart grid, new energy (renewable energy). Microgrid technology is another key technology that XIHARI is developing. **Figure 6** shows the framework of on-going microgrid project in XIHARI. This project focuses on economic operation and energy management optimization of microgrid. The microgrid consists of 20 kW PV, 2 kW wind power, 20 kWh battery, and diesel generator. Several interfaces are prepared for electrical vehicle charging system and static var generator (SVG).

6. Prospects of the Microgrid Technology

As an optimized integration of DER, microgrid has been the focus around the world. Microgrid plays an important role in the strong smart grid in China. Microgrid has many advantages; however, there are several problems that need to be solved before the large scale application in China.

Lack of national standards and specific policies of microgrid is the priority. There are standards about DER, for example, the IEEE P1547 standard. The standards of DER in China are establishing and amending, but specific standards and national policies of microgrid about the interconnection, design and planning, construction and operation, and manufacturing are still absent.

Second, the microgrid technology is not yet mature. Compare with traditional integrated power generation, the development cost and the operation cost are very high. The microgrid technology needs improving.

Third, the microgrid needs legalizing, and establishing new economic relation with bulk power system. According to the present electricity law of China, the gird-connection microgrid is illegal if it outputs electricity. Meanwhile, the output electricity of private microgrid

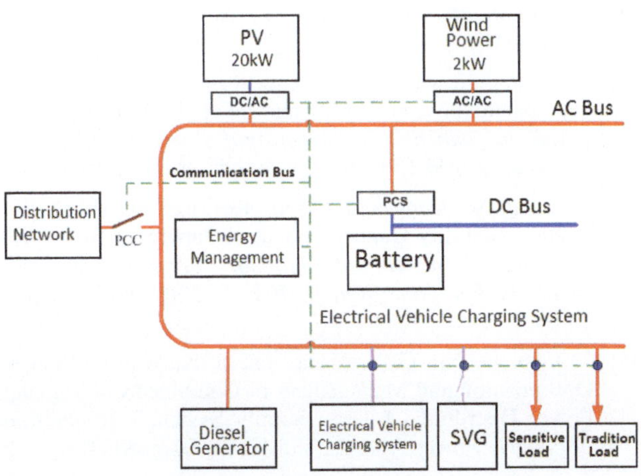

Figure 6. Microgrid in XIHARI.

conflicts with the operation of current electric power system. New economic relation should be established to ensure mutual interest.

7. Conclusions

The microgrid concept has been proposed and applied for years to solve the interconnection issues of DER to the bulk power system abroad. In China, microgrid develops with the boost of DER (with renewable energy). Though microgrid becomes more and more popular in China, the foundation of microgrid technology is weak, it needs time to realise commercial operation, especially with the support of national policies, laws and regulations. The development of microgrid technology in China is bound to be a long process.

REFERENCES

[1] B. Lasseter, "Microgrid:Distributed Power Generation," *Presented at the 2001 IEEE Power Engineering Society Winter Meeting*, Columbus, OH, USA, 2001.

[2] C. Marnay, F. Javier Rubio and A. S. Siddiqui, "Shape of the Microgrid," *In 2001 IEEE Power Engineering Society Winter Meeting*, Columbus, OH, United states, 2001, pp. 150-153.

[3] R. H. Lasseter, "MicroGrids," *In 2002 IEEE Power Engineering Society Winter Meeting*, New York, United states, 2002, pp. 305-308.

[4] R. Lasseter, A. Akhil, C. Marnay, J. Stephens, J. Dagle, R. Guttromson, A. S. Meliopoulous, R. Yinger and J. Eto, "Integration of Distributed Energy Resources," *The CERTS Microgrid Concept*, Lawrence Berkeley National Laboratory, 2002.

[5] P. Piagi and R. H. Lasseter, "Autonomous Control of Microgrids," *In 2006 IEEE Power Engineering Society General Meeting*, PES, Montreal, QC, Canada, 2006.

[6] D. K. Nichols, J. Stevens, R. H. Lasseter, J. H. Eto and H. T. Vollkommen, "Validation of the CERTS Microgrid Concept the CEC/CERTS Microgrid Testbed," *In 2006 IEEE Power Engineering Society General Meeting*, PES, Montreal, QC, Canada, 2006.

[7] S. Dai, Y. Zhang and D. K. Zhu, "Distribution Network Planning Method Containing Microgrid," *Automation of Electric Power Systems*, Vol. 22, 2010, pp. 41-45.

[8] S. W. Mei and Y. Y. Wang, "Several Basic Problems of Three-level Power Network Planning of Transmission Network-distribution Network-microgrid," *Journal of Electric Power Science and Technology*, Vol. 4, 2009, pp. 3-11.

[9] J. Shu, X. Y. Zhang,Y. L. Shen and C. H. Wu, "The Algorithm and Application in Power Sources Planning and Designing for Microgrid Based on Distributed Renewable Energy," *Control Theory & Applications*, Vol. 5, 2010, pp. 675-680.

[10] M. Wang and M. Ding, "Distribution Network Planning Including Distributed Generation," *Proceedings of Elec-*

tric Power System and Automation, Vol. 6, 2004, pp. 5-8.

[11] C. S. Wang, K. Chen, Y. H. Xie and H. F. Zheng, "Siting and Sizing of Distributed Generation in Distribution Network Expansion Planning," *Automation of Electric Power Systems*, Vol. 3, 2006, pp. 38-43.

[12] J. Wang, F. R. Li, C. Y. Feng and X. F. Wang, "Real and Reactive Power Pricing for Distribution Networks," *Automation of Electric Power Systems*, Vol. 15, 2009, pp. 29-32.

[13] C. S. Wang, L. Ma and Guo Li, "Comparison of Operation Characteristics between Two Types of Microturbines in Microgrid," *Journal of Tianjin University*, Vol.4, 2009, pp. 316-321.

[14] Y. Wang, Z. X. Lu and Y. Min, "Research and Comparison of DG Interface Simulation Models in a Microgrid," *Automation of Electric Power Systems*, Vol. 1, 2010, pp. 84-88.

[15] Y. S. Shi, Z. X. Lu, Y. Min and Y. Wang, "Analysis on Microsource Characteristics and Its Influences on Microgrid Load Voltage," *Automation of Electric Power Systems*, Vol.17, 2010, pp. 68-71.

[16] C. S. Wang, C. X. Xiao and S. X. Wang, "Synthetical Control and Analysis of Microgrid," *Automation of Electric Power Systems*, Vol. 7, 2008, pp. 98-103.

[17] W. Pei, P. S. Li, H. Y. Li, X. S. Tang, J. Z. Cheng and W. X. Zuo, "Key Technology and Testbed for Microgrid Operation Control," *Automation of Electric Power Systems*, vol.1, 2010, pp. 94-98.

[18] M. Ding, L. Zhou and R. Bi, "Optimization for Design and Application of Data Acquisition Subsystem in Micro-grid Energy Management System," *Journal of Hefei University of Technology(Natural Science)*, Vol. 12, 2010, pp. 1790-1796.

[19] S. Peng, Y. F. Cao and X. Cai, "Control of Large Scale Battery Energy Storage System Interface to Microgrid," *Automation of Electric Power Systems*, Vol. 16, 2011, pp. 38-43.

[20] Z. X. Chen, X. D. Liu and J. M. Pan, "Current Control Strategy for Grid-Connected Inverter Based on IDA-PBC," *Transactions of China Electrotechnical Society*, Vol. 8, 2011, pp. 99-105.

[21] J. J. He and W. L. Wu, "Study on Inverting Control System for Parallel Operation of Distributed Generation," *Energy Engineering*, Vol. 3, 2007, pp. 1-6.

[22] G. Wei, W. L. Wu, D. Y. Hu and Z. H. Li, "Distributed Generation and Effects of Its Parallel Operation on Power System," *High Voltage Engineering*, Vol. 1, 2007, pp. 36-40.

[23] B. B. Wu, J. H. Su, J. J. Zhang and D. Zhu, "Control Strategies of Inverter in Microgrid Island Operation," *Proceedings of the Chinese Society of Universities for Electric Power System and its Automation*, Vol. 1, 2011, pp. 1-5.

[24] F. D. Li and M. Wu, "An Improved Control Strategy of Load Distribution in an Autonomous Microgrid," *Proceedings of the CSEE*, Vol.13, 2011, pp. 18-25.

[25] X. Zhang and J. J. Liu, "A Novel Power Distribution Strategy for Parallel Inverters in Islanded Mode Microgrid," *Journal of Power Supply*, Vol.1, 2011, pp. 38-42.

[26] Q. Yang and J. H. Zhang, "Design of Microgrid Operation Model and Coordinate Protection," Shaanxi Electric Power, Vol. 1, 2011, pp. 1-5.

[27] J. H. Wang, N. L. Tai, K. Song, L. F. Gu and J. Q. Sheng, "Penetration Level Permission of for DG in Distributed Network Considering Relay Protection," *Proceedings of the CSEE*, Vol.22, 2010, pp. 37-43.

[28] B. Li and Z. Q. Bo, "Design and Research on Protection and Control of Smart Distribution Grid," *Proceedings of the CSEE*, Vol. S1, 2009, pp. 1-6.

[29] Z. Q. Yang, "The Study on Microgrid Experiment Systems," Doctor Dissertation, Tianjin University, 2010.

[30] M. Meiqin, D. Ming, S. Jianhui, L. Chang, S. Min and Z. Guorong, "Testbed for Microgrid with Multi-energy Generators," *in IEEE Canadian Conference on Electrical and Computer Engineering,* Niagara Falls, ON, Canada, 2008, pp. 637-640.

[31] M. Q. Mao, M. Ding, L. C. Zhang, J. H. Su and M. Sun, "Testbed and Information Integration of EMS for a Microgrid with Multi-energy Generation Systems," *Automation of Electric Power Systems*, Vol. 1, 2010, pp. 106-111.

[32] B. Wu, J. Liu and F. Zhuo, "The micro-grid Fast Simulation Platform Exploitation Based on PSCAD," *in 26th Annual IEEE Applied Power Electronics Conference and Exposition,* APEC 2011, TX, United states, 2011, pp. 1737-1742.

[33] L. Meng, M. J. Yang, X. W. Wang, F. Zhuo and B. Q. Liu, "A Dynamic Reactive Power Compensation Device and Its Control Method for Microgird," China, Patent for invention, 201010522245.2, 2010.

[34] H. P. Guo, L. Meng, X. W. Wang, M. J. Yang, F. Zhuo and Y. J. Zhou, "A Method to Use Virtual Impedance to Improve the Stability of Wireless Connection Between Multiple Inverters in Microgrid," China, Patent for invention, 201010522213.2, 2010.

[35] S. Zou, W. Pei and Z. Qi, "Common Interface for Interconnection between Distributed Generation and Microgrid," *Automation of Electric Power Systems*, Vol. 34, 2010, pp. 91-95.

[36] G. Zhang, X. Tang and Z. Qi, "Research on battery supercapacitor hybrid storage and its application in MicroGrid," *in Asia-Pacific Power and Energy Engineering Conference,* APPEEC 2010, Chengdu, China, 2010

[37] X. Tang, W. Deng and Z. Qi, "Research on Micro-grid Voltage Stability Control Based on Supercapacitor Energy Storage," *in 12th International Conference on Electrical Machines and Systems*, ICEMS 2009, Tokyo, Japan, 2009.

[38] L. Guo, D. Xu, C. S. Wang and S. X. Wang, "Energy Optimization and Management of Combined Cooling and Power Distributed Energy Supply System," *Automation of Electric Power Systems*, Vol.19, 2009, pp. 96-100.

Amalgamated-signal Cymometer Based on TMS320F28335

Z. Y. Xu, W. T. Lv, Z. Y. Li
College of Electrical Engineering, Zhejiang University, Hangzhou ,China

ABSTRACT

In our design, an amalgamated-signal cymometer is developed with a TMS320F28335 DSP chip ---- we use its on-chip 12-bit 16-channel AD to conduct a novel dual-sequencer synchronize sampling and averaged to improve the accuracy. Considering the algorithm, we use TI's C28x_FPU_Lib to conduct a FFT with sampled results and correct spectrum with Time-Shift Phase Difference Correcting Spectrum Method based on all-phase spectral analysis. Namely, we design a sampling-frequency self-adaption algorithm which makes operation keep on without outside-hardware command. Actually, the test result shows that our design's frequency resolution ratio is up to 0.4%, and the maximum frequency's D-value between main and minor signal is up to 29.99 kHz.

Keywords: Digital Frequency Meter; Same Precision Measurement; DSP; All-phase Spectral Analysis; Sampling-frequency Self-adaption

1. Introduction

Today's world is a digital world, in which digital signal processor playing an important role, and Ti company as the world's leading semiconductor company; its digital signal processor plays a very important role in the field [7]. Discrete Fourier transform (DFT) technology is the core of digital signal processing technology. In 1965, Cooley and Tukey proposed the Fast Fourier Transform algorithm hereinafter referred to as FFT, and it has wide application in every field of digital signal processing. Spectrum analysis is an important application of FFT, and this design is based on spectrum analysis method for signal frequency and the amplitude of the primary and secondary in the composite signal.

TMS320F28335 is TI company's new high performance 32-bit floating point digital signal processor, using 12-bit 16 channel AD to do two-channel synchronous sampling and take its average in order to improve the accuracy, and using floating-point runtime library provided by TI to do FFT computation, and using an extremely high precision time shift phase difference method based on all phase spectrum analysis to correct, getting accurate frequency and the amplitude of the primary and secondary signals, via a serial port to send to upper computer, and at the same time using the TMS320F28335's high-precision EPWM to reconstruct primary and secondary signal waveform.

2. System Solutions

2.1. The Overall Block Diagram

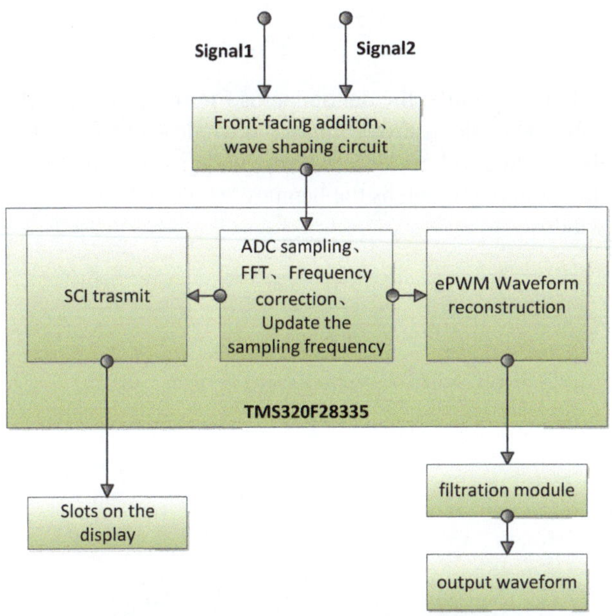

Figure 2.1. The overall block diagram..

2.2. The time Shift Phase Difference Correction Method based on all Phase FFT Spectrum Analysis

All phase FFT has better inhibiting spectrum leakage performance, which has good inhibition of spectral leakage performance of amplitude spectrum, and as well can be directly generated accurate phase spectrum [1]. Based on this, do simple arithmetic on the main spectral line with two sequences which have time shift and we can get higher accuracy of traditional phase estimate, have higher application value than phase difference correction method, and have better use value.

Description of all phase FFT spectrum analysis method:

After all phase data pretreatment and then do FFT to the data, we can get a new type of all phase of the FFT spectrum analysis as is shown in **Figure 2.2**, and its improvement in the method of truncation of data greatly improves the spectrum performance.

The process the Time-Shift Phase Difference Correcting Spectrum Method corrects Spectrum as is shown in the **Figure 2.3** below [8]:

Its Frequency estimation formula is

$$\omega^* = \frac{[\varphi_1(k^*)-\varphi_2(k^*)]}{n_L} + \frac{2k^*\pi}{N} \quad (1)$$

The deviation value of main line is

$$d\omega = \omega^* - \frac{2k^*\pi}{N} = \frac{[\varphi_1(k^*)-\varphi_2(k^*)]}{n_L} \quad (2)$$

Its Amplitude estimation formula is

$$A = \frac{|Y(k^*)|}{F_g(d\omega)} \quad (3)$$

In the formula, the denominator **Figure 2** (dω) is valuable when dω is brought into the window function of the formula of Fourier transform, and the general window function (such as the hanning window, hamming window,

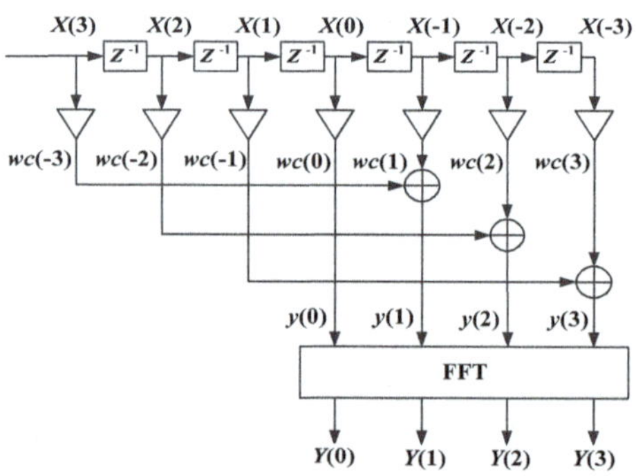

Figure 2.2. The basic block diagram of N-order all phase FFT spectrum analysis.

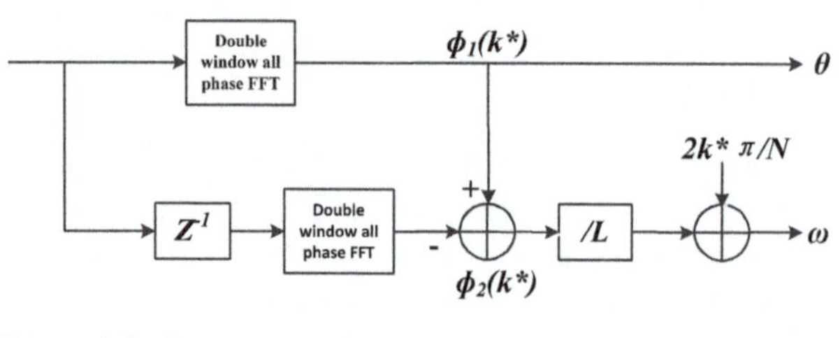

Figure 2.3. the process of spectrum correction of Time-Shift Phase Difference method

triangle window) belong to a cosine window, so their Fourier transform expression are established.

Besides, Time-Shift Phase Difference Correcting Spectrum Method has no phase difference correction of phase estimation.

Theoretically, the frequency correction precision of Time-Shift Phase Difference Correcting Spectrum Method reaches the level of $10^{-12}\Delta\omega$ degrees, and its phase correction error reaches 10^{-3}degrees, which are more precise than traditional methods.

Specific analyses on the theory of all phase spectrum analysis please refer to the reference [2], [3].

3. Design of the Hardware in the System [5]

3.1. Follower

This part of circuit mainly enhances signal, avoiding signal attenuation caused by the input and output impedance mismatch. Primary and secondary signal use the whole pole input with two followers to connect with the level after circuit, and in this not do change of amplitude but just joining a resistance whose resistance R is 1 k in the input stage.

3.2. The Summing Circuit

The module adopts the three input reverse phase summation operation circuit. because in phase end is grounding, so for the reverse side its voltage vn = 0 (according to the use of manual recommended positive input directly to the ground), and because of broken, there can be a $V_O = -R_{21}*V_1/R_{18} - R_{21}*V_2/R_{19} - R_{21}*V_3/R_{20}$, According to design, the input of two sine signal is damped to 0.5 times of original signal and then superimposed on a dc bias voltage, and the resistance for R18 = R20 = 22 k, R19 = 24 k, R21 = 22 k, and the input signal of R19 is 3.3 V dc voltage. So the superposition signal is negative, the amplitude is 0.5 times of the primary and joining U=-3.3*11/24V=-1.5125V dc offset signal.

3.3. The Second Order Butterworth Low-pass Active Filter

Due to the demand of the algorithm is higher to the signal, the module will filter the high frequency noise signal

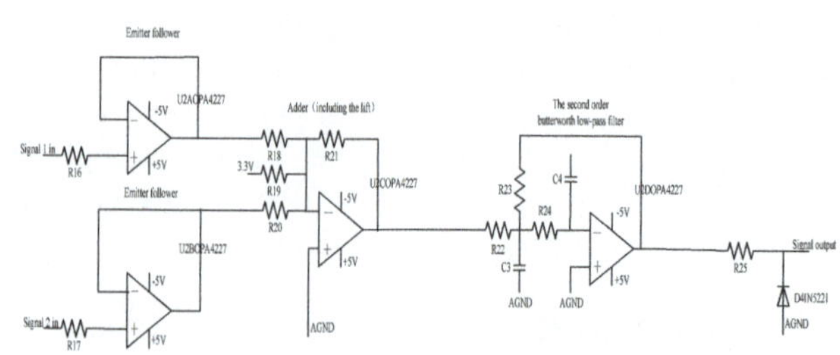

Figure 3.1. Input signal conditioning circuit.

in the input to the DSP signal, to reduce the interference signal on the calculation of FFT, and use the negative input at the same time, will be through the adder originally negative signal into a positive.

OPA4227 precision itself introduces less noise, consider attenuation rate of a first order butter worth filter for each frequency doubling is 6 db, every ten times as much as 20 db; attenuation rate of second order butterworth filter for each frequency doubling is 12 db, and attenuation rate of the third order butter worth filter for frequency doubling is each 18 db and so on, so we only need the second order to meet the demand. In this, R22 = R23 = 100 k, R24 = 22 k, C3 = 180 pF, and cutoff frequency is 33 KHZ, phase reversal 180 degrees when in 120 KHZ.

3.4. Limiter Circuit

This module uses the diode and a resistor in parallel, R25 = 10 k, diode using 1 n5221, positive pressure limiting at 2.4 V, negative for 0.6 V, and according to the calculation, after the amplitude limiting, output positive maximum voltage limit is 0.75 + 1.5125 = 2.2625 V, the limit of minimum forward voltage is 1.5125-0.75 = 0.7625 V, which are within the control of clipping. While DSP AD requests input amplitude is in 0 ~ 3.3 V, the design meets the requirements.

4. The System Software Design

System software design adopts modular design, and it combines subroutine which completes the specific functions into a functional module, invoked by the main monitoring program invocation. Software overall block diagram is as shown in the figure below.

System software consists of main functional modules: initialization module, AD sampling module, FFT calculation module, spectrum correction module, SCI sending module, ePWM waveform reconstruction module, the watchdog module and interrupt module.

Specific operation modes of System software design please refer to the reference [4].

Main computational program flow is as shown in the figure below [6]:

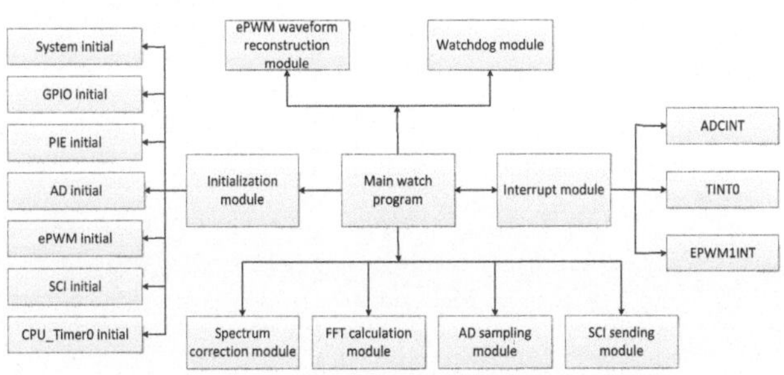

Figure 4.1.general program of the software.

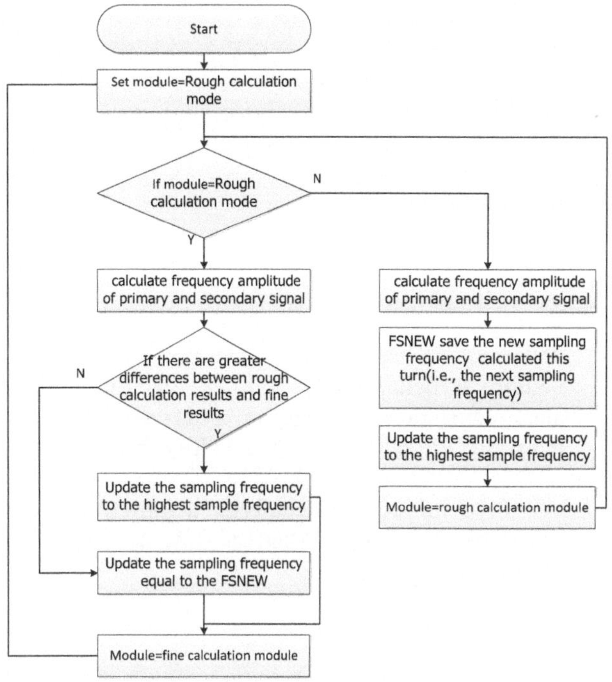

Figure 4.2. Main computational program flow chart.

5. The Core Code

Important structure definition

```
Type def struct {
  Uint16 Main Index;
  Uint16 Minor Index;
} SIGSEARCH_F32_STRUCT;

Type def struct {
  Float 32 Freq;
  Float 32 Mag;
} APCM_F32_STRUCT;

Type def struct {
  APCM_F32_STRUCT Main;
  APCM_F32_STRUCT Minor;
  Uint16 Flag;
} CORRECTION_F32_STRUCT;
```

2. Spectrum correction function

```
//===================================
// Function: APCM_F32_STRUCT RFFT_f32_apcm
(RFFT_F32_STRUCT *fft1, RFFT_F32_STRUCT
*fft2,float32 Fs,Uint16 index)
//===================================
APCM_F32_STRUCT RFFT_f32_apcm
(RFFT_F32_STRUCT *fft1, RFFT_F32_STRUCT *fft2,
float32 Fs,Uint16 index)
{
    APCM_F32_STRUCT ans;
    float32 phi1, phi2;
float32 deltaphi, delta, deltafreqk;
```

```
    float 32 temp, h;

    phi1=*(fft1->PhaseBuf+index);
    phi2=*(fft2->PhaseBuf+index);
    deltaphi=phi2-phi1;
    delta=mod(deltaphi,2*PI);
    if (delta<-PI)
        delta=delta+2*PI;
    else if(delta>PI)
        delta=delta-2*PI;

    deltafreqk=delta/(2*PI);
    deltafreqk=deltafreqk+(deltafreqk==0)*EPS;
    ans.Freq=(index+deltafreqk)*Fs/fft1->FFTSize;
    temp=sin(PI*deltafreqk)+(sin(PI*deltafreqk)==0)*
EPS;
    h=2*PI*deltafreqk*(1-deltafreqk*deltafreqk)/temp;
    ans.Mag=h*h*(*(fft1->MagBuf+index))/2;
    return ans;
}
//====================================
```

REFERENCES

[1] K. J. Xu, "Theory and Application of TMS320X281X DSP," Beijing University of Aeronautics and Astronautics Press, 2006.

[2] K. Ding, "Discrete Spectrum Correction Theory and Technology," Science Press, 2007.

[3] RIFEDG, VINCENTGA. Use of the discrete fourier transform in the measurement of levels and tones[J]. Bell Syst. Tech. J. , 1970, 49(2): 197- 228.

[4] Z. H. Wang, "Digital Signal All Phase Spectrum Analysis and Filtering Technology," Electronic Industry Press, 2009.

[5] "TMS320F28335 Data Manual," Texas Instruments.

[6] G. Z. Zhao, "Signal Analysis and Processing," Mechanical Industry Press, 2005.

[7] Z. Z. Li, "Digital Signal Processing and Application with MATLAB," Tsinghua University Press, 2008.

[8] "Assembly Good Paper of C2000 Grand Prix of TI C2000 Senior Embedded Controlled in 2008," Electronic Industry Press, 2008.

Development of High Torque and High Power Density Hybrid Excitation Flux Switching Motor for Traction Drive in Hybrid Electric Vehicles

Erwan Sulaiman[1,2*], **Takashi Kosaka**[2]

[1]MyEV, Department of Electrical Power Engineering, University Tun Hussein Onn Malaysia, Johor, Malaysia

[2]Department of Electrical & Computer Science Engineering, Nagoya Institute of Technology, Nagoya, Japan

ABSTRACT

This paper presents design feasibility study and development of a new hybrid excitation flux switching motor (HEFSM) as a contender for traction drives in hybrid electric vehicles (HEVs). Initially, the motor general construction, the basic working principle and the design concept of the proposed HEFSM are outlined. Then, the initial drive performances of the proposed HEFSM are evaluated based on 2D-FEA, in which the design restrictions, specifications and target performances are similar with conventional interior permanent magnet synchronous motor (IPMSM) used in HEV. Since the initial results fail to achieve the target performances, deterministic design optimization approach is used to treat several design parameters. After several cycles of optimization, the proposed motor makes it possible to obtain the target torque and power of 333 Nm and 123 kW, respectively. In addition, due to definite advantage of robust rotor structure of HEFSM, rotor mechanical stress prediction at maximum speed of 12,400 r/min is much lower than the mechanical stress in conventional IPMSM. Finally, the maximum torque and power density of the final design HEFSM are approximately 11.41 Nm/kg and 5.55 kW/kg, respectively, which is 19.98% and 58.12% more than the torque and power density in existing IPMSM for Lexus RX400h.

Keywords: Hybrid Excitation Flux Switching Machine (HEFSM); Field Excitation Coil (FEC); Permanent Magnet (PM); Hybrid Electric Vehicle (HEV)

1. Introduction

Hybrid excitation machine (HEM) which consists of permanent magnet (PM) and field excitation coil (FEC) as their combined flux sources, has several unique features that can be applied in HEV drive system. In general HEM can be classified into four categories based on the location of PM and FEC such as 1) both PM and FEC are located at rotor side [1-3], 2) the PM is in the rotor while the FEC is in the stator [4], 3) the PM is in the rotor while the FEC is in the machine end [5,6], and 4) both PM and FEC are located in the stator [7-9]. All HEMs mentioned in the first three consist of a PM in the rotor and can be categorized as "hybrid rotor-PM with FEC machines" while the final machine can be referred as "hybrid stator-PM with FEC machines". Based on its principles of operation, in which the fluxes sources are generated in stator side and moved into the rotor, the fourth machine is also known as "hybrid excitation flux switching machine" (HEFSM) which is getting more popular recently [10,11]. With all active parts located on the stator, HEFSM has the advantages of 1) robust rotor structure which is becoming more suitable to be applied for high-speed drive applications, 2) due to the fact that the all major heats are accumulated in stator part, a simple cooling system can be applied compared with a complex water jacket system used in IPMSM for Lexus RX400h, and 3) the additional FEC can be used to control flux with variable flux capabilities.

Various combinations of stator slot and rotor pole for HEFSM have been developed for high speed applications. For example, 12Slot-10Pole HEFSM has been proposed such as in [12,13]. However, the machine in [12] has a separated PM and C-type stator core that makes it difficult to manufacture, and the design is not yet optimized

*Corresponding author.

for HEV applications while the machine in [13] has a limitation of torque and power production in high current density condition. This is due to insufficient stator yoke width between FEC and armature coil slots resulting in magnetic saturation and negative torque production. To reduce the supply frequency of inverter, 6Slot-5Pole HEFSM has been proposed by the authors. Although the proposed machine has met the target performances, the problem of unbalanced pulling force due to odd number of poles is difficult to overcome [14]. In addition, some researchers have proposed 6Slot-8Pole machines but these types of machines have problems of high torque ripple and back-emf waveforms, which are usual concerns for this type of eight pole machine [15,16].

In this paper, design feasibility and optimization studies are conducted to 12Slot-10Pole HEFSM in effort to achieve the target performances for HEV applications. **Figure 1** illustrates the cross-sectional view of the main machine part of the initial HEFSM. The motor is composed of 12 PMs and 12 FECs distributed uniformly in the midst of each armature coil while the three-phase armature coils are accommodated in the 12 slots for each 1/4 stator body periodically. In this motor, the PMs and FECs produce six north poles interspersed between six south poles. The flux paths caused by both PM and mmf of FEC under open circuit condition are demonstrated in **Figure 2**. The term, "flux switching", is created based on the changes in polarity of each flux in each stator tooth, depending on the motion of the rotor. When the rotor rotates, the fluxes generated by PM and FEC link with the armature coil flux alternately. For the rotor rotation through 1/10 of a revolution, the flux linkage of the armature coil has one periodic cycle and thus, the frequency of back-emf induced in the armature coil becomes ten times of the mechanical rotational frequency.

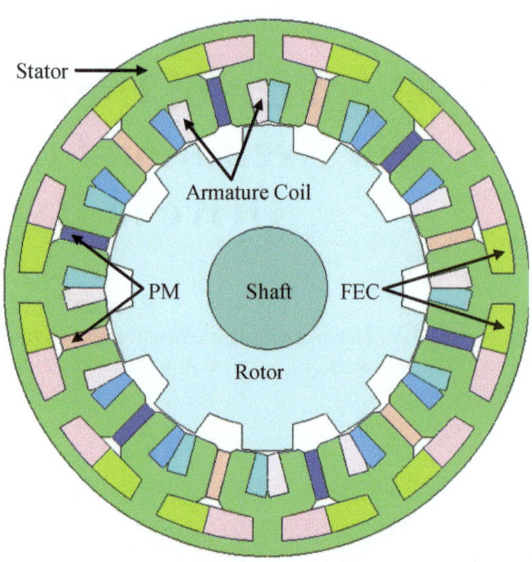

Figure 1. 12Slot-10Pole HEFSM.

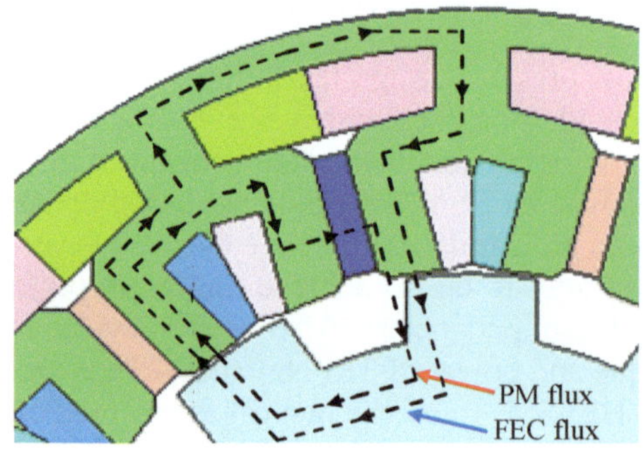

Figure 2. Flux paths of PM and FEC in 12Slot-10Pole HEFSM.

Table 1. HEFSM design restrictions and specifications.

Items	IPMSM	HEFSM
Maximum DC-bus voltage inverter (V)	650	650
Maximum inverter current (A_{rms})	360	360
Maximum J_a (A_{rms}/mm^2)[a]	31	30
Maximum J_e (A/mm^2)[b]	NA	30
Stator outer diameter (mm)	264	264
Motor stack length (mm)	70	70
Shaft radius (mm)	30	30
Air gap length (mm)	0.8	0.8
PM weight (kg)	1.1	1.1
Maximum speed (r/min)	12,400	12,400
Maximum torque (Nm)	333	333
Reduction gear ratio	2.478	2.478
Max. axle torque via reduction gear (Nm)	825	825
Maximum power (kW)	123	>123
Power density (kW/kg)	3.5	>3.5

J_a = current density in armature coil; J_e = current density in FEC.

2. Design Requirements, Restrictions and Specifications for HEV Applications

The design requirements, restrictions and specifications of the proposed HEFSM for HEV applications are similar with IPMSM for Lexus RX400h listed in **Table 1** [17]. The electrical restrictions related with the inverter such as maximum 650 V DC bus voltage and maximum 360 V inverter current are set. The limit of the armature coil current density, J_a and the FEC current density, J_e is set to 30 A_{rms}/mm^2 and 30 A/mm^2, respectively. The weight of the PM is 1.1kg similar with PM volume in IPMSM. The target torque of 333 Nm with reduction gear ration of 2.478 is set, hence, realizing the maximum axle torque via reduction gear of 825 Nm. The maximum operating speed is set to 12,400 r/min and the target power is set to be more than 123 kW. As the proposed HEFSM consists of very simple structure with concentrated winding in all coils, the target motor weight to be designed is set to be less than 35 kg, resulting in that the proposed machine promises to achieve the maximum power density of more than 3.5 kW/kg. Commercial FEA package, JMAG-Studio ver.10.0, released by Japanese Research Institute (JRI) is used as 2D-FEA solver in this design.

3. Initial Performances of the Proposed HEFSM Based on 2D-FEA

Initially, performances of the proposed HEFSM in open circuit condition such as back-emf and cogging torque are analyzed as shown in **Figures 3** and **4**. In **Figure 3**, the amplitude of the fundamental component in which the induced voltage is generated from the flux of PM only is 123.6 V. The back-emf is slightly sinusoidal which results a small amount of cogging torque of approximately 1.06 Nm peak-to-peak. However, when J_e is set to 15 A/mm^2, the induced voltage is slightly distorted and the amplitude is increased to 263.2 V which is more than double of that under no FEC current. This is due to the field strengthening effect by the additional FEC. For load analysis, performances of the machine at maximum J_a and J_e are analyzed. The torque and power obtained at base speed 5731.4 r/min are 175.9 Nm and 105.6 kW, respectively, which is less than the target value. To investigate this issue, the torque versus J_e at various J_a is plotted as depicted in **Figure 5**. It is obvious that the torque is increased with increasing J_e up to certain J_e and begins to decrease when higher J_e is applied as shown in red circle. For instance, at J_a of 30 A_{rms}/mm^2, the maximum torque of 181.66 Nm is obtained when J_e is set to 20 A/mm^2. However, the torque starts to reduce when J_e is set higher than this value. For J_a of 20 A_{rms}/mm^2 and 25 A_{rms}/mm^2, the maximum torque obtained are 152.72 Nm and 170.94 Nm respectively, when J_e is set to 15 A/mm^2. The torque also starts to reduce when J_e is set higher than this value. Similarly, the same phenomenon occurs at the condition of J_a of 5 A_{rms}/mm^2, 10 A_{rms}/mm^2 and 15 A_{rms}/mm^2 where the torque starts to reduce when J_e is set higher than 10 A/mm^2.

To explain this phenomenon, further investigation is examined on the flux density distribution at three conditions 1) before maximum torque, 2) at maximum torque, and 3) after maximum torque. For example, at maximum J_a of 30 A_{rms}/mm^2, flux distribution at J_e of 10 A/mm^2, 20 A/mm^2 and 30 A/mm^2 are investigated as shown in **Figure 6**. It can be seen that for low J_e of 10 A/mm^2, the flux can easily flow to the direction according to its principle as shown in **Figure 6(a)**. Nevertheless, the flux flow to the left part starts to saturate between armature coil upper slot and FEC lower slot marked in blue circle when J_e is set to 20 A/mm^2 as shown in **Figure 6(b)**. In this case, some of the flux from FEC which flow to the right side

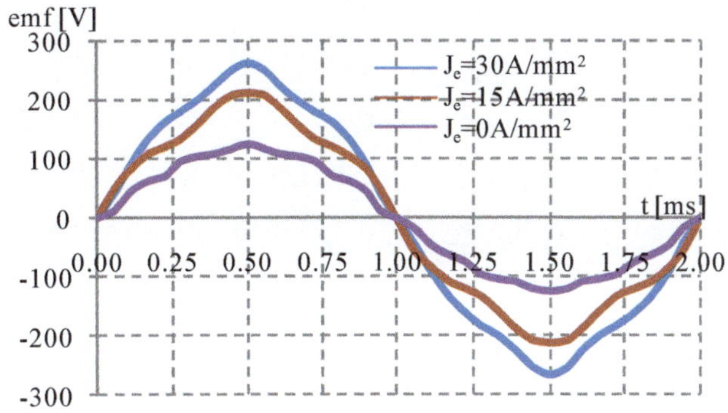

Figure 3. Back-emf at 3000 r/min.

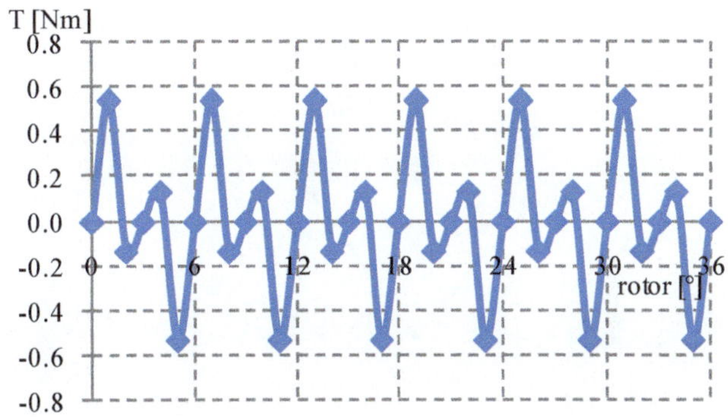

Figure 4. Cogging torque of the original design HEFSM.

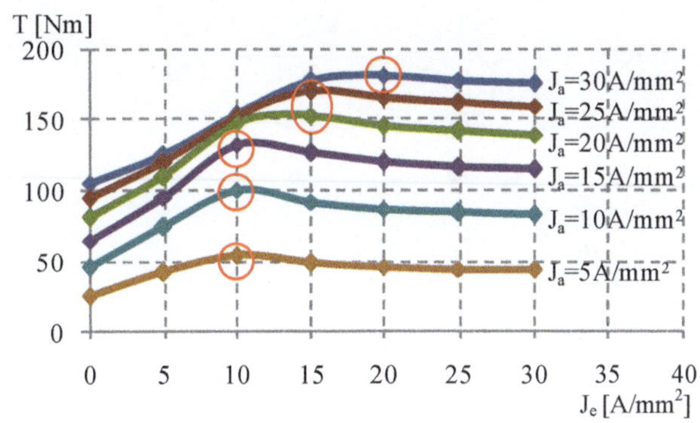

Figure 5. Torque versus J_e at various J_a.

cancelled the flux from PM as shown in red circle which results in reducing the torque production. Hence, for the maximum J_e of 30 A/mm^2 where much FEC flux is generated, the flux flow to the left part is totally saturated between armature coil upper slot and FEC lower slot marked in blue circle. Therefore, much higher flux from the stator outer yoke passes the FEC pitch move towards the PM in the right side. This flux also cancelled the PM flux and some of the flux is forces to flow into the rotor side producing much negative torque as shown in **Figure 6(c)**, hence reducing the torque production. Thus, one of the methods that can be used to overcome this problem is by investigating the suitable length between armature coil upper slot and FEC lower slot to avoid flux saturation.

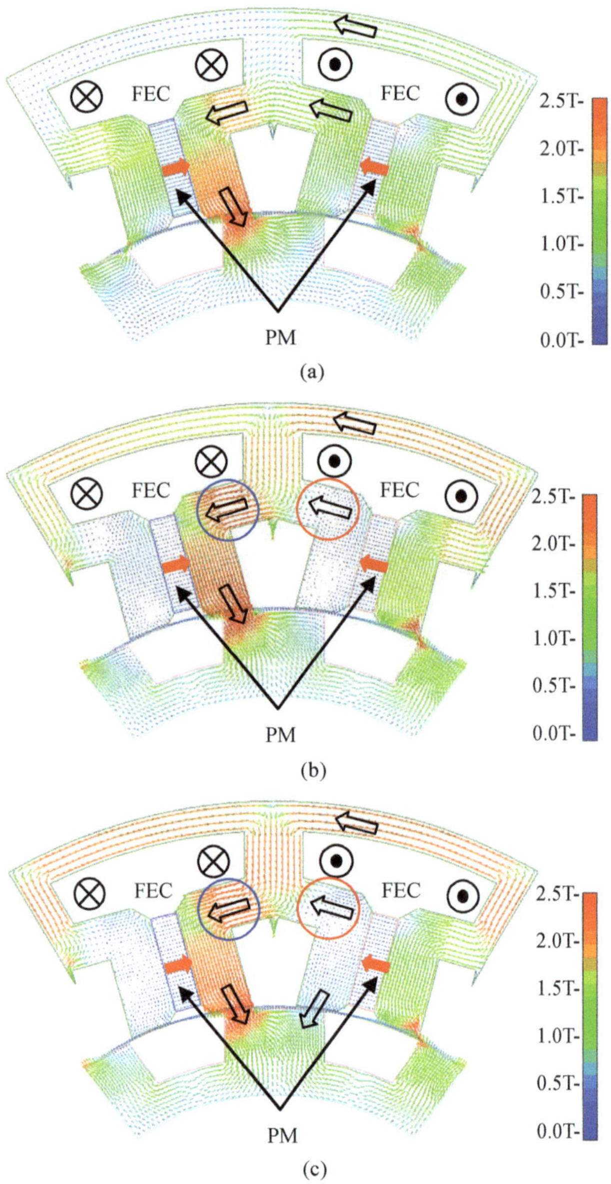

Figure 6. Flux vector diagram of various J_e at maximum J_a of 30 A_{rms}/mm^2 (a) J_e = 10 A/mm², T = 153.4 Nm (b) J_e = 20 A/mm², T_{max} = 181.7 Nm (c) J_e = 30 A/mm², T = 175.9 Nm.

4. Design Methodology for Improvements

To make a simple design, the shape of armature coil slot and air gap between the inner and outer PM are redesign, so that design free parameters of D_1 to D_{10} can be defined as illustrated in **Figure 7**. The first step is carried out by updating the rotor parameters, D_1, D_2 and D_3 while keeping D_4 to D_{10} as constant. As the torque increases with the increase in rotor radius, D_1 which is considered as the dominant parameter that can improve the torque is firstly treated. In this condition, D_4, D_6, D_8, D_9 and D_{10} are simply shifted to the new position by following the movement of D_1, while D_5 and D_7 are kept constant. Then, by selecting D_1 at its maximum performance, both rotor pole width D_2 and rotor pole depth D_3 are varied. Once the maximum performance from the combination of D_2 and D_3 is determined, the second step is carried out by changing the FEC slot parameters D_4, D_5 and D_6 while keeping the other parameters constant. Then, by using the combination of D_4 to D_6 that bring out the maximum performance at the second step, the third step is carried out by varying the armature coil slot parameters D_7 and D_8 with keeping other parameters constant. The necessary armature coil slot area, S_a is determined by varying armature coil depth, D_7 and armature coil width, D_8 to accommodate natural number of turns, N_a for armature coil. Furthermore, to ensure the PM is not demagnetized at temperatures as high as 180˚C, D_9 and D_{10} are adjusted with keeping the same PM volume. The method of changing D_1 to D_{10} is treated repeatedly until the target performances are achieved.

All design parameters are adjusted with keeping air gap length of 0.8 mm constant under maximum J_a and J_e. In addition, at the final design, the corners circled in **Figure 7** are designed as a curve to ensure all flux at the edge of the shape flow more smoothly, hence increases the performance of the machine. For the rotor inner pole, the curve designed not only increased the flux flows but also increase the rotor mechanical strength of the machine, make it more robust to work in high speed condition. Finally, after few cycle of optimization, the machine satisfied the target requirements and performances for HEV applications. The cross sectional views of the final design HEFSM is depicted in **Figure 8**, while details of final parameters are listed in **Table 2**. The differences between the initial and the final design HEFSM are 1)

Table 2. Initial and final design parameters.

	Details	Initial	Final
	PM volume (kg)	1.1	1.1
D_1	Rotor radius (mm)	80.2	88.2
D_2	Rotor pole width (mm)	12.5	9.5
D_3	Rotor pole depth (mm)	12.2	23.2
D_4	Permanent magnet height (mm)	24.0	15.0
D_5	FEC slot pitch (mm)	7.4	20.0
D_6	Stator outer core thickness (mm)	7.4	8.0
D_7	Armature coil width (mm)	8.0	6.0
D_8	Armature coil depth (mm)	19.1	32.71
D_9	Distance between air gap and PM (mm)	2.0	0.5
D_{10}	Distance between FEC and PM (mm)	4.0	0.5
N_a	No. of turns of armature coil	7	9
T	Torque (Nm)	175.86	334.5
N	Speed (r/min)	5731.4	3701.2
P	Power (kW)	105.55	129.6
pf	Power factor	0.368	0.452

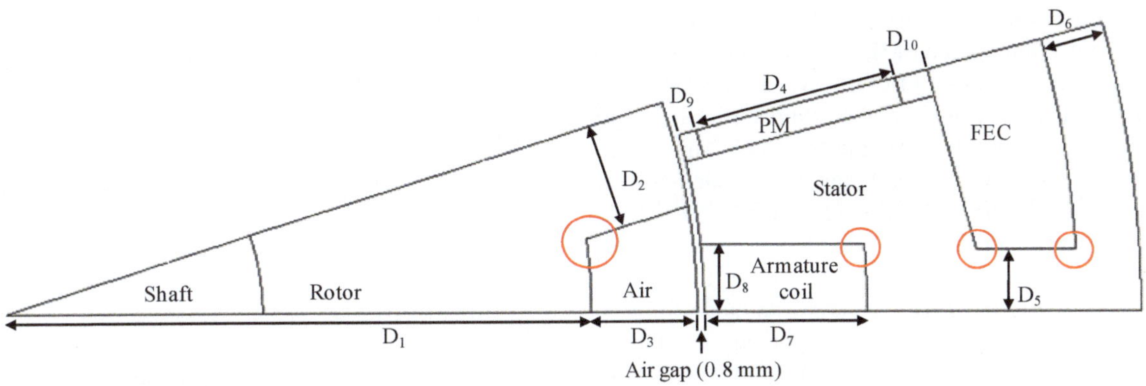

Figure 7. Design parameter defined as D_1 to D_{10}.

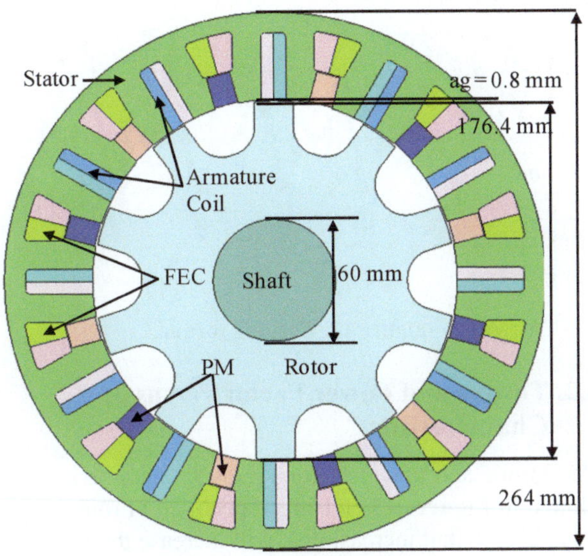

Figure 8. Final Design HEFSM.

the rotor radius of the final design is longer than the initial rotor radius which gives more torque as had been expected, 2) the armature coil width of the final design is less than the initial design, but has high armature coil depth which results in more number of turns, 3) the FEC slot area is reduced approximately 40% from the initial design to cover some volume of stator yoke used for armature coil, 4) the final design has less PM depth with high PM width to keep the same PM volume of 1.1 kg, 5) the final design has no gap between armature coil upper slot and FEC lower slot which solved the flux saturation problem, 6) the stator outer core thickness is higher than the initial design to allow more flux to flow smoothly, and 7) the final design has a stator yoke with a straight "I shape" that makes the flux flow more easily into the rotor. As a proof, **Figure 9** illustrates the flux distribution of the final design HEFSM for J_e of 20 A/mm^2 and 30 A/mm^2 with maximum J_a of 30 A_{rms}/mm^2. As the gap between armature coil upper slot and FEC lower slot in the final design HEFSM is expanded and considered

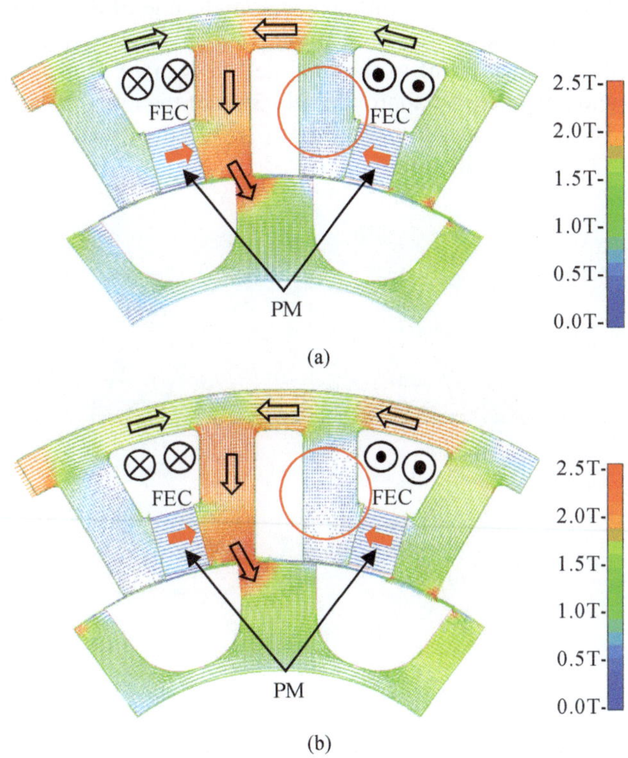

Figure 9. Flux vector diagram of various J_e at maximum J_a of 30 A_{rms}/mm^2 (a) J_e = 20 A/mm^2, T = 280.4 Nm (b) J_e = 30 A/mm^2, T = 334.4 Nm.

negligible, the magnetic saturation caused by higher FEC current is relaxed. Thus, the improved design maintains the same torque for both current density conditions and enables to extract higher power factor.

5. Results and Performances of the Final Design HEFSM

5.1. Flux Path at Open Circuit Condition

The open circuit field distribution for PM and FEC of the final design HEFSM are investigated based on 2D-FEA as illustrated in **Figure 10**. **Figure 10(a)** illustrates the flux path due to mmf of PM only, while **Figure 10(b)** represents the combination of flux line from both PM and mmf of FEC at maximum FEC density, J_e of 30 A/mm^2. In **Figure 10(a)**, it is obvious that almost 100% flux of PM flow in the stator iron around the FEC. This yields negligible cogging torque and almost no back-emf at open-circuit condition under the maximum speed operation, which makes it easy to protect the switching devices when the inverter is shut down due to some failures. In contrast, from **Figure 10(b)**, a large amount of fluxes flow to the rotor side by field strengthening excitation, resulting in the maximum torque production with the aid of hybrid excitation.

Furthermore, **Figures 11** and **12** illustrate the comparison between back-emf and cogging torque of the initial and final design at 3000 r/min, respectively. It is clear that, the back emf of the final design is more sinusoidal and much less by approximately 36% of the initial design. The final cogging torque is also reduced by more than 50% of the initial design.

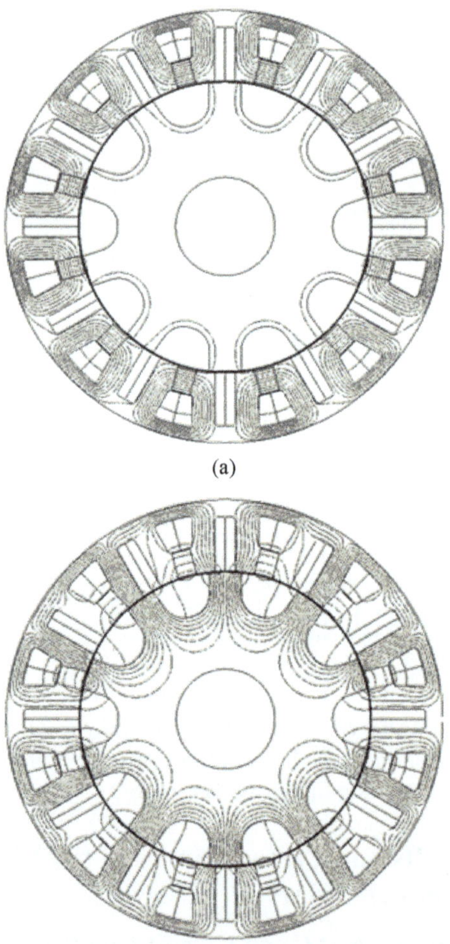

(a)

(b)

Figure 10. Flux path of the final design HEFSM at open circuit condition (a) PM only (b) PM and maximum FEC current density of 30 A/mm^2.

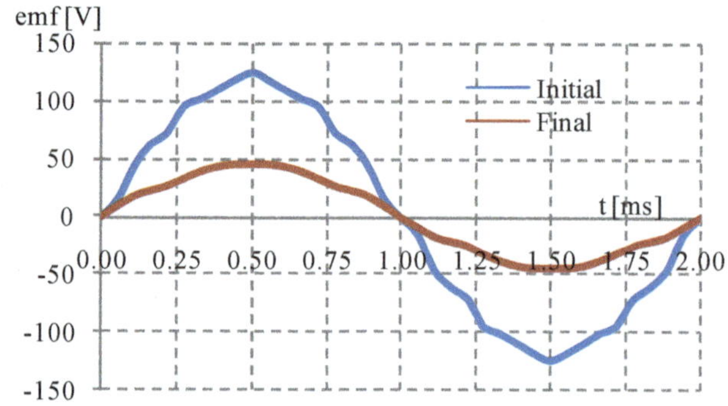

Figure 11. Back-emf.

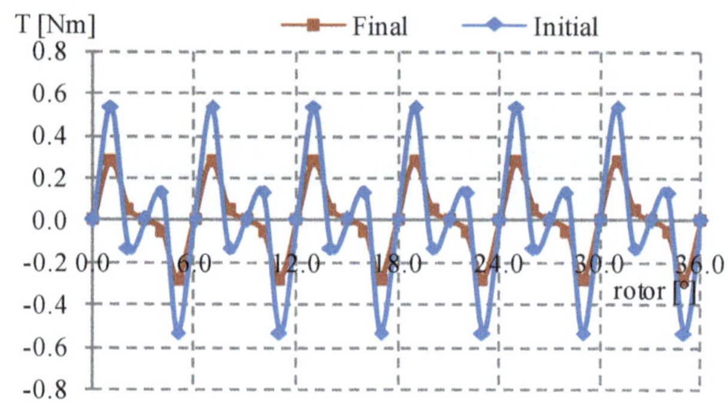

Figure 12. Cogging torque.

5.2. Torque and Power Factor versus J_e Characteristics

The torque and power factor versus J_e characteristics are plotted in **Figures 13** and **14**, respectively. From the plot, it is obvious that increasing J_a will increase the torque but will reduce the power factor. To equilibrate this situation, J_e is increased so that the power factor can be improved and kept constant even if J_a is very high. The plots clearly show that maximum torque of 334.4 NM is obtained when J_a and J_e are set to 30 A/mm^2 as their maximum with the power factor of 0.452. However, for low J_a of less than 15 A_{rms}/mm^2, the torque are slightly reduced with the increasing of J_e of more than 20 A/mm^2. This situation occurs due to excessive FEC flux that generates negative torque thus reducing the performances, similar with the original HEFSM discussed previously. The comparison between the torque and power factor versus J_e under maximum J_a for the initial and final design HEFSM is depicted **Figure 15**. The torque and power factor of the final design HEFSM increased approximately 50% and 23%, respectively compared with the original design.

5.3. Torque and Power versus Speed Characteristics

The torque versus speed characteristics of the IPMSM and the final design HEFSM are plotted in **Figure 16**.

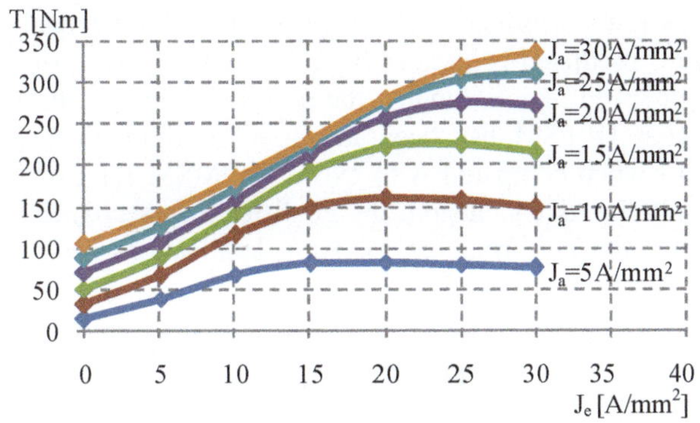

Figure 13. Torque vs J_e.

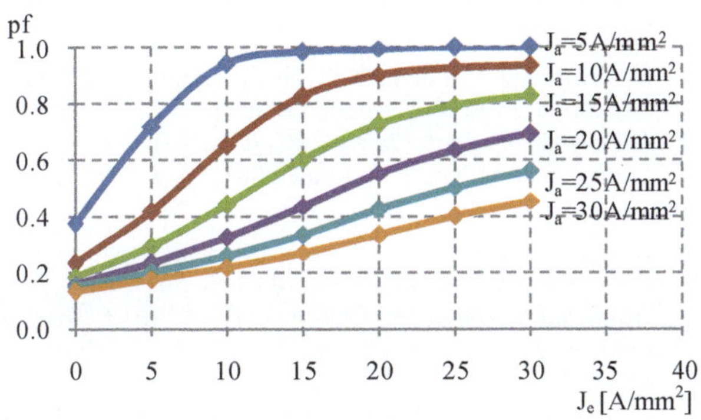

Figure 14. Power factor vs J_e.

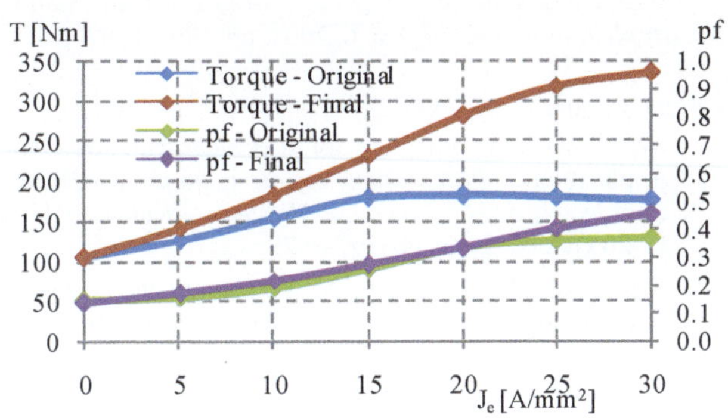

Figure 15. Torque and power factor vs J_e.

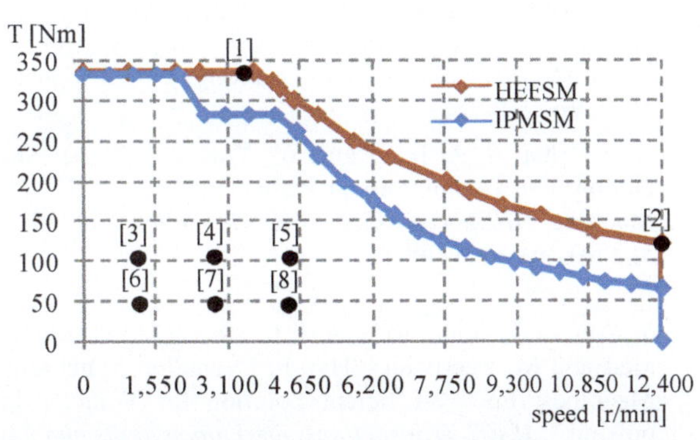

Figure 16. Torque vs speed characteristics.

The maximum torque obtained for IPMSM and HEFSM are 333 Nm and 334.4 Nm, respectively. It is obvious that the HEFSM has better torque condition and produced much higher torque capability in high speed region. Meanwhile, **Figure 17** illustrates the power versus speed characteristics of the IPMSM and the final design HEFSM. From the graph, 1) at maximum torque, the power achieved for IPMSM and the final design HEFSM is 72.1 kW and 129.6 kW, at the speed of approximately 2100 r/min and 3701 r/min, respectively, 2) the maximum power obtained is 123 kW for IPMSM and 162.8 kW for HEFSM 3) the average power of the IPMSM and the HEFSM at normal driving mode of 3000 - 6000 r/min are 113.7 kW and 133.7 kW, respectively, which proves that the HEFSM has better performance than IPMSM in frequent driving condition. The total weight of the final design HEFSM including stator iron, rotor iron, PM, armature coil, FEC, and estimation of both coil ends is 29.3 kg, which is 16.2% less than the estimated of 35 kg for IPMSM. Thus, the maximum torque density and maximum power density are 11.41 Nm/kg and 5.55 kW/kg, respectively, which is much higher than the target requirements for HEV applications. The maximum torque and power density of the final design HEFSM are increased approximately by 20.0% and 36.9%, respectively compared to 9.51 Nm/kg and 3.51 kW/kg of existing IPMSM.

5.4. Motor Loss and Efficiency

The motor loss and efficiency are calculated considering iron losses in all laminated cores, and copper losses in armature coil and FEC. The detailed loss and motor efficiency of the final design HEFSM at maximum torque, maximum power, and frequent operating point under light load driving condition noted as No. 1 to No. 8 in **Figure 16** are illustrated in **Figure 18**. At high torque operating points No. 1, the efficiency is slightly degraded due to increase in copper loss while at high-speed operating point No. 2, the motor efficiency is degraded due to increase in iron loss. Furthermore, at frequent driving operation No. 3 to No. 8 under low load condition, the proposed machine achieves relatively high efficiency of more than 93%.

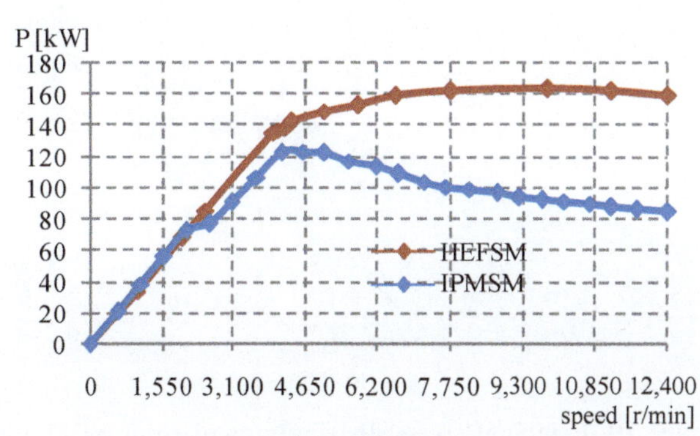

Figure 17. Power versus speed characteristics.

5.5. Rotor Mechanical Strength

The mechanical stress prediction of rotor structure is calculated by centrifugal force analysis based on 2D-FEA as depicted in **Figure 19**. The maximum stress at 12,400 r/min reaches 46 MPa and 28 MPa for the original and final design, respectively, which is much smaller than allowable maximum stress of 300 MPa in conventional electromagnetic steel. This is a great advantage of the final design HEFSM that makes it more applicable and suitable to operate in high-speed application compare to conventional IPMSM.

6. Conclusion

In this paper, design feasibility studies and performance analysis of 12Slot-10Pole HEFSM for traction drive in the target HEV have been presented. The design refinement has been clearly demonstrated and finally achieved the target performances. In addition, the rotor mechanical stress predicted is good enough for the machine to run in high-speed region. Finally, the power density of the final design HEFSM has been increased of more than one third when compared with existing IPMSM for LEXUS RX400h. As conclusion, the goal of this research to get maximum performances for HEV applications has been successfully achieved.

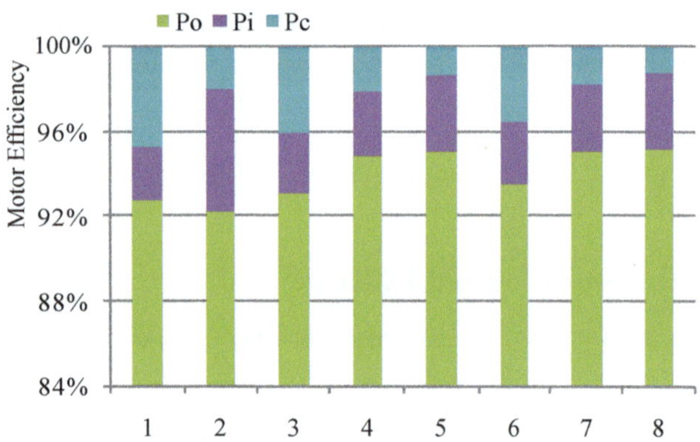

Figure 18. Motor efficiency.

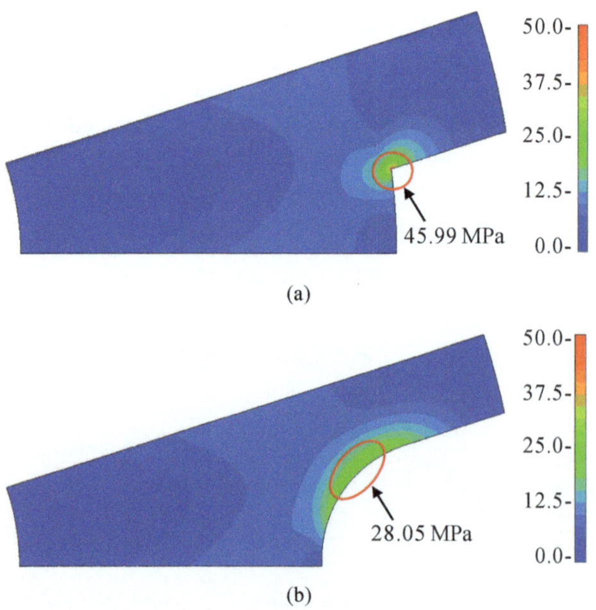

Figure 19. Principal stress distributions of rotor at 12,400 r/min (a) initial design (b) final design.

REFERENCES

[1] X. G. Luo and T. A. Lipo, "A Synchronous/Permanent Magnet Hybrid AC Machine," *IEEE Transactions on Energy Conversion*, Vol. 15, No. 2, 2000, pp. 203-210.

[2] N. Naoe and T. Fukami, "Trial Production of a Hybrid Excitation Type Synchronous Machine," *Proceedings of IEEE International Electric Machines and Drives Conference*, Cambridge, 17-20 June 2001, pp. 545-547.

[3] D. Fodorean, A. Djerdir, I. A. Viorel and A. Miraoui, "A Double Excited Synchronous Machine for Direct Drive Application-Design and Prototype Tests," *IEEE Transactions on Energy Conversion*, Vol. 22, No. 3, 2007, pp. 656-665.

[4] J. A. Tapia, F. Leonardi and T. A. Lipo, "Consequent-Pole Permanent-Magnet Machine with Extended Field-Weakening Capability," *IEEE Transactions on Industry Applications*, Vol. 39, No. 6, 2003, pp. 1704-1709.

[5] T. Kosaka and N. Matsui, "Hybrid Excitation Machines with Powdered Iron Core for Electrical Traction Drive Applications," *Proceedings of International Conference on Electrical Machines and Systems*, Wuhan, China, 17-20 October 2008, pp. 2974-979.

[6] T. Kosaka, M. Sridharbabu, M. Yamamoto and N. Matsui, "Design Studies of Hybrid Excitation Motor for Main Spindle Drive in Machine Tools," *IEEE Transactions on Industrial Electronics*, Vol. 57, No. 11, 2010, pp. 3807-3813.

[7] Z. Chen, Y. Sun and Y. Yan, "Static Characteristics of A Novel Hybrid Excitation Doubly Salient Machine," *Proceedings of the Eight International Conference on Electrical Machines and Systems*, Nanjing, 27-29 September 2005, pp. 718-721.

[8] K. T. Chau, J. Z. Jiang and W. Yong, "A Novel Stator Doubly Fed Doubly Salient Permanent Magnet Brushless Machine," *IEEE Transactions on Magnetics*, Vol. 39, No. 5, 2003, pp. 3001-3003.

[9] Y. Amara, L. Vido, M. Gabsi, E. Hoang, A. H. Ben Ahmed and M. Lecrivain, "Hybrid Excitation Synchronous Machines: Energy-Efficient Solution for Vehicles Propulsion," *IEEE Transactions on Vehicular Technology*, Vol. 58, No. 5, 2009, pp. 2137-2149.

[10] D. Fodorean, A. Djerdir, I. A. Viorel and A. Miraoui, "A Double Excited Synchronous Machine for Direct Drive

Application-Design and Prototype Tests," *IEEE Transactions on Energy Conversion*, Vol. 22, No. 3, 2007, pp. 656-665.

[11] E. Sulaiman, T. Kosaka and N. Matsui, "Design Improvement and Performance Analysis of 12S-10P Permanent Magnet Flux Switching Machine with Field Excitation Coil," *Journal of Electrical System*, Vol. 8, No. 4, 2012, pp. 425-432.

[12] H. Wei, M. Cheng and G. Zhang, "A Novel Hybrid Excitation Flux-Switching Motor for Hybrid Vehicles," *IEEE Transactions on Magnetics*, Vol. 45, No. 10, 2009, pp. 4728-4731.

[13] E. Hoang, M. Lecrivain and M. Gabsi, "A New Structure of a Switching Flux Synchronous Polyphased Machine with Hybrid Excitation," *Proceedings of European Conference on Power Electronics and Applications*, Aalborg, 2-5 September 2007, pp. 1-8.

[14] E. Sulaiman, T. Kosaka and N. Matsui, "Design Optimization and Performance of A Novel 6-Slot 5-Pole PMFSM with Hybrid Excitation for Hybrid Electric Vehicle," *IEEJ Transactions on Industry Applications*, Vol. 132, No. 2, 2011, pp. 211-218.

[15] A. Zulu, B. C. Mecrow and M. Armstrong, "A Wound-Field Three-Phase Flux-Switching Synchronous Motor with All Excitation Sources on The Stator," *IEEE Transactions on Industry Applications*, Vol. 46, No. 6, 2010, pp. 2363-2371.

[16] E. Sulaiman, T. Kosaka and N. Matsui, "High Power Density Design of 6Slot-8Pole Hybrid Excitation Flux Switching Machine for Hybrid Electric Vehicle," *IEEE Transactions on Magnetics*, Vol. 47, No. 10, 2011 pp. 4453-4456.

[17] M. Kamiya, "Development of Traction Drive Motors for the Toyota Hybrid Systems," *IEEJ Transactions on Industry Applications*, Vol. 126, No. 4, 2006, pp. 473-479.

A Financial Approach to Evaluate an Optimized Combined Cooling, Heat and Power System

Shahab Bahrami*, **Farahbakhsh Safe**
Electrical Engineering Department, Sharif University of Technology, Tehran, Iran

ABSTRACT

Iran's removing subsidy from energy carrier in four years ago leads to spike electricity price dramatically. This abrupt change increases the interest on distributed generation (DG) because of its several benefits such as lower electricity generation price. In Iran among all type of DGs, because of wide natural gas network infrastructure and several incentives that government legislated to support combined cooling, heat and power (CCHP) investors, this type of technology is more prevalent in comparison with other technologies. Between existing CCHP technologies, certain economic choices are to be taken into account. For different buildings with different load curves, suitable size and operation of CCHP should be calculated to make the project more feasible. If CCHP does not well suited for a position, then the whole energy efficiency would be plunged significantly. In this paper, a model to find the optimal size and operation of CCHP and auxiliary boiler for any users is proposed by considering an integrated view of electricity and natural gas network using GAMS software. Then this method is applying for a hospital in Tehran as a real case study. Finally, by applying COMFAR III software, useful financial parameters and sensitivity analysis are calculated.

Keywords: Combined Cooling Heat and Power (CCHP); Energy Hub; Optimal Size; Financial Analysis

1. Introduction

The development of poly-generation smart grids represents an interesting solution to satisfy electricity and heat demand and emission reduction [1,2]: poly-generation smart grids generate electricity, heating and cooling thermal power close to end users, solving the main disadvantages of the centralized generation approach, due to energy transmission [3]. In fact, the distributed generation approach has several benefits over the others, such as: 1) reduction of transmission and distribution costs: about the 30% of the costs related to electricity supply relates to these costs. Local connections do not generate high capital costs and energy losses for long distances to be wired with overhead facilities; 2) decrease of energy dissipation: piping and conversion devices dissipate almost 6% of produced energy, [4] increasing costs and emissions; in a smart grid, these kinds of losses are avoided; 3) increase of energy efficiency: the simultaneous supply of electrical and thermal demand allow to reduce energy waste, improving system global efficiency; since thermal energy is less easily transported than electricity, distributed generation approach (production close to users) is essential. Because of mentioned benefits deregulation has evolved in all three sectors of the power system (*i.e.* generation, transmission, and distribution) from centralized to a decentralized status. One of the main concepts in deregulation is Micrograds which are used at the distribution level [1]. Microgrid, with its decentralized electricity generation, combined with on site production of heat, could provide reliable and electric power as well as heat and cooling to its consumers at an economic cost. Nowadays, following the expansion of natural gas networks and also benefits of this energy carrier such as lower emission level and prices, CCHP technologies have attained unprecedented level of popularity as one of the most important distributed energy resources. The US Environmental Protection Agency (EPA) found that the primary applications for CHP in the commercial sectors are those building types with relatively high and coincident electric and hot water demands such as hotels, universities and hospitals. Use of CHP thermal output for absorption cooling could increase the efficiency of CHP system in commercial sectors, this integrated system which provides these demands named

*Corresponding author.

combined cooling heat and power (CCHP).

One of the major factors for users to choose a CCHP system is the overall cost of CCHPs which is largely dependent on its size [1]. Hence finding the optimized size of a CCHP is economically important.

Generally, an optimized CCHP can be evaluated by analyzing two main factors: costs and benefits. Cost is one of the main components in nearly all DG financial analysis, but is inadequate for complete evaluations. Furthermore, reliability enhancements [5], power cost saving, power loss and emission reduction [6] are also key elements in deciding which CCHP should be installed.

The cost of generation of electricity, heat and cooling from a CCHP can be classified into capital investment cost, operation and maintenance (O&M) costs, fuel cost and depreciation cost. On the other hand, the benefits from the CCHP placement can be classified into power loss reduction, significantly decreasing the expected energy not supplied which is a favorable effect in a power system.

CHP can inject its power directly into distribution feeders and by alleviating transmission losses the benefits of power loss reduction become quite clear [6]. Moreover, reliability enhancement has received substantial attention as it reduces the costs of losses incurred by utility customers as a result of power failures [7].

All of these costs and benefits are calculated in terms of Present Value Factor (PVF), accumulated over the economic life of the respective equipment. It is a common practice for a decision maker to translate future cash flows into their present values.

From a number of recent publications [1-9], it can be seen that in a deregulated power system, each individual distribution company may wish to determine the costs and benefits of DG planning under different circumstances. It is difficult to find a single planning method that satisfies all objectives simultaneously. In this paper, a value-based planning method for CCHP placement based on the energy hub concept is proposed. The proposed method takes the benefits and costs of CCHP placement into account and determines the optimal sizing for CCHP placement. Test results show that with proper size selection, CCHP placement can be used to improve service reliability, and reduce power loss and emission costs [8].

The survey of previous literature on DER (Distributed Energy Resource) planning as well as optimal DER deployment in the radial (conventional) as well as meshed-type distribution systems indicates that a number of similar studies [9-16], encompass sensitivity analysis to modern soft computing techniques, such as genetic algorithms (GAs), evolutionary programming (EP), DER-CAM, etc. Special mention can be made to [9] and [14]. Reference [9] proposes a method for distributed generator planning based on GAs and considers customer interruption cost (CIC) as the benefit of distributed generators placement but the benefit of waste heat recovery is not considered. Sheikhi *et al.* [10] proposed an optimization model to find the optimal size and operation for combined cooling, heat and power systems, in order to reduce power loss and enhance service reliability of the system. Reference [11] gives the economic analysis of the microgrid, which evolves from the existing low-voltage (LV) network, on the basis of cost and benefit of potential reliability improvements. Reference [12] examines some successful experiences of DG in Central Virginia Electric Cooperative. References [13,14] use the DER-CAM technique to minimize the cost and proposes a model for location and optimal selection of DER system, which could work parallel to the macrogrid. Reference [15] uses the evolutionary-algorithmic (EA) approach to optimize placement of DG in a meshed microgrid. Reference [16] focuses on the optimal distributed generator placement in a conventional radial distribution system.

The contents of this paper are organized into the following six sections. Determination of the optimum operational point, the energy hub concept and a brief overview of the energy hub modeling are presented in Section 2. Section 3 discusses the potential benefits of deploying energy hubs. Section 4 provides detailed formulations of the problem and case studies are debated in detail in section 5. Finally, conclusions are drawn in "List of Symbols and Abbreviations".

2. Energy Hub Concept and Modeling

Some conceptual approaches for an integrated view of transmission and distribution systems with distributed generation have been published. Besides "energy-services supply systems" [17], "basic units" [18], and "micro grids" [19], so-called "hybrid energy hubs", are suggested, where the term "hybrid" implies the use of multiple energy carriers [20,21]. An energy hub is considered a unit where different energy carriers can be converted, conditioned, and maybe stored. It represents an interface between different energy infrastructures and/or loads. Some worth mention works in this field are listed in **Table 1**. **Figure 1** demonstrates an example of an energy hub.

The CHP device couples the three energy systems at the same time that produces electricity, heat and cooling from natural gas. The absorption chiller converts wasted heat from CHP or produced one from boiler to cooling power.

Consider a converter device as depicted in **Figure 2** converts an input energy carrier α into β. Input and output power flows are not independent; they are considered to be coupled,

$$L_\beta = c_{\alpha\beta} \times P_\alpha \quad (1)$$

where P_α and L_β are the steady state input and output energy flows respectively, $c_{\alpha\beta}$ is the coupling factor which defines the coupling between input and output energy flow. For a simple converter device with one input and one output, the coupling factor corresponds to the converter's steady state energy efficiency.

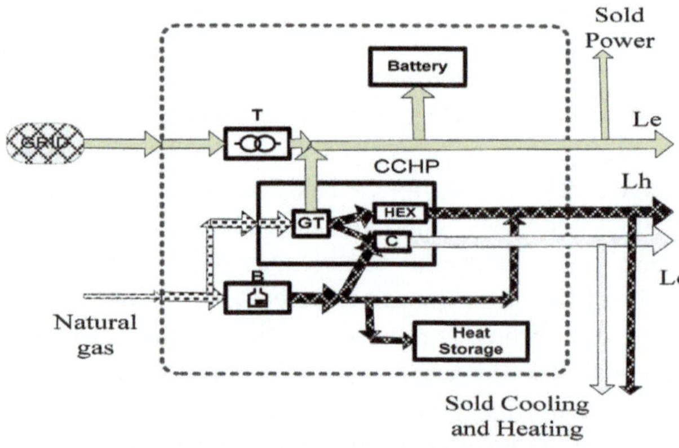

Figure 1. An energy hub containing an electric transformer, a CCHP, a boiler (B), battery and heat storage.

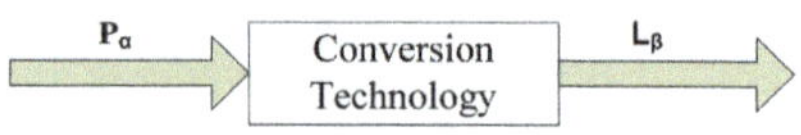

Figure 2. Model of energy converter.

A general model covering all types of couplings can be stated all power inputs $P_\alpha, P_\beta, \cdots, P_\omega$ and outputs $L_\alpha, L_\beta, \cdots, L_\omega$ in vector form and enables the formulation of a multi-input, multi-output power conversion as follow [7]:

$$\begin{bmatrix} L_\alpha \\ L_\beta \\ \vdots \\ L_\omega \end{bmatrix} = \begin{bmatrix} C_{\alpha\alpha} & C_{\beta\alpha} & \cdots & C_{\omega\alpha} \\ C_{\alpha\beta} & C_{\beta\beta} & \cdots & C_{\omega\beta} \\ \vdots & \vdots & \ddots & \vdots \\ C_{\alpha\omega} & C_{\beta\omega} & \cdots & C_{\omega\omega} \end{bmatrix} \begin{bmatrix} P_\alpha \\ P_\beta \\ \vdots \\ P_\omega \end{bmatrix} \quad (2)$$

3. Proposed Optimization Method

Investigating the best size of CCHP, auxiliary boiler, heating and electricity storage devices of an energy hub system have a substantial effect on the users' benefits. This section proposes an analytical method to determine the most advantageous selection.

To find the best energy hub elements between existing

Table1. A summary of major works on Energy Hub (EH).

Author	Approach	Publish year
Andersson [22]	A modeling framework for future energy systems	2007
Favre-Perrod [23]	VoFEN (EH concept)	2005
Hemmes [24]	EH concept	2006
Geidl [25]	EH modeling	2007
Kopple [26]	EH reliability	2012
Galus [27]	PHEV in EH	2011
Kienzle [28]	DSM &Uncertainty in EH	2011
Arnold [29]	Renewable &MPC in EH	2011
Kraus [30]	A country as EH	2011
Schuzle [31]	Renewable modeling &Pricing in EH	2011
Parisio [32]	A robust optimization on uncertainty environment in EH	2012
Carradore [33]	Voltage regulation in EH	2009
Robertson [34]	EH for multiple energy carrier	2009
Ramiraz-elizendo [35]	Unit commitment in EH	2011
Velez [36]	Control strategy for EH	2011
Chehreghani [37]	Mathematical optimization modeling of EH	2008
Syed [38]	EH for plug-in hybrid fuel cell vehicle	2010
Sheikhi [1]	Financial ana lysis &optimal size & operation of EH	2012
Haghifam [39]	Multi objective electric distribution system expansion planning	2009

choices the value based planning is employed. The costs of energy hub placement include investment, operation and maintenance cost (O & M) of CHP, absorption chiller, auxiliary boiler and heat and electricity storage devices. To find the benefit term for CCHP placement, it is assumed that the outputs of CCHP are fully sold. Emission reduction and reliability enhancement are other major terms that would be added to form the total benefit. This planning method attempts to realize the minimum cost solution where the overall benefits can be maximized.

The total input flow P_g splits up to different converters such as CCHP and auxiliary boiler in **Figure 1**. Dispatch factor, γ, specifies how much of the total input power P_g flows through the CCHP. At the same time, absorption chiller uses heating power to generate cooling. α and β show the proportion of heating power that is produced by auxiliary boiler and CHP, consumed by the chiller. To investigate optimal value of this parameter, an appropriate objective function which is considered the net benefit for the energy hub system has to be formed.

For modeling CO_2 emission effects on power generation two factors, χ_e and χ_g, are introduced. These parameters show how many dollars you have to pay as penalty more than the electricity and gas price to compensate their harmful results for emitting greenhouse gases.

Based on the social costs of carbon emissions, it is assumed that the price of carbon is \$30 per ton (\$0.03 per kg) which needs to increase with inflation rates [40]. With these extensions, multi-period multi-carrier optimal power flow and limitations can be stated as nonlinear programming (NLP) structure:

$$L_e(N)+KK(N)+SE_{in}(N)-SE_{out}(N) = \eta_{ee}P_e(N)+\gamma_N\eta_{ge}P_g(N) \quad (3)$$

$$L_h(N)+H(n) = \left(\eta_{gf}(1-\gamma(N))\alpha(N)+\beta(N)\gamma(N)\eta_{gh}\right)P_g(N) + S_{outh}(N)-S_{inh}(N) \quad (4)$$

$$L_C(N)+CC(n) = \left(\eta_{gf}(1-\gamma(N))(1-\alpha(N))+(1-\beta(N))\gamma(N)\eta_{gh}\right) \times\eta_{ac}P_g(N)+S_{outc}(N)\times\eta_{ac} \quad (5)$$

Equality constraints which describe the electricity, heat and cooling flow through the hub are given in (3 - 5) *KK*(*N*), *CC*(*N*) and *H*(*N*) indicates the amount of exported electricity, cooling and heat in each hour.

$$SE(N)=SE(N-1)+SE_{in}(N)-SEout(N) \quad (6)$$

$$S(N)=0.985\times S(N-1)+\sinh(N) - south(N)-soutc(N) \quad (7)$$

$$SE_{in}(N)\le 0.25SE_m \quad (8)$$

$$SE_{out}(N)\le 0.25SE_m \quad (9)$$

$$SE(N)\le SE_{\max} \quad (10)$$

$$\sinh(N)\le 20\ \text{kw}\cdot\text{h} \quad (11)$$

$$soutc(N), south(N)\le 15\ \text{kw}\cdot\text{h} \quad (12)$$

$$S(N)\le S_m \quad (13)$$

The amount of heat in the heating storage devices and electricity storage units are as in (6, 7). The coefficient 0.985 in Equation (7) shows 1.5% of stored heat gets lost per hour. Constrains which describe in (8-12) indicate the maximum and minimum input and output of battery and heat storage. S_m in (13) is the maximum capacity of heat storage.

$$P_g(N)\le P_{gm} \quad (14)$$

$$P_e(N)\le P_{em} \quad (15)$$

$$\forall N: 0\le\alpha(N),\beta(N),\gamma(N)\le 1 \quad (16)$$

Limitation of the dispatch factors in (16) has to be regarded as well.

$$Z_2=\Sigma\left(\sinh(N)+south(N)+soutc(N)\right)P_{hT} \quad (17)$$

$$Z_1=\Sigma P_g(N)(g+x_g)+P_e(N)(e(N)+X_e) \quad (18)$$

Equation (17) is the price of pumping heat auxiliary boiler or CHP in to the heat storage and from heat storage to the heating load. P_{hT} is the price of pumping heat for each kWh.

$$\begin{aligned}\text{Benfit} &= \sum_N L_e(N)(e(N)+X_e)+\Sigma kk(N)(e(N)+X_e+b) \\ &+\Sigma\gamma_{(N)}P_g(N)\left(\frac{g}{\eta_{gf}}\eta_{gh}\right)+\Sigma(1-\gamma(N))P_g(N)g \\ &+\Sigma(1-\gamma(N))\eta_{gf}(1-\alpha(N)) \\ &+(1-\beta(N))\gamma(N)\eta_{gh}\eta_{ac}P_g(N)C(N)30\times CPVF \\ &+\left(\left[EENS^{old}-EENS^{OPT}\right]\times VOLL\right)CPVF\end{aligned} \quad (19)$$

$$\text{cost} = \text{capital cost}+(\text{running cost})CPVF \quad (20)$$

$$\begin{aligned}\text{capital cost} &= \text{costCHP}[\$/\text{kw}]\times\text{cap } CHP \\ &+\text{costBoiler}[\$/\text{kw}]\times cap\ Boiler \\ &+\text{costChiller}[\$/\text{kw}]\times cap\ chiler \\ &+\text{Heating storoge cost}\times S_m \\ &+\text{Electircity storog cost}\times SE_M \\ &+\text{Fixed Cost}\end{aligned} \quad (21)$$

$$\text{runing cost} = (Z_1 + Z_2) \times 30 + \text{maintenance Cost} \quad (22)$$

$$C_m = \frac{360}{N} (\text{maintenance cost per kW} \cdot \text{h}) \sum P_g(N) \times \gamma(N) \quad (23)$$

C_m is the maintenance cost per year. The main part of maintenance cost is related to CHP then other maintenance costs have been ignored. Equations (17)-(23) formulate the benefit and cost in details.

$$\eta_{ge} \gamma(N) P_g(N) \le \text{cap}CHP \quad (24)$$

$$\eta_{gf} (1 - \gamma(N)) P_g(N) \le CapBoiler \quad (25)$$

$$\begin{aligned} &(\eta_{gf} (1 - \gamma(N))(1 - \alpha(N)) \\ &+ (1 - \beta(N)) \gamma(N) \eta_{gh} P_g(N)) \eta_{ac} \le \text{capChiller} \end{aligned} \quad (26)$$

$$CC(N) \le C_M \quad KK(N) \le P_{seM} \quad H(N) \le H_M \quad (27)$$

Inequalities (24-26) show constrains for maximum capacity of CHP, chiller and auxiliary boiler. Maximum allowable values of heating, cooling and electrical power for sale are given by (27).So the objective function of the problem can be expressed as:

$$\text{BMC} = \{\text{Benefit} - \text{Cost}\} \quad (28)$$

Using CHP to produce electricity eliminates the cost of transmission and this is one of the important factors that make CHP as a financially attractive option to produce electricity. Decreasing the cost of transmission dose not benefit the end users directly and is a beneficial factor for governments. To make end users share this benefit, governments provide some bonus schemes for electricity producers by distributed generation. This bonus has been added to the base price of electricity exported to the grid.

Finding this added value needs some calculations and suppositions .The effective efficiency considered as follows and it must be more than the mean efficiency of conventional power generation.

$$\eta_E = \frac{\eta_e}{1 - (\eta_t \times \mu)} \quad (29)$$

η_E: Effective efficiency.

η_e: Electrical efficiency.

η_t: Thermal efficiency.

μ: Percentage of the used heat of CHP.

The amount of saving natural gas when CHP is used to produce 1 kW·h of electricity is calculated as follow:

$$SG = \left[\frac{860}{HV_g} \times \left(\frac{100}{\eta_{ave}(1 - L)} - \frac{100}{\eta_E} \right) \right] \quad (30)$$

where:

SG: saving natural gas when 1kWh of electricity generated by CHP instead of conventional power generation system [m^3].

HV_g: Heating value of natural gas [kcal/m^3].

η_{avg}: mean efficiency of conventional power generation.

L: percentage of transmission loss of electrical grid.

Multiplying the natural gas price by the above value results in the bonus that would be added to the base price of electricity:

$$b = SG \times \text{natural gas price} \quad (31)$$

Obviously, efficiency and size of energy hub elements considerably affect the optimum value of parameters (P_e, P_g, γ, α, β).

The main objective in this paper is to calculate the optimum size of CHP, absorption chiller and auxiliary boiler in an energy hub.

$$\text{PVF} = \frac{(1 + if)}{(1 + ir)} \quad (32)$$

$$\text{CPVF} = \frac{(\text{PVF})^{EL} - 1}{\text{PVF} - 1} \quad (33)$$

CPVF is the compounding present value factor. *ir, if* and *EL* are respectively the per unit (p.u) interest rate, p.u inflation rate and economic life of the equipment.

Fixed cost term in (22) consists of the cost of the central controller, load controllers, interfacing equipment and low voltage circuit breaker [41,42].

By introducing a CHP system, the reliability of supplyingan electrical load increases substantially. This increase depends on the size of CHP. To explain the effect of CHP on power system reliability, Expected Energy Not Supplied (EENS) shows up in the (21). EENS is one of the most important indices in generating capacity adequacy evaluation. The EENS for one year can be calculated using the following equation:

$$\text{EENS} = \sum_{k=1}^{n} P_k A_k \quad \text{MW} \cdot \text{h/year} \quad (34)$$

where P_k is the probability of having a capacity outage equal to O_k; O_k, the magnitude of the capacity outage; and A_k, the energy not supplied because of the capacity outage O_k.

The Value of Lost Load (VoLL) in (19) is the estimated amount that customers receiving electricity with bilateral contracts would be willing to pay to avoid a disruption in their electricity service [41].

4. Financial Analysis

The financial analysis of investment projects is typically carried out using the technique of discounted cash flow (DCF) analysis. This section introduces concept of DCF analysis for the derivation of project performance criteria such as net present value (NPV) and internal rate of return (IRR).

Discounted Cash Flow (DCF) analysis is the technique used to derive financial performance criteria for investment projects. Cash flow analysis is simply the process of identifying and categorizing of cash flows associated with a project or proposed course of action, and making estimates of their values. Discounted cash flow analysis is an extension of simple cash flow analysis and takes into account the time value of money. A number of criteria are used in DCF to estimate project performance including Net Present Value (NPV), and Internal Rate of Return (IRR) and Dynamic Payback Period (DPP) [43].

4.1. Net Present Value

The net present value (NPV) is the sum of the discounted annual cash flows.

$$\text{NPV} = \sum_{T=1}^{EL} \frac{a_T}{(1+ir)^T} \quad (35)$$

A project is regarded as financially desirable if the NPV is positive [43].

4.2. Internal Rate of Return (IRR)

The internal rate of return (IRR) is the interest rate such that the discounted sum of net cash flows is zero. Generally speaking, the higher a project's internal rate of return, the more desirable it is to undertake the project. As such, IRR can be used to rank several prospective projects a firm is considering. Assuming all other factors are equal among the various projects, the project with the highest IRR would probably be considered the best and undertaken first [44].

The value of IRR such that

$$\sum_{T=1}^{EL} \frac{a_T}{(1+IRR)^T} = 0 \quad (36)$$

4.3. Dynamic Payback Period

The dynamic payback period (DPP) is the number of years for the projects to break even, *i.e.*, the number of years for which discounted annual net cash flows must be summed before the sum becomes positive (and remains positive for the remainder of the project's life). The dynamic payback period indicates the number of years until the investment in a project is recovered. It is a useful criterion for a firm with a short planning horizon, but does not take account of all the information available, *i.e.*, the net cash flows for years beyond the payback period [45].

5. Case Study

The high thermal to electrical ratio of hospitals shows that hospitals have the potential to capture the waste heat generated by a CCHP system. Healthcare inpatient facilities can use the thermal energy for space heating and hot water for laundry and kitchen facilities. The high values of many key criteria show that healthcare inpatient facilities offer great potential for large capacity CCHP systems. Many hospitals also proactively look for cost effective energy solutions because of their energy costs. The potential to meet the high power quality and reliability needs with a CHP system is also of great interest to hospitals.

The hospital which is considered as case study, **Figure 3**, has 10,000 square-meters comprised of three main sections. The hospital operates 24 hours a day all year-round (8760 hours per year). To increase the energy efficiency of this hospital, a detailed energy audit has been done for this hospital and installing a CCHP system was introduced as one of the key solutions to decrease energy price effectively because of its location, it is not possible to sell heat and cooling energy and also there is no storage device in this case study.

Energy load profiles (**Figures 4-6**) and energy price (**Figure 7**) of this hospital are depicted as follow. The price of cooling is considered to be 1.2 times more than electricity price. Note that in **Figures 5** and **6** there are two load profiles. One of them denotes winter and autumn day load sample and the other indicates load profile of summer and spring days. Note that prices of electricity,

Figure 3. Hospital main building.

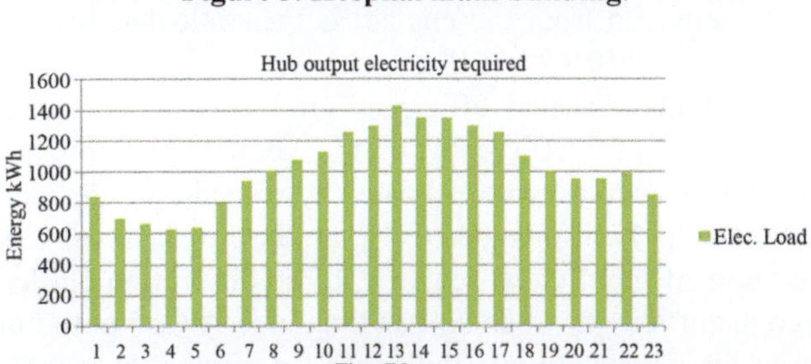

Figure 4. Electricity consuption in a normal day for the hospital [summer and winter].

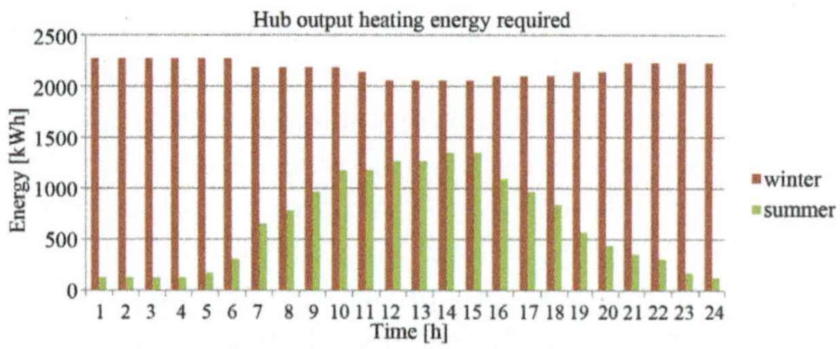

Figure 5. Heating energy consuption in a normal day for the hospital [summer and winter].

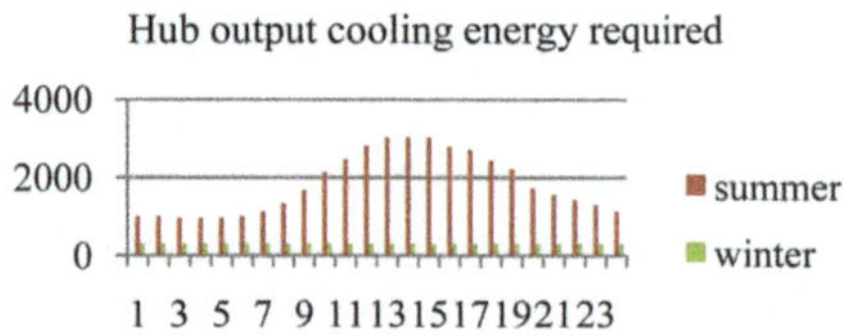

Figure 6. Cooling energy consuption in a normal day for the hospital [summer and winter].

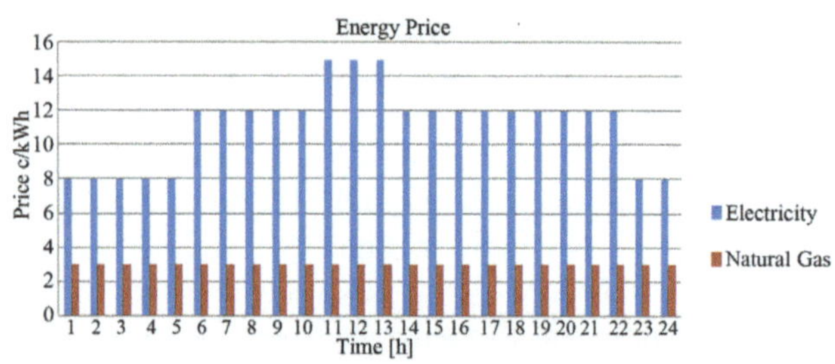

Figure 7. Energy price.

cooling and heat have an escalation rate of 14% per year. In this case, since extra recovered heat could not be sold, the extra heat is passed to a heat dump radiator. The heat dump radiator is cooled by electrically driven fans.

In this study, all efficiencies are independent of power and have a constant value. The typical energy distribution for internal combustion engines is provided [46]. It shows that 30% of the fuel energy is converted to heat energy rejected through the coolant and another 30% of the fuel energy is rejected as heat through the exhaust gas. The total efficiency of heat exchangers for the coolant and exhaust gas is estimated to be 0.85, and the total fuel-to-thermal-energy conversion efficiency (*i.e.*, total heat recovered from the engine) is then calculated to be $(30\% + 30\%)\cdot(0.85) = 51\%$.

The boiler thermal efficiency (η_{gf}) is assumed to be 90%. The total efficiency of the cooling components (chiller efficiency) was estimated by considering the Coefficient of Performance (CoP), amount of heat moved per unit of input work required, of an absorption chiller and the efficiency of an air handling unit. A CoP of 0.7 is used for the absorption chiller and an efficiency of 0.85 is used for the air handling unit. The total efficiency of cooling components is then calculated to be $(0.7) \times (0.85) \times 100 = 60\%$. The total efficiency of the heating components is estimated at 85% which is the efficiency of the air handling unit.

The thermal energy losses due to energy transport/transmission in the network are neglected in this simulation because the pipes are well insulated in the facility. CHP and boiler costs depend on their sizes. **Figures 8** and **9** depict these relations.

A summary of energy hub elements' efficiency information for the algorithm and the data needed for optimization problem are listed in **Table 2**. In **Table 3**, fix and variable cost of absorption chiller are shown [30]. Bonus for selling electricity to the grid is calculated by (22). Furthermore, the boundary conditions are shown in **Table 4**. Note that the all parameters are positive.

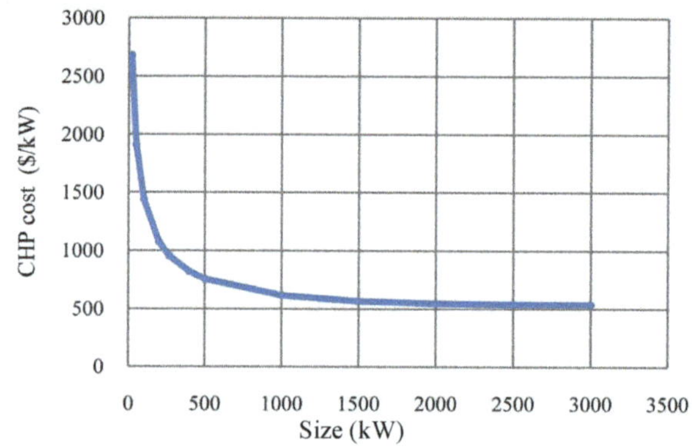

Figure 8. CHP cost.

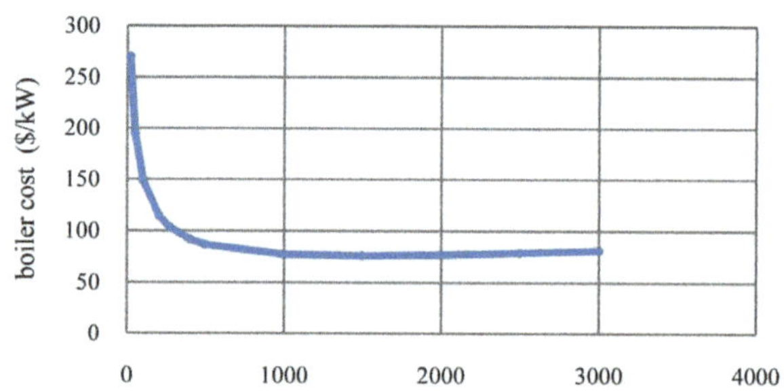

Figure 9. Boiler cost.

Table 2. Performance characteristics of CCHP and auxiliary boiler.

Maintenance Cost ($/kWh)	η_{ge}^{CHP}	η_{gh}^{CHP}	η_{gh}^{B}	η_{ee}	$\eta^{chiller}$
0.01	35%	40%	90%	98%	60%

Table 3. Cost of heating storage devices and absorption chiller.

	Absorption chiller
Fixed cost ($)	20,000
Variables cost ($/kW)	115

The interest rate (*ir*) is 0.20 p.u., the inflation rate (*if*) is 0.12 p.u., the economic life cycle (*EL*) of all equipment is considered to be 15 years [47]. For Tehran X_e= 1.32 $/kWh and X_g=0.6 $/kWh [48,49]. In this case study, analysis on Tehran Natural Gas shows that it has the Higher Heating Value (HHV) of 10.35 kW·h/N·m^3.

As it was described in section 3 installing CHP as a secondary supply for electricity, decreases EENS. In this case study, since the hospital equipped with two standbys 1 MW diesel generator, this effect could be overlooked [5]. To solve the above problem, GAMS software is used and the best size of energy hub elements is evaluated. **Table 4** demonstrates optimal size of energy hub elements and **Table 5** summarizes some important financial parameters.

Figures 10 and **11** depict imported and exported electricity in a winter and summer days. Between 10:00 a.m. and 1:00 p.m. as electricity price is too high, CHP works full loaded and exports electricity to the grid. **Figure 12** shows the installed CHP system in the hospital which is connected to the electricity network. **Figures 13** and **14** depict how the payback period and IRR will be affected according to variation in interest rates and component prices. **Figure 15** implies that if discount rate is less than 20%, NPV would soar dramatically.

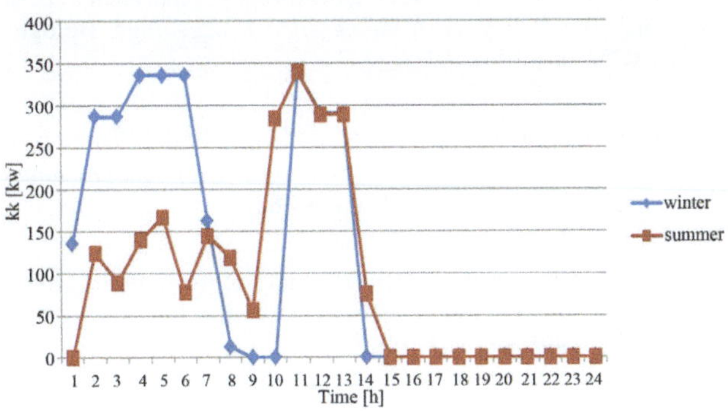

Figure 10. Exported electricity to the grid.

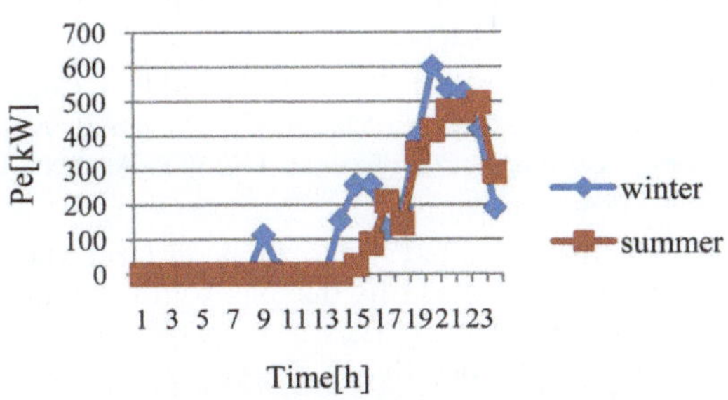

Figure 11. Imported electricity from the grid.

Table 4. Optimized value of energy hub elements.

CHP Capacity (kW)	Auxiliary Boiler Capacity (kW)	Absorption Chiller Capacity
1750	4400	3000 kW = 850 refrigeration tons

Figure 12. Installed CHP system in the hospital.

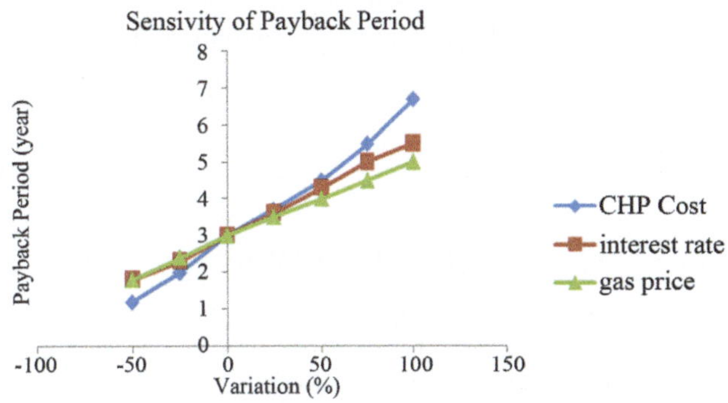

Figure 13. Sensitivity of payback period.

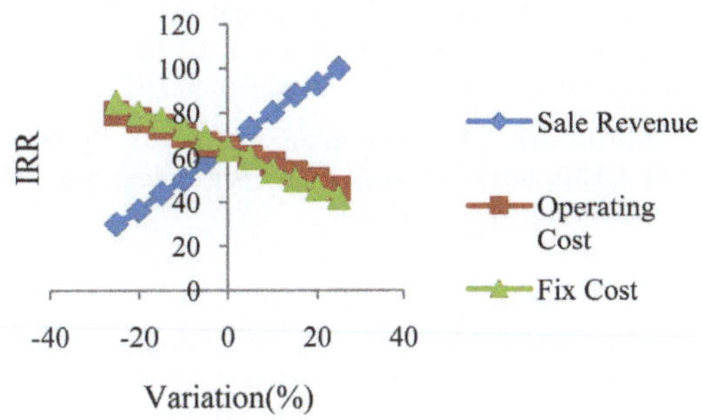

Figure 14. Sensitivity of IRR.

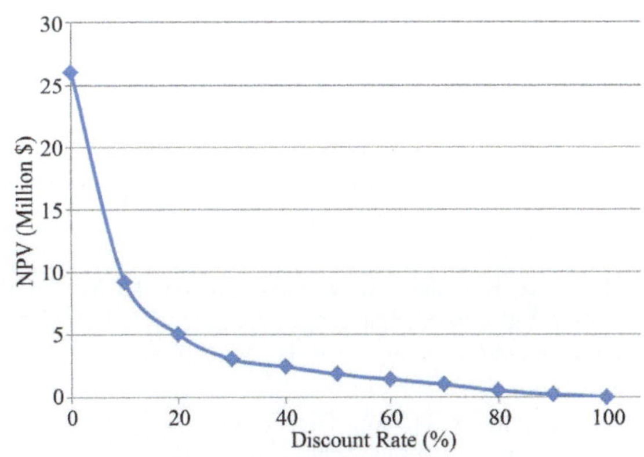

Figure 15. Sensitivity of NPV.

Table 5. Financial parameters.

Internal Rate of Return (IRR(%))	Net Present Value (NPV(million $))	Dynamic Payback Period (DPP(year))
64%	9.22	3

Finally, the net price of 1 kWh electricity which is generated by CCHP calculates as follow:

$$\begin{aligned}&\text{Electricity price by CCHP}\\&=\frac{\text{Natural gas price}}{\text{Total electricity generated by CCHP}}\\&+\frac{\text{O \& M Cost}+\text{Depreciation Cost}}{\text{Total electricity generated by CCHP}}\\&+\frac{\text{CCHP Cost}\times ir}{\text{Total electricity generated by CCHP}}\\&-\frac{\text{Benefit of heat recovery for cooling and heating}}{\text{Total electricity generated by CCHP}}\\&=4.36\ \text{Cent/kWh}\end{aligned} \quad (37)$$

REFERENCES

[1] A. Sheikhi, A. M. Ranjbar, M. Mahmoody and F. Safe, "CHP Optimize Selection Methodology for an Energy Hub System," *10th International Conference on Environment and Electrical Engineering*, Rome, 8-11 May 2011, pp. 1-5.

[2] US EPA, "Inventory of US Greenhouse Gas Emissions and Sinks: 1990-2001," EPA 430-R-03-004, US Environmental Protection Agency, Washington DC, 2003.

[3] J. McDonald, "Adaptive Intelligent Power Systems: Active Distribution Networks," *Energy Policy*, Vol. 36, No. 12, 2008, pp. 4346-4351.

[4] V. H. Mendez, J. Rivier, J. I. De La Fuente, T. Gomez, J. Arceluz and J. Marin, "Impact of Distributed Generation on Distribution Investment Deferral," *International Journal of Electrical Power & Energy Systems*, Vol. 28, No. 4, 2006, pp. 244-252.

[5] R. Billinton and R. N. Allen, "Reliability Evaluation of Engineering Systems, Conceptand Techniques," 2nd Edition, Plenum Press, New York, 1992.

[6] R. Graham and W. Chow, "Technical and Economic Assessment of Combined Heat and Power Technologies for Commercial Customer Applications," EPRI Project Manager, 2003.

[7] H. Cho, R. Luck, S. D. Eksioglu and L. M. Chamra, "Cost Optimized Real Time Operationof CHP System," *Energy and Buildings*, Vol. 41, No. 4, 2009, pp. 445-451.

[8] A. Sheikhi, B. Mozafari and A. M. Ranjbar, "CHP Optimize Selection Methodology for a Multicarrier Energy System," *Power Tech Conference*, Trondheim, 19-23 June 2011, pp. 1-7.

[9] J. Teng, Y. Liu, C. Chen and C.-F. Chen, "Value-Based Distributed Generator Placements for Service Quality Improvements," *International Journal of Electrical Power and Energy Systems*, Vol. 29, No. 3, 2007, pp. 268-274.

[10] A. M. Sheikhi and R. H. Oraee, "Financial Analysis and Optimal Size and Operation for a Multicarrier Energy System," *Energy and Buildings*, Vol. 48, 2012, pp. 71-78.

[11] A. Sheikhi and A. M. Ranjbar, "Optimal Dispatch of a Multiple Energy Carrier System Equipped with a CCHP," *ICREPQ Conference*, Las Palmas de Gran Canaria, 2011.

[12] P. J. Mago and L. M. Chamra, "Analysis and Optimization of CCHP Systems Based on Energy, Economical, and Environmental Considerations," *Energy and Building*, Vol. 41, No. 10, 2009, pp. 1099-1106.

[13] R. Hashemi, "A Developed Offline Model to Optimal Operation of Combined Cooling, Heating and Power Systems," *IEEE Transaction on Energy Conversion*, Vol. 24 No. 1, 2009, pp. 222-229.

[14] P. M. Costa and M. A. Matos, "Economic Analysis of Microgrids including Reliability Aspect," 9th International Conference on Probabilistic Methods Applied to Power System, Stockholm, 11-15 June 1, 2006, pp. 1-15.

[15] H. Asano and S. Bando, "Operational Planning Method and Economic Analysis of Microgrid with Intermittent Renewable Energy and Battery Storage," *29th IAEE International Conference*, Potsdam, 7-10 June 2006.

[16] A. R. Abdelaziz and W. M. Ali, "Dispersed Generation Planning Using a New Evolutionary Approach," *IEEE Bologna Power Tech Conference Proceedings*, Bologna, 23-26 June 2003, 5 Pages.

[17] R. Dugan, T. E. McDermott and G. J. Ball, "Planning for Distributed Generation," *IEEE Industry Applications Magazine*, Vol. 7, No. 2, 2001, pp. 80-88.

[18] H. M. Groscurth, T. Bruckner and R. Kümmel, "Modeling of Energy Service Supply System," *Energy*, Vol. 20, No. 9, 1995, pp. 941-958.

[19] I. Bouwmans and K. Hemmes, "Optimising Energy Systems—Hydrogen and Distributed Generation," *2nd International Symposium on Power System Market Aspects*, Stockholm, 2-4 October 2002.

[20] R. H. Lasseter and P. Piagi, "Microgrid: A Conceptual solution," *Proceedings of IEEE 35th Annual Power Electronics Specialists Conference (PESC)*, Aachen, 20-25 June 2004, pp. 1-6.

[21] R. Frik and P. Favre-Perrod, "Proposal for a Multifunctional Energy Bus and Its Interlink with Generation and Consumption, Diploma Thesis," Power Systems and High Voltage Laboratories, Zurich, 2004.

[22] G. Andersson, "A Modeling Framework for Future Energy Systems," Project Home Page, 2006, http://www.eeh.ee.ethz.ch/psl/research/vofen.html

[23] P. Favre-Perrod, M. Geidl, G. Koeppel and B. Klockl, "A Vision of Future Energy Network," IEEE Inaugural Conference and Exposition in Africa, Durban, 11-15 July 2005, pp. 13-17.

[24] K. Hemmes, J. L. Zachariah-Wolf and G. Andersson, "Towards Multi-Source Multi-Product Energy Systems," *Hydrogen Energy*, Vol. 32, No. 10-11, 2007, pp. 1332-1338.

[25] M. Geidl, "Integrated Modeling and Optimization of Multi-Carrier Energy Systems," PhD Thesis, Eidgenössische Technische Hochschule Zürich, Zürich, 2007.

[26] G. Koppl and G. Andersson, "Reliability Modeling of Multicarrier Energy Systems," *Energy*, Vol. 34, No. 3, 2009, pp. 235-244.

[27] M. D. Galus and G. Andersson, "Power System Considerations of Plug-In Hybrid Electric Vehicles Based on a Multi Energy Carrier Model," *IEEE Power & Energy Society General Meeting*, Calgary, 26-30 July 2009, pp. 1-8.

[28] F. Kienzle, P. Ahcin and G. Andersson, "Valuing Investments in Multi-Energy Conversion, Storage, and Demand-Side Management Systems under Uncertainty," *IEEE Transaction on Sustainable Energy*, Vol. 2, No. 2, 2011, pp. 194-202.

[29] M. Arnol and G. Andersson, "Investigating Renewable Infeed in Residential Area Applying Model Predictive Control," *IEEE Power & Energy Society General Meeting*, Minneapolis, 25-29 July 2010, pp. 1-8.

[30] T. Krause, F. Kienzle, Y. Liu and G. Anderssson, "Inter Connected National Energy Systems Using an Energy Hub Approach," *IEEE PowerTech*, Trondheim, 19-23 June 2011, pp. 1-7.

[31] M. Schuzle, L. Friedrich and M. Gautschi, "Modeling and Optimization of Renewables: Applying the Energy Hub Approach," *IEEE Conference on Sustainable Energy Technologies*, Singapore City, 24-27 November 2008, pp. 83-88.

[32] A. Parisio, C. D. Vecchio and A. Vaccaro, "A Robust Optimization Approach to Energy Hub Management," *Electrical Power and Energy Systems*, Vol. 42, No. 1, 2012, pp. 98-104.

[33] L. Carradore and R. Turri, "Modeling and Simulation of Multi-Vector Energy Systems," *IEEE Conference on Power Tech*, Bucharest, 28 June-2 July 2009, pp. 1-7.

[34] E. M Robertson, *et al.*, "Outline for an Integrated Multiple Energy Carrier Model of the UK Energy Infrastructure," *44th International Universities Power Engineering Conference* (*UPEC*), Glasgow, 1-4 September 2009, pp. 1-5.

[35] L. Forbes, S. T. Galloway and G. W. Ault, "An Approach for Modeling a Decentralized Energy System," *45th International Universities Power Engineering Conference* (*UPEC*), Cardiff, 31 August-3 September 2010, pp. 1-5.

[36] L, M, Ramirez-Elizondo and G. C. Paap, "Unit Commitment in Multiple Energy Carrier Systems," *IEEE Conference*, Starkville, 4-6 October 2009, pp. 1-6.

[37] V. Velez, L. Ramirez-Elizendo and G. C. Paap, "Control Strategy for an Autonomous Energy System with Electricity and Heat Flows," *IEEE Conference*, Hersonissos, 25-28 September 2011, pp. 1-6.

[38] F. Syed, "Analysis of a Clean Energy Hub Interfaced with a Fleet of Plug-In Fuel Cell Vehicles," Master Thesis, University of Waterloo, Waterloo, 2011.

[39] M. Setayeshnazar and M. R. Haghifam, "Multi Objective Electric Distribution System Expansion Planning Using Hybrid Energy Hub Concept," *Electric Power Systems Research*, Vol. 79, No. 6, 2009, pp. 899-911.

[40] M. Geidl, "A Green Field Approaches for Future Power Systems," Cigre Session 41, Paris, 2006.

[41] A. Sheikhi, A. M. Ranjbar, H. Oraee and A. R. Moshari, "Optimal Operation and Size for an Energy Hub with CCHP," *Energy and Power Engineering Journal*, Vol. 3, No. 5, 2011, pp. 641-649.

[42] A. K. Basu and S. Chowdhury," Impact of Strategic Deployment of CHP-Based DERs on Microgrid Reliability," *IEEE Transactions on Power Delivery*, Vol. 25, No. 3, 2010, pp. 1697-1705.

[43] D. G. Newnan, T. G. Eschenbach and J. P. Lavelle, "Engineering Economic Analysis," 9th Edition, Oxford University Press, Oxford, 2004.

[44] M. Y. Khan, "Theory & Problems in Financial Management," McGraw Hill Higher Education, New York, 1993.

[45] M. A. Main, "Project Economics and Decision Analysis, Volume I: Deterministic Models," 2nd Edition, PennWell Publishing, Tulsa, 2002.

[46] American Society of Heating Refrigerating and Air-Conditioning Engineers, "Ashrae Handbook—Hvac Systems and Equipment," 2009.

[47] A. Sheikhi, *et al.*, "CHP Optimized Selection Methodology for a Multi-Carrierenergy System," *International of Review Electrical Engineering* (*I.R.E.E*), Vol. 6, No. 4, 2011, pp. 1-7.

[48] H. Cho, "Evaluation of CCHP Systems Performance Based on Operational Cost Primary Energy Consumption, and Carbon Dioxide Emission by Utilizing an Optimal Operation Scheme," *Applied Energy*, Vol. 86, No. 12, 2009, pp. 2540-2549.

[49] B. Rolfsman, "CO_2 Emission Consequences of Energy Measures in Buildings," *Building and Environment*, Vol. 37, No. 12, 2002, pp. 1421-1430.

List of Symbols and Abbreviations

N Time intervals of optimization

$L_e(N)$ Electrical energy demand in the time interval [kW·h]

$L_h(N)$ Heating energy demand in the time interval [kW·h]

$L_c(N)$ Coolingenergy demand in the time interval [kW·h]

$e(N)$ Electricity price $\$/\text{kw} \cdot \text{h}$

g Gas price $\$/\text{kw} \cdot \text{h}$

ir Interest rate

iff Inflation rate

EL Economic life of the equipment [year]

MC Maintenance cost $\$/\text{kw} \cdot \text{h}$

C_m Yearly maintenance cost $\$/\text{year}$

η_E Effective efficiency of CHP

η_{ee} Transformer efficiency 20kV/440kV

η_{ge} Electrical efficiency of CHP

η_{gf} Auxiliary boiler efficiency

η_{gh} Thermal efficiency of CHP

η_{ac} Absorption chiller efficiency

χ_e Emission factor for average power generation $\$/\text{kw} \cdot \text{h}$

χ_g Emission factor for gas power generation $\$/\text{kw} \cdot \text{h}$

$VOLL$ Value of lost load $\$/\text{kw} \cdot \text{h}$

HHV Higher heating value

CoP Coefficient of performance

P_{hT} Pumping heat price [cent/kWh]

Variables:

$P_e(N)$ Purchased electricity [kW·h]

$P_g(N)$ Purchased natural gas [kW·h]

P_{gM} Maximum purchased natural gas [kW]

P_{gm} Minimum purchased natural gas [kW]

P_{eM} Maximum purchased electricity [kW]

P_{em} Minimum purchased electricity [kW]

$\gamma(N)$ Dispatched factor for natural gas inlet

$\alpha(N)$ Dispatch factor for auxiliary boiler

$\beta(N)$ Dispatch factor for CHP

$S_{inh}(N)$ Input rate of heat storage kW·h

$S_{outh}(N)$ Output rate of heat storage kW·h

S_M Nominal capacity of heat storage kW

$SE_{in}(N)$ Input rate of battery kW·h

$SE_{out}(N)$ Output rate of battery kW·h

SE_M Nominal capacity of battery kW

Benefit(X) Benefit of using the element X [$]

Cost(X) Cost of using element X [$]

Cap(X) Capacity of element X [kW]

BMC Benefit minus cost [$]

$KK(N)$ Exported electricity to the grid [kW·h]

Z_1 Price of heat transfer from CHP, heat storage and auxiliary boiler to the load [$]

Z_2 Price of purchased electricity and gas [$]

$H(N)$ Exported heat [kW·h]

H_M Maximum heat exported [kW·h]

$CC(N)$ Cooling exported [kW·h]

C_M Maximum cooling exported [kW·h]

P_{SeM} Maximum exported electricity from CHP to the grid [kW·h]

EENS Expected energy not supplied

P_k The probability of having a capacity outage equal to O_k

O_k Outage capacity

A_k The energy not supplied because of the capacity outage O_k

a_k Annual net cash flow

DCF Discounted cash flow

IRR Internal rate of return

DPP Dynamic payback period

NPV Net present value

Impact Analysis of Off-board Charger to Power Quality

Yubo Fan, Chunlin Guo, Wenbo Qi, Zheci Tang
State Key Laboratory for Alternate Electrical Power System with Renewable Energy Sources,
North China Electric Power University, Beijing, China

ABSTRACT

In this paper, we tested the entire charging process of a single off-board charger in one charging station in Beijing. Among the testing data, we chose the typical power quality parameters and compared them with national standard. Then we drew conclusions as follows: 1) Electric vehicle battery is the capacitive load. It can export reactive power when charging. 2) In the charging process of the off-board charger, indicators of voltage deviation, frequency deviation, power factor, and voltage distortion rate are qualified. 3) Off-board charger produces odd harmonics in the charging process, and with increasing harmonic order, harmonic content reduces. There is a certain amount of high-order harmonic in off-board charger, mainly distributing around 6650 kHz. 4) Generated harmonics of the actual device, the harmonic is mainly reflected in the current, voltage, only a small distortion.

Keywords: Electric Vehicles; Charger; Power Quality

1. Introduction

With the development of electric vehicles, charging infrastructure also advanced. However, due to the high power charger for charging stations like switching power supply, rectifier, inverter, etc, are usually using power electronic technology, which is a highly non-linear electrical equipment, the large-scale use of it will cause serious power quality problems on the grid. For example, harmonic problem, we have known it for a long time. The flowing harmonic in the grid will result in distortion of the grid voltage and it also limited the application of non-linear system.

At present, some scholars have conducted researches on electric vehicles charging impact on grid, including research on the impact of the increasing electric vehicles on medium or low voltage power network. Issues [1-6] like load, voltage, loss, three-phase unbalance, harmonics etc are included.

But these studies only analyzed the minor part of the charging station about power quality[7-11], there is no comprehensive analysis about all the indicator of power quality. Besides, most research results are concluded from simulation, and they have simplified real operating system. In this case, we can say that their conclusions are questionable.

*This work is supported by: National High Technology R&D Program of China (863Program) (2012AA050804).Key Project of the National Research Program of China (2011BAG02B14), National High Technology R&D Program of China (863 Program) (2011AA05A109)

Therefore, based on the testing data of the single off-board charger charging process, we have systematically analyzed the power quality problems including voltage deviation, frequency deviation, voltage fluctuations and flickers, three-phrase unbalance and harmonics etc. caused by off-board charger systematically and drawn conclusions.

2. Power Quality Indicators

2.1. Chinese National Standard Limit

1) Voltage deviation

According to GB/I2325-2008 Power quality-Deviation of supply voltage, the allowable deviation of three -phrase supply power that below 20 kV is ±7% of system rated voltage.

2) Frequency deviation

The nominal frequency of the power system in China is 50 Hz. According to GB/T15945-1995 Quality of electric energy supply-Permissible deviation of frequency for power system, the allowable frequency deviation of the power system is ±0.2 Hz

3) Voltage fluctuation

According to GB12326-2008 Power quality--Voltage fluctuation and flicker, the voltage fluctuations of common connection points ranked 0.38 kv and 10 kv caused by impact load should below 2%.

4) Three-phrase unbalance

Three-phrase unbalance refers to three-phase voltage (or current) amplitude is inconsistent and exceed the specified range in the power system. The unbalance factor expression,

$$\varepsilon_u = \frac{U_2}{U_1} \times 100\% \qquad (1)$$

where, U_1—the R.M.S. of three-phrase voltage positive-sequence component, V;

U_2—the R.M.S. of three-phrase voltage negative-sequence component, V.

According to GB/T15543-2008 Power quality-Admissible three-phrase voltage unbalance factor, the common voltage unbalance limit of every user is 1.3%, *i.e.*, the side normal voltage unbalance of 0.38 kv caused by charger should below 1.3%.

5) Harmonic voltage

The total ratio of voltage harmonic distortion refers to the percentage of higher harmonic voltage RMS and the fundamental voltage RMS. China have set a standard for allowable values of harmonic voltage of the power supply.

According to GB/T14549-93 Quality of electric energy supply-Harmonics in public supply network, the total harmonic distortion rate limit is 5.0% and 4.0% , the corresponding harmonic voltage is 0.38 kv and 10 kv. The following table shows that harmonic voltage in the in- point should below values in **Table 1**.

6) Harmonic current

As to public electric network of 0.38 kv, when the reference short-circuit capacity is 10 MVA, Harmonic current permissible value shown in **Table 2**.

When the reference short-circuit capacity is not 10 MVA, harmonic current allowed values are permitted in accordance with the size of the actual minimum short-circuit capacity scaling.

$$I_{hi} = h(h_T / h_I)^{1/\partial} \qquad (2)$$

where, I_{hi} means the RMS of each harmonic current converted; h represents the allowed values of harmonic current acquired from power system; h_T means the minimum short-circuit capacity of single set of devices; h_I means the reference short-circuit capacity of point of common coupling; ∂ represents phase superposition coefficients, ∂'s value of each harmonic current shown in **Table 3**:

With 0.38 kV electric utility network, grid system short-circuit capacity of 10 MVA (the benchmark short circuit capacity), according to *GB/T14549—93* quality of electric energy supply--harmonics in public supply network, the utility grid injected harmonic current component (root mean square value) should not exceed in exemplar 4-2 (harmonic current permissible value of injected into the common connection point) in the allowable value (as shown in **Table 4**).

Since the Test points of 0.4 kV bus at a minimum short-circuit current of 10 kA, so the minimum short-circuit capacity of the low-voltage side of the transformer is [11]:

$$S_{sc} = \sqrt{3} \times 0.4 \times 10 = 6.928\ (MVA) \qquad (3)$$

The actual minimum of short circuit capacity is $S_T = 6.928MVA$, Which is use to compute harmonic finite value of the common points, and from that we got **Table 5**.

Table 1. Harmonics voltage limit value in public supply network (phrase voltage).

System Rated Voltage (KV)	THD Urel to h1 (%)	Harmonic Voltage Ratio (%)	
		Odd	Even
0.4	5.0	4.0	2.0
10	4.0	3.2	1.6

Table 2. IEC-3-4 Harmonic current permissible value.

Harmonic Order n	5	7	11	13	17	19
Permissible Harmonic Current (I_n/I_1)/%	9.5	6.5	3.1	2.0	1.2	1.1

Table 3. ∂'s value of each harmonic current.

Harmonic Order n	5	7	11	13	>13
∂	1.2	1.4	1.8	1.9	2.0

Table 4. Harmonic current permissible value of injected into the point of common coupling.

Rated Voltage /kV	Reference short-circuit capacity / MVA	Harmonic order and harmonic current permissible value (A)											
		2	3	4	5	6	7	8	9	10	11	12	13
0.38	10	78	62	39	62	26	44	19	21	16	28	13	24
		Harmonic order and harmonic current permissible value (A)											
		14	15	16	17	18	19	20	21	22	23	24	25
		11	12	9.7	18	8.6	16	7.8	8.9	7.1	14	6.5	12

Table 5. Measured limit value of harmonic current.

Harmonic Order n	3	5	7	11	13	17	19
Harmonic Current /A	4.03	5.06	5.14	5.27	4.93	4.00	3.56

2.2. Standard Comparison between Chinese and IEC's

Table 6 shows a comparison of power quality between national and IEC standards. The table shows that the national standards allowable values of power quality below the IEC standard allowable value, so if they meet the requirements of the national standards of China, they must meet the IEC standard.

3. Test Description

The test is aimed at a single off-board charger (parameters in **Table** 7) whose rated power is 15 kW in an EV charging station of Beijing. A power quality testing is carried out through the entire charging process of 8T sanitation truck battery power (200 AH/96V) in the charging station.

The three-stage charging method is used in the charging process of the 8T sanitation truck battery power (200 AH/96V). The first stage is the constant current limit pressure mode, the charging current I = 60 A, the limit voltage U = 103 V; the second stage is a constant current limiting mode, the charging voltage U = 103 V, the limit current I = 20 A; the third stage is charging stopped, the charge cut-off current is 0.1C (20 A). The monitor analyzer of the test is American FLUKE three-phase power quality analyzer (Model: Fluke1760 Basic). Continuous recording of three-phase data is on the AC side of charger for the whole process of charging.

Test point is selected in the AC input side of the charger; the testing diagram is shown in **Figure 1**.

Table 6. A comparison of power quality between national and IEC standards.

Power quality indicators	Chinese national standard	IEC
Voltage deviation	the rated voltage is ± 7% of the high-voltage power supply and low voltage power users of 10 kV and below	No standard
Frequency deviation	the allowable value of frequency deviation of power system is ± 0.2 Hz; when the compatible value is larger, the deviation value can be as large as ± 0.5 Hz	Short-term changes is ± 1 Hz,and the steady-state is smaller
Voltage harmonics	When the nominal voltage of power grid is 0.38 KV ， correspondingly, Odd harmonic is 4%， Even harmonic is 2%， Total harmonic distortion is 5%； When the nominal voltage of power grid is 10kv, correspondingly , Odd harmonic is 3.2%， Even harmonic is 1.6%， Total harmonic distortion is 4%；	Third harmonic 5%， Fifth harmonic 6%， Seventh harmonic 5%， Eleventh harmonic 3.5%， Thirteenth harmonic 3%， Total harmonic distortion 8%； As to very short effect(within 3 s seconds)，compatible level equals to K* the above values, K=1.3+0.7/45*（h-5）,THD=11%。
Voltage fluctuations and flickers	The voltage fluctuations are not larger than3% of supply voltage in common conditions; short term flicker（10 min）p_{st}=1,long term flicker（2 h）p_{lt}= 0.8	The voltage fluctuations are not larger than 3% of the nominal supply voltage;short term flicker（10 min）p_{st}=1,long term flicker（2 h）p_{lt}=0.8
Three-phrase unbalance	The amissible factor is 2% in points of common coupling in the common conditions of power system, short time is no longer than 4%, unbalance factor caused by users is 1.3%	negative-sequence component is 2% of positive-sequence component, it can reach 3% in certain place

Table 7. Off-board charger technical parameters.

Heading	off-board charger
Technical Specifications	HEV-Z-ER125 A/120 V
Input voltage	AC380 V±15%
AC input frequency	45-65 Hz
Output voltage	DC120V（Adjustable）
Output current	125 A
Maximum output power	15 kW

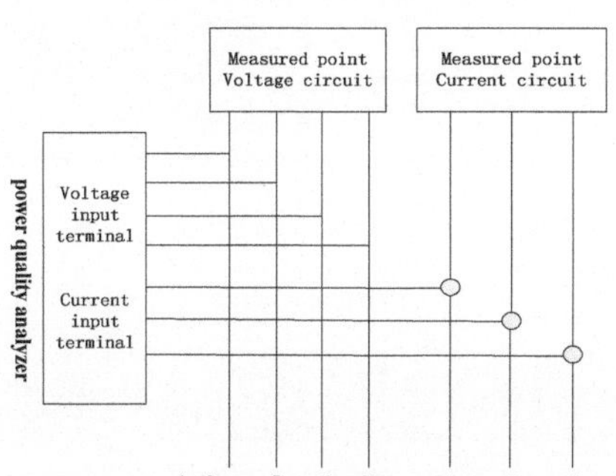

Figure 1. Test wiring diagram.

4. Test Outcome

1) Startup surge

It takes the charger about 32 seconds from start to the stage of constant-current limit voltage charging, but great surge current will be produced in the moment of starting. The above figure indicates that the current surge produced by single charger starting is very large: the amplitude can up to 20.3 A. After the chargers are constructed in large scale, the current surge produced by many chargers starting at the same time will have very great impact on the power grid (as shown in **Figure 2**).

2) Voltage RMS and current RMS

When charger provides normal charging to the power battery, the voltage measured by power quality analyzer is 237 V, and the current 9.2 A.

3) Power and power factor

Conclusion: Over 95% of the charging is constant power charging. Reactive powers of the test points are negative, indicating the presence of reactive power backfeed and the capacitive load. The charger is stable, the power factor when charger stably running is almost 1, which is in line with the requirements of the relevant national standards (as shown in **Figure 3**).

4) Grid frequency deviation

Figure 4 shows the measured grid frequency trends, and also indicates that the maximum frequency deviation of charger network side is -0.018, which is within the allowable value of GB (± 0.2 Hz) and so in line with national standards.

5) Voltage Deviation

Taking phase A as an example. As can be seen from **Figure 5**, the maximum of voltage upper deviation is of about 4% and that of the voltage lower deviation is zero. Both of them are less than the limit of 7%. So a single charger charging will not cause the AC side voltage deviation exceeded.

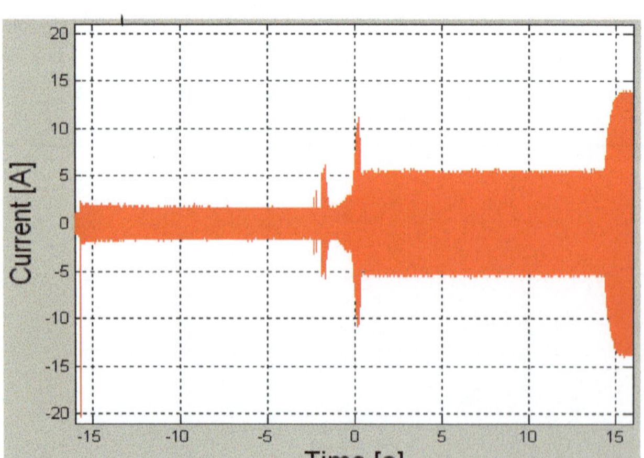

Figure 2. The surge current waveform when charger starting.

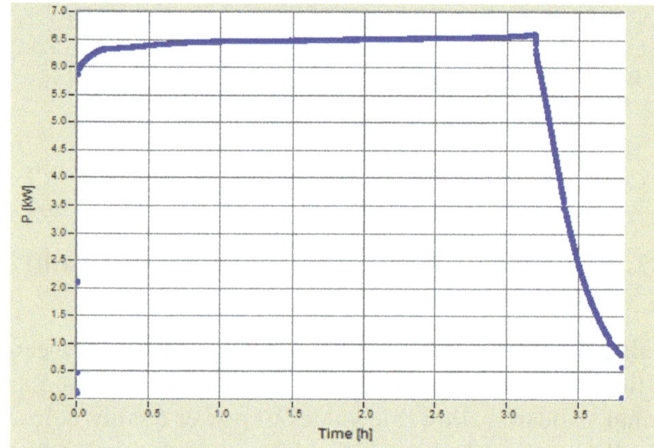

(a) Charger input active power curve

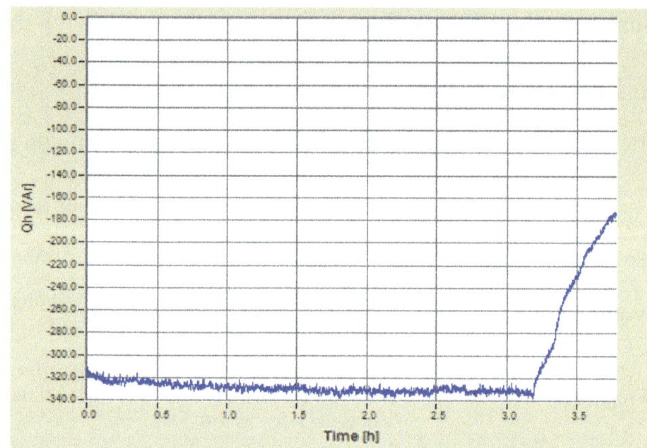

(b) Charger input reactive power curve

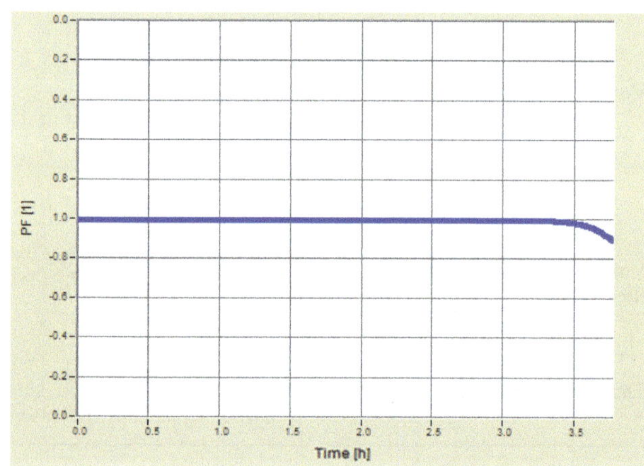

(c) Charger power factor curve

Figure 3. Power and power factor.

6) Voltage fluctuation

Figure 6 shows the maximum voltage fluctuation produced by a single charger AC side is 2.4%, which exceeds the GB limit of 2%. However, the maximum variation in voltage produced in more than 95% of the whole charging process is within limits. So the single charger charging substantially will not cause excessive voltage fluctuations in the AC side.

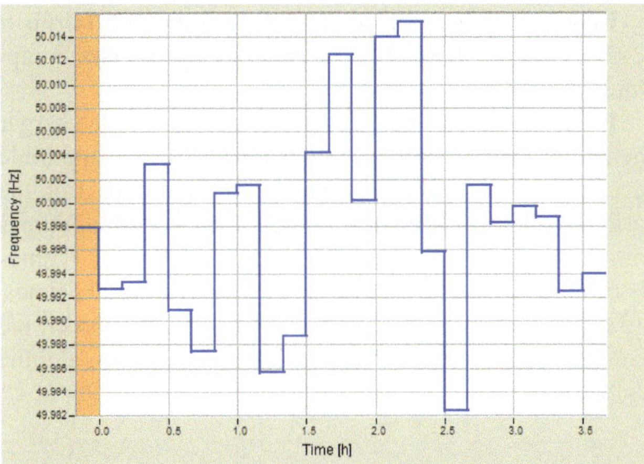

Figure 4. The charger network side frequency curve.

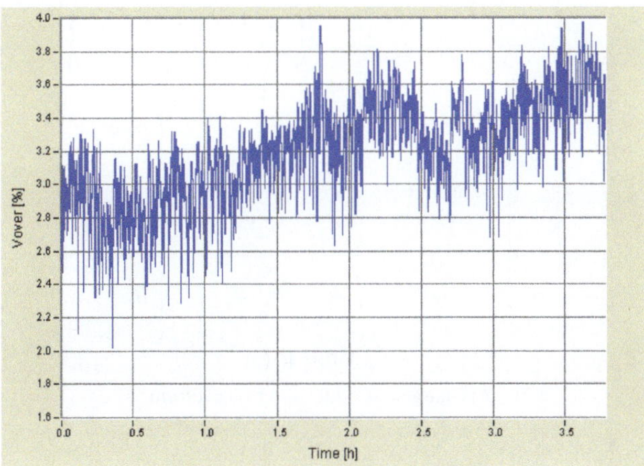

(a) The voltage upper deviation

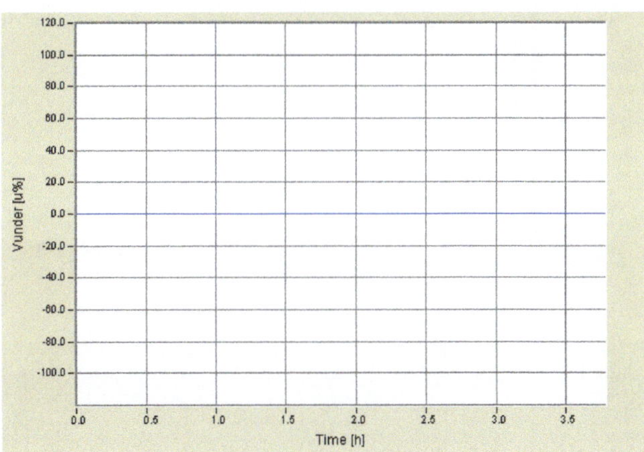

(b) The voltage lower deviation

Figure 5. Voltage deviation.

7) Three-phase unbalance

Figure 7 shows that the three-phase voltage unbalance factor is below 0.4%, far less than the GB 1.3%. So the three-phase unbalance of charger network side is in line with the provisions of the relevant national standards.

8) Harmonic voltage

Figure 8(a) is a period of voltage waveform (Urms = 242 V, Irms = 9.86 A) in power stationary phase captured by the oscilloscope **Figure 8** shows that a single charger running has inconspicuous impact on the network side voltage. THD_u is less than 2%, and the odd harmonics and even harmonics ratios are both no more than 2%. So the charger network side harmonic voltage is in line with the provisions of the relevant national standards.

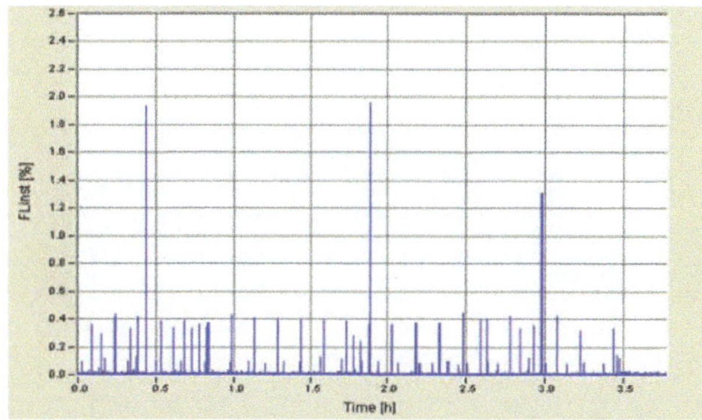

Figure 6. The charger AC side voltage fluctuations diagram of the entire charging process.

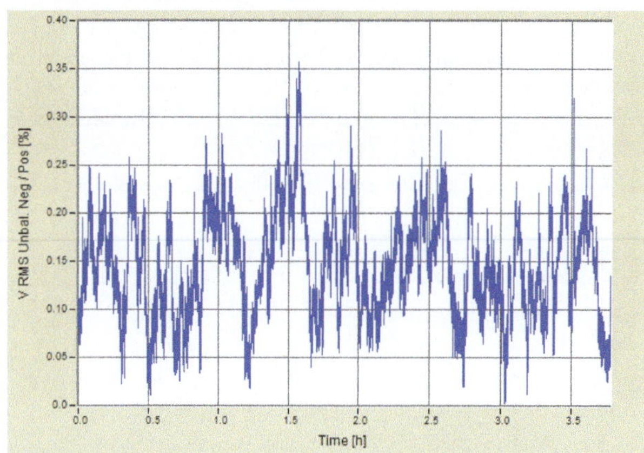

Figure 7. Time-varying trend of three-phase voltage unbalance factor.

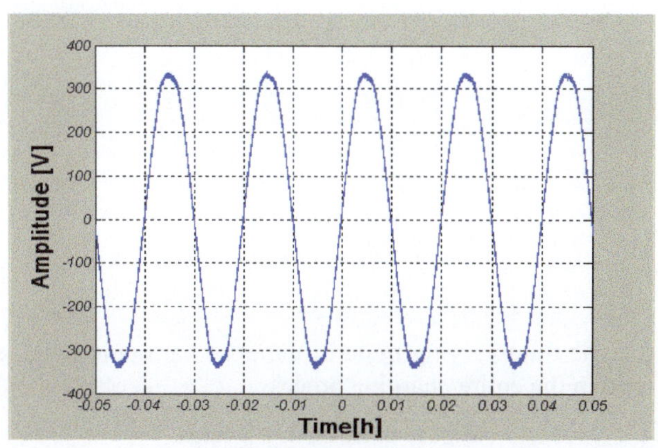

(a) The voltage waveform of charger AC side

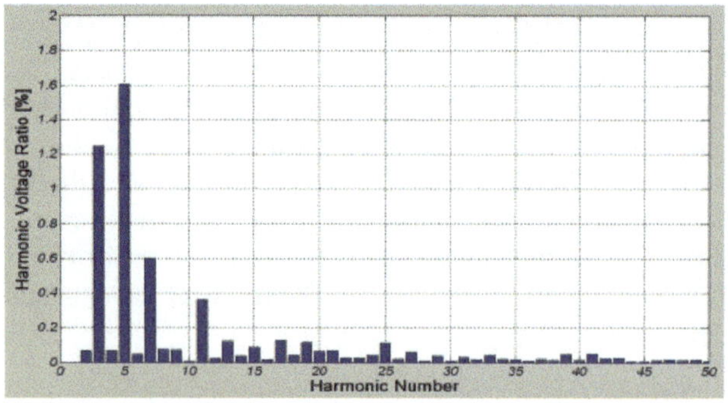

(b) Charger AC side harmonic voltage spectrum

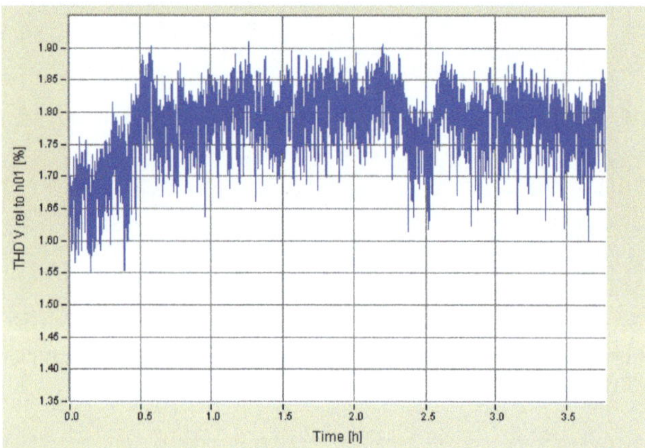

(c) The total harmonic voltage distortion rate of the completed charging process

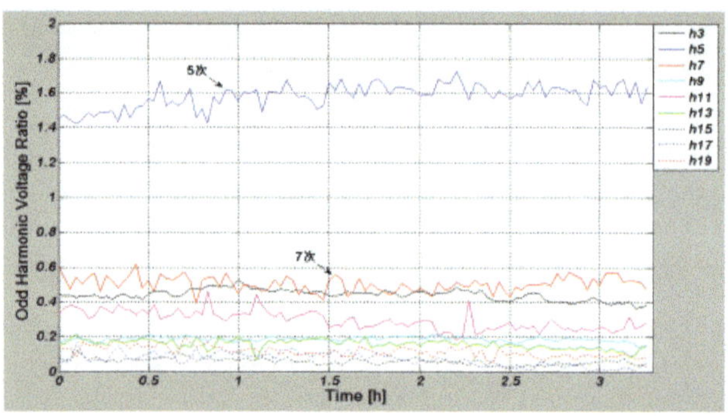

(d) The time-varying trend of the odd harmonic voltage ratio in the entire charging process

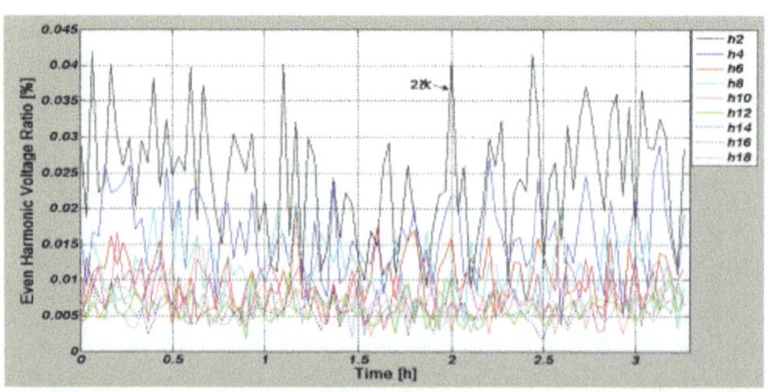

(e) The time-varying trend of the even harmonic voltage ratio in the entire charging process

Figure 8. Harmonic voltage.

9) Harmonic Current

Figure 9(a) is a period of voltage waveform (Urms = 242 V, Irms = 9.86 A) in power stationary phase captured by the oscilloscope.

Figure 9 shows that in even harmonics, single charger generates twice maximum I_{h2} sometimes the amplitude even ups to 0.11 A, which may have some impact on the grid; In odd harmonics, single charger generates 3 times minimum harmonic, mainly generating 6 k ± 1 times harmonic, k = 1,2,3, ..., that is, 5 times, 7 times, 11 times, 13 times, 17 times, 19 times, ... The higher the order, the smaller the amplitude, wherein the I_{h5} maximum, and the I_{h5} second.

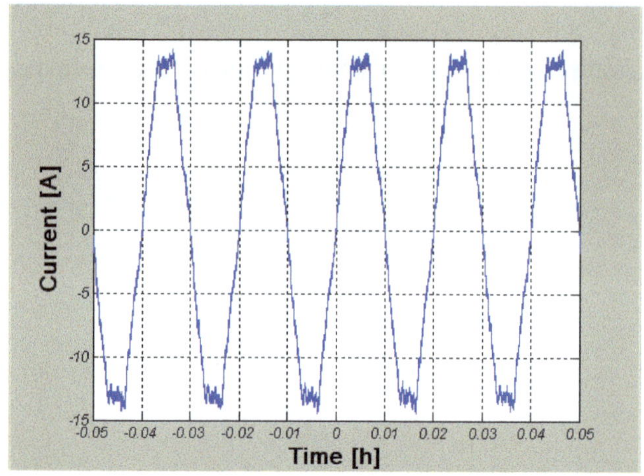

(a) Charger AC side current waveform

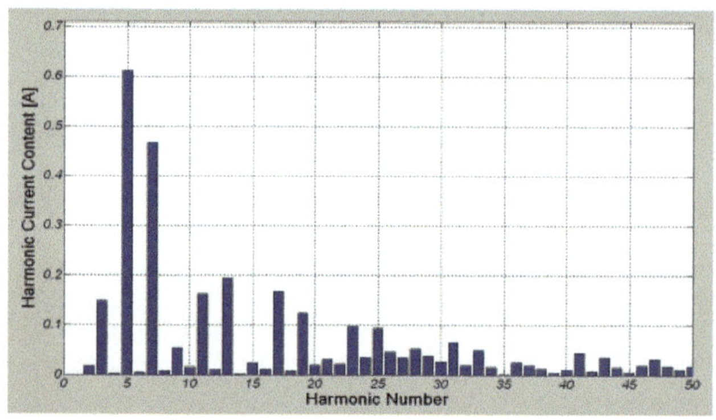

(b) Charger AC side harmonic current spectrum

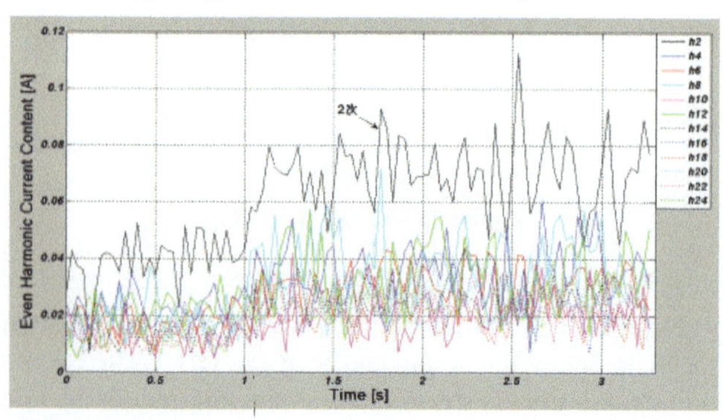

(c) Time-varying trend of the even harmonic current content throughout the charging process

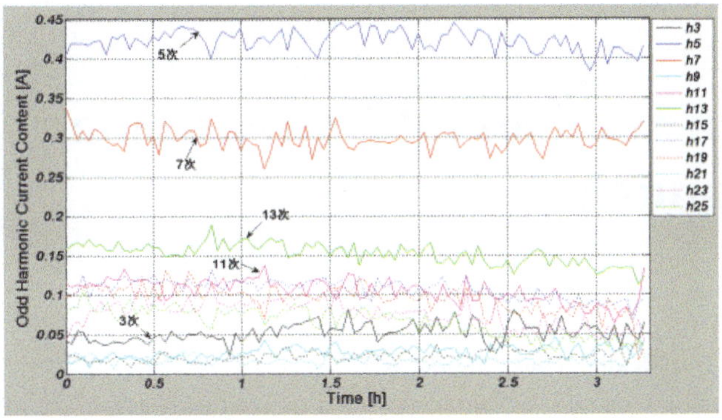

(d) Time-varying trend of the odd harmonic current content throughout the charging process

Figure 9. Harmonic Current.

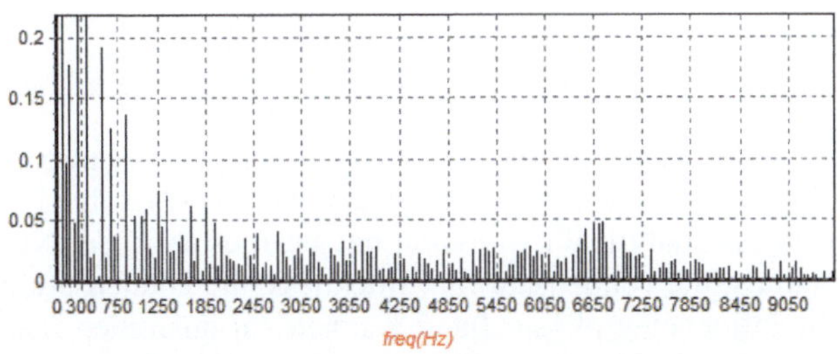

Figure 10. Charger AC side 0-10 kHz current harmonic spectrum.

10) Higher Harmonic Phenomenon

Conclusion: The preceding analysis shows that the low-order harmonics of the off-board charger is very low, which is in line with relevant international standards. But **Figure 10** shows that there is also higher harmonic, mainly distributing around 6650 kHz

5. Conclusions

After monitoring the charging process of single-board charger in electric vehicles charging station, and evaluating the testing results, we can draw conclusions as follows:

1) Electric vehicle battery is the capacitive load. It can export reactive power when charging.

2) In the charging process of the off-board charger, indicators of voltage deviation, frequency deviation, power factor, and voltage distortion rate are qualified.

3) Off-board charger produces odd harmonics in the charging process, and with increasing harmonic order, harmonic content reduces. There is a certain amount of high-order harmonic in off-board charger, mainly distributing around 6650 kHz.

4) Generated harmonics of the actual device, the harmonic is mainly reflected in the current, voltage, only a small distortion.

In fact it also shows that in the actual grid operation, to ensure that the grid power quality standards, vehicle charging station will consider providing compensation device and filtering device in associated distribution system.

REFERENCES

[1] K. Clement Nyns, E. Haesen and J. Drisen, "The Impact of Charging Plug-in Hybrid Electric Vehicles on a Residential Distribution Grid," *IEEE Transaction on Power System,* Vol. 25, No. 1, 2010, pp. 371-380.

[2] G. A. Putrus, P. Suwanapingkaral, D. Johnston, *et al.,* Impact of Electric Vehicles on Power Distribution Networks, *Proceeding of IEEE Vehicle Power and Propulsion Conference*, September 7-10, 2009, Dearborn, MI, USA, 827-831.

[3] S. N. Shao, M. Pipattanasomporn and S. Rahman, "Challenges of PHEV Penetration to the Residential Distribution Network," *Proceeding of IEEE Power & Energy Society General Meeting*, Calgary, Canada, July 26-30, 2009.

[4] J. Taylor, A. Maitra, M. Alexander, *et al.*, "Evaluation of the Impact of Plug-in Electric Vehicle Loading on Distribution System Operation,"*Proceeding of IEEE Power & Energy Society General Meeting*, Calgary, Canada, July 26-30, 2009.

[5] C. Roe, F. Evanelos, J. Melsel, *et al.*, "Power System Level Impacts of PHEVs," *Proceeding of the 42nd Hawaii International Conference on System Sciences,*Hawaii, HI,USA, January 5-8, 2008.

[6] M. Basu, K. Gaughan and E. Coyle, "Harmonic Distortion Caused by PHEV Battery Chargers in the Distribution Systems Network and Its Remedy," *Proceeding of the 39th International University Power Engineering Conference*,Bristol, UK, September 6-8, 2004.

[7] M. Y. Xu, X. H. Mu, H. Zhang, Y. He, M. J. Zhang and X. G. Chen, "Analysis of the Influence the Electrical Vehicle Charging Station to the Grid Harmonics," *Heilongjiang Electric Power*, No. 1, 2012.

[8] L. L. Ma and J. Q. Zhang, "Study on Electric Vehicle Charging Device Model Based on PSCAD and Harmonic Analysis," *Shanxi Electric Power*, Vol. 7, 2012, pp. 28-32.

[9] M. Huang, S. F. Huang and J. C. Jiang, "Harmonic Study of Electric Vehicle Chargers," *Journal of Beijing Jiaotong University*, Vol. 5, 2008.

[10] Q. Zhang, W. J. Han, J. H. Yu, C. Y. Li and L. F. Shi, "Simulation Model of Electric Vehicle Charging Station and the Harmonic Analysis on Power Grid," *Transactions of China Electrotechnical Society*, Vol. 2, 2012.

[11] Q. Liu, "Charging Modes of Electric Vehicle and the Impact Analysis of Charging Station to Power Quality on Power Grid," China High-Tech Enterprises, Vol. 27, 2011.

23

Condensation Heat Transfer in Horizontal Non-Circular Microchannels

Hicham El Mghari[1,2], Mohamed Asbik[1], Hasna Louahlia-Gualous[2]
[1]LP2MS, URAC08, Université Moulay Ismaïl, Faculté des Sciences, Meknès, Maroc
[2]LUSAC, Université de Caen Basse Normandie, Saint Lô, France

ABSTRACT

This investigation contributes to a better understanding of condensation heat transfer in horizontal non-circular microchannels. For this purpose, the conservation equations of mass, momentum and energy have been numerically solved in both phases (liquid and vapor), and all the more so the film thickness analytical expression has been established. Numerical results relative to variations of the meniscus curvature radius, the condensate film thickness, the condensation length and heat transfer coefficients, are analyzed in terms of the influencing physical and geometrical quantities. The effect of the microchannel shapes on the average Nusselt number is highlighted by studying condensation of steam in-square, rectangular and equilateral triangular microchannels with the same hydraulic diameter of 250 μm.

Keywords: Condensation; Microchannel; Numerical Simulation; Capillary Regime; Heat Transfer

1. Introduction

Understanding the heat transfer behavior of condensation flow in microchannels is important for a broad variety of engineering applications. Although there have been a number of investigations on boiling flow in microchannels, there are relatively few experimental data and theoretical analyses relative to condensation processes available in the literature, especially, for condensation inside a noncircular microchannel. In the chapter 6 of the reference [1], an overview of minichannels and microchannels condensation has been exposed.

Most of the physical and mathematical models that focused on annular condensation heat transfer in circular channel were developed in the previous works. Begg *et al.* [2] studied annular film condensation in a small circular tube to predict the shape of the liquid-vapor interface along a miniature tube leading to the complete condensation phenomena in small diameter tubes. Louahlia-Gualous and Asbik [3] conducted a numerical model predicting heat transfer for condensation of pure refrigerant and binary mixture in a mini-tube. Miscevic *et al.* [4] developed a stationary condensation capillary flow model based on the separate flow approach by taking into account the coupling between a cylindrical interface and a hemispherical interface. Recently, Ribeiro *et al.* [5] experimentally investigated the thermal-hydraulic performance of microchannel condensers using three different copper metal foams structures with distinct pore densities and porosities (0.893 and 0.947) as enhanced surfaces on the air-side. Their results are compared with the conventional condenser surface. El Achkar *et al.* [6] investigated the experimental heat transfer in the isolated bubbles zone of a transparent circular cross-section micro condenser. The evolution of vapor quality was experimentally determined by using the image processing. The energy balance was then used to calculate the temperature of the liquid in the isolated bubbles zone, showing that liquid and vapor were not in thermal equilibrium. The sensible heat transfers and latent heat transfers were then compared. However, a fundamental understanding of local mechanisms of heat and mass transfer cannot be accomplished using multichannels because the mass flow rates for each microchannel and condensate flow regime are unknown. Also, condensation heat transfer and pressure drop depend on the corresponding structure of the two phase flow (mist flow, annular flow, bubbly flow, or slug/plug flow) as shown by Odaymet *et al.* [7]. Three main condensation flows were identified in a small circular tube: the annular flow, the intermittent or elongated bubbles flow, and the spherical bubbles flow by Louahlia-Gualous and Mecheri [8]. Annular flow is especially found to be one of the dominant condensation flows in microchannelsas shown by Odaymet and Louahlia-Gualous [9] and Quan *et al.* [10].

On the other hand, various theoretical models have been proposed to predict the local heat transfer related to the condensation annular flow in non-circular channel where the surface tension plays a predominant effect on the condensate flow, more specifically, in the channel corners. Indeed, Zhao and Liao [11] analyzed annular film condensation heat transfer inside vertical mini triangular channel using three zones: the thin liquid film flow on the sidewall, the condensate flow in the corners, and the vapor core flow in the center. Wu and Cheng [12] carried out a simultaneous visualization and measurement experiment to perform condensation flow patterns of steam flowing through an array of trapezoidal silicon microchannels with a hydraulic diameter of 82.8 μm. Wang *et al.* [13] and Wang & Rose [14] proposed a theoretical model for condensation annular flow in a horizontal square and equilateral triangular channel with hydraulic diameter ranging from 0.5 to 5 mm by taking into account the effects of gravity, surface tension, and interfacial shear stress. They obtained the local heat transfer coefficient for refrigerants R134a, R22, R152a, CO2, propane, ammonia, and R410a by assuming that the channel wall temperature is uniform. Furthermore, they proposed one correlation for condensation heat transfer in the square and triangular microchannels in which surface tension and viscosity are the predominant parameters controlling condensate film thickness. Additionally, in references ([15-18]) for which the contents are not detailed here, readers could find more information about the use of various non-circular microchannel shapes.

The main purpose of the present work is to determine the heat transfer coefficient during the steam condensation inside horizontal various non-circular microchannels (rectangle, square, or equilateral triangle). Indeed, the classical mathematical model of the annular condensation flow is retained in both phases (liquid and vapor). An appropriate numerical method is used to solve the differential equations system obtained from different conservation equations (mass, momentum and energy). To compute the heat transfer coefficient, a new and simple geometrical method is used to express the condensate film thickness.

2. Physical Model

The physical model investigated in this paper is illustrated on the **Figure 1**. It concerns, a horizontal non circular microchannel with a hydraulic diameter D, which is cooled with a uniform wall heat flux density q. At the microchannel inlet, the vapor (steam) mass flux, the pressure and temperature of the steam is given at the saturated state. The vapor condenses inside the microchannel having wall temperature lower than vapor saturation temperature. The film thickness on the heat exchange surface varies along the axial direction with the vapor quality.

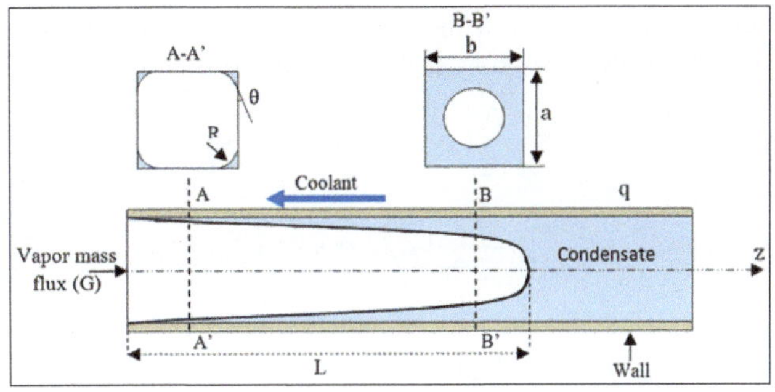

Figure 1. Schematic representation of the annular condensation in a rectangular microchannel.

When total condensation occurs, the end of the condensation zone has a hemispherical meniscus. Its length L is one of the unknown parameters in the physical model. The condensate film flows along the axial direction under effects of the pressure, surface tension, and shear stress.

3. Mathematical Formulation

3.1. Hypotheses

The mathematical formulation of the problem is based on the following principal assumptions:

- For liquid and vapor phases, thermophysical properties are assumed to be constant.
- In the microchannel, the flow is supposed to be steady-state, laminar, one-dimensional and axis-symmetrical.
- The free surface of the condensate film is smooth.
- Gravity forces are negligible compared to the effects of surface tension.
- The heat transfer from the cooling fluid to the condensate flow is assumed to be one-dimensional.
- In the condensate, the temperature profile is supposed linear.
- The saturation temperature of the vapor is assumed to remain constant along the microchannel.

3.2. Conservation Equations

The modeling approach developed here describes the liquid and vapor phases separately. The governing equations are used in the Cartesian coordinates as shown in the **Figure 1**.

3.2.1. Mass Conservation

The average parameters over a cross-section are used in liquid and vapor phases of the condensation flow respecting continuity conditions at the liquid-vapor interface. The equation of the mass conservation can be written for each local cross section as follows:

$$\frac{1}{A}\frac{\partial}{\partial z}\left(\rho_l U_l (1-\alpha) A\right) = -\Gamma \tag{1}$$

$$\frac{1}{A}\frac{\partial}{\partial z}\left(\rho_v U_v \alpha A\right) = \Gamma \tag{2}$$

where Γ represents the volumic rate of phase change. By convention, its sign is negative for condensation and positive for evaporation. So, we can write:

$$\Gamma = -\mathrm{d}\dot{m}_L/\mathrm{d}z = \mathrm{d}\dot{m}_v/\mathrm{d}z \tag{3}$$

A_ξ, A_ξ and U_ξ are respectively the microchannel cross section, the density and the axial velocity. The subscript ξ refers to the considered phase (ξ = v or L).

3.2.2. Momentum Conservation

For a fixed position z along the microchannel, the condensate film thickness in noncircular microchannels is much thicker into the microchannel corners than elsewhere, especially; at the internal considered circumference because of the surface tension effect. This is the reason that the axial flow in the film region between the corners is neglected.

From the forces balance illustrated on the **Figure 2**, the momentum conservation equations in the control volume of the length dz, are given by the Equations (4) and (5). These relationships are essentially expressed in terms ofthe interfacial shear at the liquid-vapor interface, the shear wall friction at the liquid-wall contact surface, and the pressure forces on the liquid area. In both phases, all the physical properties are assumed to be constant, and the influences of the gravity and the buoyancy forces are neglected. So, in the liquid and vapor phases, the momentum conservation equations are:

- In the liquid phase:

$$-A_L\frac{\mathrm{d}P_L}{\mathrm{d}z} + \tau_{vL}\frac{\mathrm{d}S_{vL}}{\mathrm{d}z} - \tau_w\frac{\mathrm{d}S_{Lw}}{\mathrm{d}z} - U_L\frac{\mathrm{d}\dot{m}_L}{\mathrm{d}z} = \dot{m}_L\frac{\mathrm{d}U_L}{\mathrm{d}z} \tag{4}$$

- In the vapor phase:

$$-A_v\frac{\mathrm{d}P_v}{\mathrm{d}z} + \tau_{vL}\frac{\mathrm{d}S_{vL}}{\mathrm{d}z} - U_v\frac{\mathrm{d}\dot{m}_v}{\mathrm{d}z} = \dot{m}_v\frac{\mathrm{d}U_v}{\mathrm{d}z} \tag{5}$$

A_ξ, A_ξ and U_ξ are respectively the microchannel cross-section, the pressure and the axial velocities in the phase $\xi(v \text{ or } L)$. Furthermore, we indicate that $\mathrm{d}S_{vL}$ is the liquid–vapor interface surface along $\mathrm{d}z$, $\mathrm{d}S_{Lw}$ is the wet heat exchange surface along $\mathrm{d}z$, and τ_{vL} is the shear stress at liquid-vapor interface, τ_w is the shear stress at the microchannel heat exchange surface.

3.2.3. Energy Conservation

The local energy equation in the liquid phase can written as:

$$q\mathcal{P}z = \dot{m}_L(z)h_{fg} \tag{6}$$

The total energy equation defined in the length of the total condensation zone as:

$$q\mathcal{P}L = \dot{m}_{v,in}h_{fg} \tag{7}$$

where q is the heat flux density, $\dot{m}_L(z)$ is the local liquid mass flow rate $\mathcal{P}$ is the microchannel perimeter, and z is the microchannel abscissa.

3.2.4. Curvature Radius Expression

To express the curvature radius derivative, we need to use the Laplace-Young equation:

$$\frac{\mathrm{d}P_L}{\mathrm{d}z} = \frac{\mathrm{d}P_v}{\mathrm{d}z} + \frac{\sigma}{R^2}\frac{\mathrm{d}R}{\mathrm{d}z} \tag{8}$$

Combining Equations (1)-(8), we get:

$$\frac{\mathrm{d}R}{\mathrm{d}z} = \frac{0.5 f_{vL}\rho_v \mathcal{P}_{vL} U_v^2\left[1+\frac{A_L}{A_v}\right] - 0.5 f_L \rho_L \mathcal{P}_{Lw} U_L^2 - \frac{2q\mathcal{P}A_L}{h_{fg}}\left[\frac{U_v}{A_v}+\frac{U_L}{A_L}\right]}{\sigma\frac{A_L}{R^2} + 2\frac{\rho_v U_v^2 A_L^2}{A_v R} - \frac{2\rho_L A_L}{R}U_L^2} \tag{9}$$

3.2.5. Dimensionless Equations

To establish the dimensionless equations, the following variables are used:

$$R^* = \frac{R}{D_h}, z^* = \frac{z}{D_h}, L^* = \frac{L}{D_h}, \bar{\rho} = \frac{\rho_v}{\rho_L}, \mathcal{P}^* = \frac{\mathcal{P}}{D_h}, P^* = \frac{PD_h}{2\sigma}, U_V^* = \frac{A\rho_v U_v}{\dot{m}_{v,in}}, U_L^* = \frac{A\rho_L U_L}{\dot{m}_{v,in}} \tag{10}$$

Then the dimensionless form of the derivative curvature radius, the velocity gradients and the pressure gradients in both phases, are given by:

$$\frac{\mathrm{d}R^*}{\mathrm{d}z^*} = \frac{0.5 f_{vL}\mathcal{P}_{vL}^* U_v^{*2}\left[1+\frac{A_L^*}{A_v^*}\right] - 0.5 f_L \rho^* \mathcal{P}_{Lw}^* U_L^{*2} - 2Bo\mathcal{P}^* A_L^*\left[\rho^*\frac{U_L^*}{A_L^*}+\frac{U_v^*}{A_v^*}\right]}{\frac{A_L^*\rho^*}{Re\,Ca} + \frac{2A_L^{*2}U_v^{*2}}{A_v^* R^*} - \frac{2\rho^* A_L^* U_L^{*2}}{R^*}} \tag{11}$$

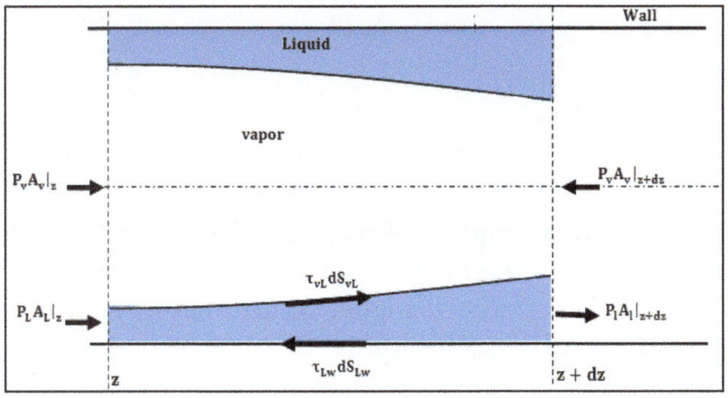

Figure 2. Forces balance for two-phase flow in a horizontal microchannel.

$$\frac{\mathrm{d}U_L^*}{\mathrm{d}z^*}=\frac{Bo\mathcal{P}^*}{A_L^*}-2\frac{U_L^*}{R^*}\frac{\mathrm{d}R^*}{\mathrm{d}z^*} \tag{12}$$

$$\frac{\mathrm{d}U_v^*}{\mathrm{d}z^*}=-\frac{Bo\mathcal{P}^*}{A_v^*}-2\frac{A_L^*}{A_v^*}\frac{U_v^*}{R^*}\frac{\mathrm{d}R^*}{\mathrm{d}z^*} \tag{13}$$

$$\frac{\mathrm{d}P_L^*}{\mathrm{d}z^*}=\frac{1}{4}\frac{f_{vL}Ca\,Re\mathcal{P}_{vL}^*U_v^{*2}}{\bar{\rho}A_L^*}-\frac{1}{4}\frac{f_LCa\,Re\mathcal{P}_{Lw}^*U_L^{*2}}{A_L^*}-\frac{Bo\,Ca\,Re\mathcal{P}^*U_L^*}{A_L^*}+\frac{Ca\,ReU_L^{*2}}{R^*}\frac{\mathrm{d}R^*}{\mathrm{d}z^*} \tag{14}$$

$$\frac{\mathrm{d}P_v^*}{\mathrm{d}z^*}=-Ca\,ReU_v^*\frac{\mathrm{d}U_v^*}{\mathrm{d}z^*}+\frac{Ca\,Re}{A_v^*}\frac{\mathrm{d}A_v^*}{\mathrm{d}z^*}+\frac{Ca\,ReU_v^{*2}}{R^*}\frac{A_L^*}{A_v^*}\frac{\mathrm{d}R^*}{\mathrm{d}z^*}-\frac{1}{4}\frac{f_{vL}Ca\,Re\mathcal{P}_{vL}^*U_v^{*2}}{\bar{\rho}A_v^*} \tag{15}$$

$$Bo=q/Gh_{fg},\quad Ca=\frac{\mu_lG}{\sigma\rho_L},$$

$$Re=\frac{\dot{m}_{v,in}D_h}{A\mu_L},\quad \dot{m}_{v,in}=GA$$

are respectively the boiling number, the capillary number, the vapor Reynolds number and the vapor mass flux.

Definitions of other parameters appearing in the above relations (11-15) are such that:

1) The liquid friction factor for laminar is given by:

$$f_L=CRe_L^{-1} \tag{16}$$

C is the Poiseuille number given in [19].

For turbulent flow, the liquid friction coefficient is determined from the Blasius equation [1]:

$$f_L=0.0791\mathrm{Re}_L^{-0.25} \tag{17}$$

In this expression, the liquid Reynolds number Re_L is calculated assuming the liquid single phase:

$$Re_L=\frac{\rho_LU_LD_L}{\mu_L},\quad D_L=\frac{4A_L}{\mathcal{P}_{Lw}} \tag{18a-b}$$

2) The interfacial frictional coefficient taking into account the effect of the condensation process on the interfacial shear stress is defined by [20]:

$$f_{vL}=f_{v,0}\frac{\varnothing e^{\varnothing}}{e^{\varnothing}-1} \tag{19a}$$

where $f_{v,0}$ is the friction factor for single phase vapor flow defined for laminar flow as :

$$f_{v,0}=CRe_v^{-1} \tag{19b}$$

For turbulent flow, $f_{v,0}$ is determined from:

$$f_{v,0}=0.0791Re_v^{-0.25} \tag{20a}$$

$$Re_v=\frac{\rho_vU_vD_v}{\mu_v},\quad D_v=\frac{4A_v}{\mathcal{P}_{vL}}, \tag{20b}$$

The factor $\varnothing$ is defined as the ratio of the local condensation mass flow rate to the vapor mass flow rate rebounding from the liquid-vapor interface [2], which is approximated by:

$$\varnothing=\frac{\mathrm{d}q}{\mathcal{P}\mathrm{d}z}\frac{2}{h_{fg}U_v\rho_vf_{v,0}} \tag{21}$$

3.2.6. Dimensionless Boundary Conditions

Equations (11) to (15) are solved using the following dimensionless boundary conditions:

1) At the microchannel inlet ($z^*=0$):

- the flow mass flux G is imposed;
- the temperature and pressure at the saturated state are given;
- the non-dimensional vapor velocity is

$$U_v^*(z^*=0)=1;$$

- the non-dimensional liquid velocity is

$$U_L^*(z^*=0)=0;$$

- the non-dimensional curvature radius is

$$R^*(z^*=0)=\frac{\sigma}{P_{v0}D_h}$$

2) At the position $z^*=L^*$ corresponding to the end of the condensation zone:

- the non-dimensional outlet liquid pressure is:

$$P_l(z^*=L^*)=\frac{P_0D}{2\sigma};$$

- for the rectangular cross section, the dimensionless curvature radius is [22]:

$$R^*(z^*=L^*)=\varepsilon\sin(\beta)/(2\cos(\theta+\beta)D_h)$$

where $\varepsilon=\min(a,b)$;

- for the equilateral triangle cross section, the dimensionless curvature radius a is [18] :

$$R^*(z^*=L^*)=\mathcal{P}\sin(\beta)/(6\cos(\theta+\beta)D_h)$$

4. Solution Procedure

The above boundary conditions combined with non- dimensional Equations (11)-(15) constitute the mathematical model of two-phase flow in capillary regime with a vapor-liquid phase change. Solution of this mathematical model is not trivial since one of the limit positions remains to be found, the value of L being one of the unknowns of the problem. So, the iteration process is used to solve the mathematical model. To start the calculations, the saturated vapor at the inlet of the tube is assumed having the known mass flux, temperature, pressure, and vapor quality. For the next axial step $\left(z^* + \Delta z^*\right)$, the following steps are executed:

1) To start calculation, an arbitrary total condensation length is assumed,

2) Calculation of the curvature radius from Equation (11),

3) Calculation of the local liquid velocity from Equation (12);

4) Calculation of the local vapor velocity from Equation (13);

5) Calculation of the liquid pressure from Equation (14);

6) Calculation of the local vapor pressure from Equation (15);

7) Calculation of the total condensation length from Equation (26).

Steps 2) to 7) are repeated until the value of the condensation length obtained at the iteration number "it" is approximately equal to the one determined at iteration number "it-1".

Knowing the total condensation length, calculations are then made for the next values of z locations. For each *z* location, steps 2) to 7) are made and calculation is stopped when *z* value is approximately equal to the total condensation length L.

5. Computation of the Heat Transfer Coefficient and the Condensation Length

5.1. Calculation of the Heat Transfer Coefficient

Assuming that the temperature profile is linear in the condensate film, the local heat transfer is expressed by:

$$h(z) = \frac{\lambda_L}{\delta(z)} \tag{22a}$$

and hence the average heat transfer coefficient in the microchannel condensation length *L* is performed by the relationship:

$$h_m = \frac{1}{L}\int_0^L h(z)\,\mathrm{d}z \tag{22b}$$

Two last quantities are used to determine the average and the local Nusselt numbers respectively:

$$Nu_{av} = \frac{h_m D}{\lambda_L} \tag{23a}$$

$$Nu_{loc} = \frac{h(z) D}{\lambda_L} \tag{23b}$$

$\delta(z)$ is the average condensate film thickness for each z location along the microchannel. Using the geometrical considerations from the **Figure 3**, the following analytical expression is established:

$$\delta(z) = \frac{1}{\frac{\pi}{2}}\int_0^{\frac{\pi}{2}} R\left[\frac{\cos(\theta)}{\cos(\Omega)} - 1\right]\mathrm{d}\Omega \tag{24}$$

5.2. Calculation of the Condensate Length

The parameter d*R*/d*z* given by the Equation (9) is expressed in terms of the limit curvature radius R_L.

At *z* = *L* location, d*R*/d*z* is to be infinite which causes a calculation problem. To avoid this complication, we set $z = L - \varepsilon$ where ε is an infinitesimal parameter. The limit values of the curvature radius and its first derivative are defined as [22]:

$$R_L = R\big|_{z=L-\varepsilon} = \frac{\varepsilon \sin\beta}{2\cos(\theta+\beta) D_h} \tag{25}$$

$$\left.\frac{\mathrm{d}R}{\mathrm{d}z}\right|_{z=L-\varepsilon} = \frac{\varepsilon\cos\theta - R_L}{R_L\cos\theta} \tag{26}$$

To estimate the numerical value of the condensation length a dichotomy method was performed using Equation (26), between the inlet of the microchannels which corresponds to $(z = 0$ and $R = 0)$ and the end of the condensation zone where $(z = L - \varepsilon$ and $R = R_L)$.

6. Results and Discussions

6.1. Numerical Results Validation

The validation of the results obtained in the present work for condensation of water in a square microchannel, are compared to the predictions of various correlations available in the literature. These correlations are proposed for condensation heat transfer in microchannels and macrochannels. Among these predictive correlations, those of condensation in microchannels are defined by Wang *et al.* [26], Koyama *et al.* [23], and Wang & Rose [14]. **Figure 4** shows the comparisons between the average Nusselt number obtained from the present numerical model and those predicted by seven correlations: Dobson *et al.* [22], Wang *et al.* [26], Koyama *et al.* [23], Wang & Rose [14], Traviss *et al.* [24], Shah [19], and Ackers *et al.* [25]. Results presented in the **Figure 4** are those obtained for condensation of steam in a horizontal square microchannel with hydraulic diameter of 110 μm, for various inlet-

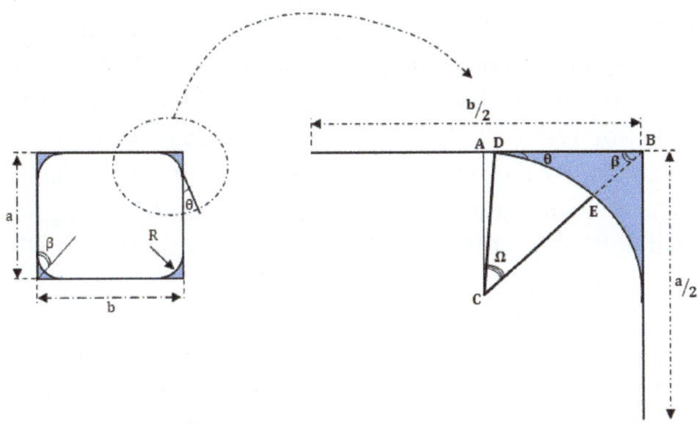

Figure 3. Distribution of the condensate film in the microchannel corners.

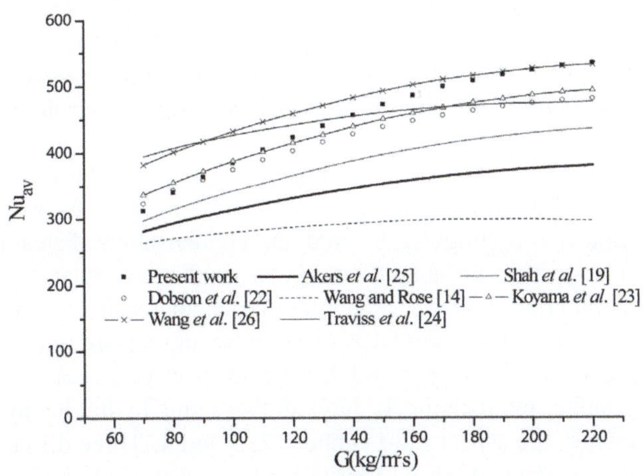

Figure 4. Comparison of the present work with predictions of literature correlations.

vapor mass fluxes ranging from 70 to 220 kg/m²s. It is found that the best predictions of the present average Nusselt number are obtained by the correlations of Dobson *et al.* [22] and Koyama *et al.* [23] for which the mean relative deviation is about 7%. Both of these correlations are developed in the case of an annular condensation heat transfer of several pure refrigerants which is carried out in the channels with small hydraulic diameter. The correlations of Traviss *et al.* [24] and Ackers *et al.* [25] devoted to the convective condensation heat transfer in the macro-scale channels have been also evaluated to the micro-scale. It can be seen that the correlation of Traviss *et al.* [24] over predicts highly the present results for low mass fluxes whereas the Ackers's correlation [25] under predicts them highly for all the range of the tested mass fluxes. Adding up, the correlation of Shah [19] is proposed for condensation heat transfer in channels with large hydraulic diameter from 7 to 40 mm and Wang *et al.* [26] correlation is defined for condensation heat transfer inside a horizontal small rectangular channel with hydraulic diameter of 1.46mm. Average Nusselt numbers obtained by both of these correlations give the same trend and the reasonable results compared to those predicted numerically. The Shah's correlation under predicts numerical results with a maximum average deviation of about 16%. As for the correlation of Wang *et al.* [26], it over predicts the present results with a maximum deviation of 22% obtained at low mass fluxes. Moreover, correlation of Wang & Rose [14] based on the Nusselt theory including the effects of interfacial shear stress and surface tension on condensation heat transfer is also evaluated. Predictions of Nusselt number from the correlation of Wang & Rose [14] are not in accordance with those of this study since they gave the highest deviation of 37% form the numerical results.

6.2. Peripheral Condensate Film Thickness and Local Heat Transfer

Numerical results are given in the **Figures 5(a) and (b)** and the **Figures 6(a) and (b)** for steam condensation in square section microchannel with hydraulic diameter of 110 and 250 µm. Computations were conducted for vapor mass flux G = 90 kg/m²s, contact angle θ = 15°, heat flux density q = 100 kW/m². For the conditions used in the present computation, the boiling number, the capillary number, the inlet steam temperature and pressure are maintained constant.

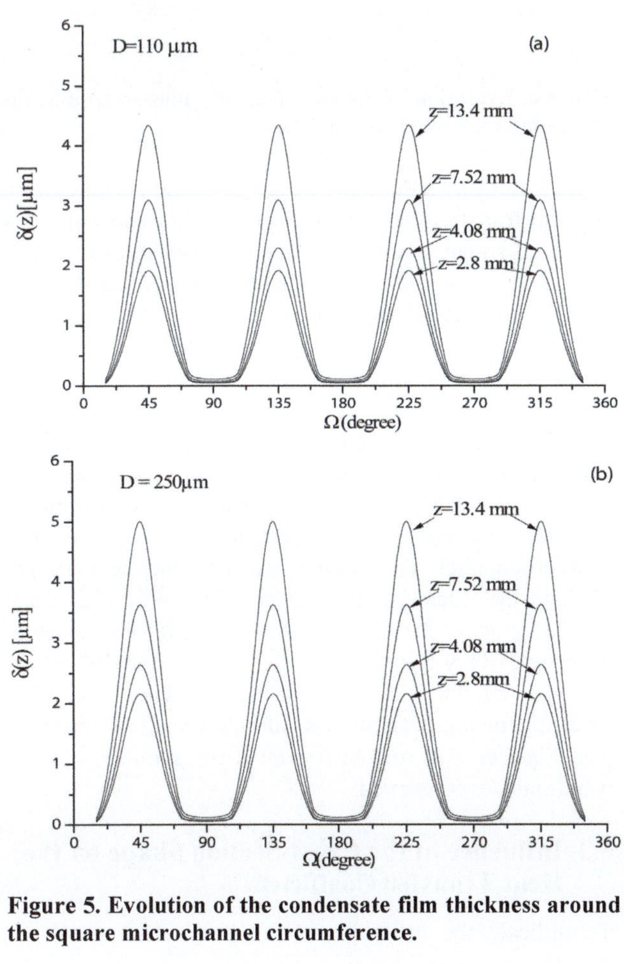

Figure 5. Evolution of the condensate film thickness around the square microchannel circumference.

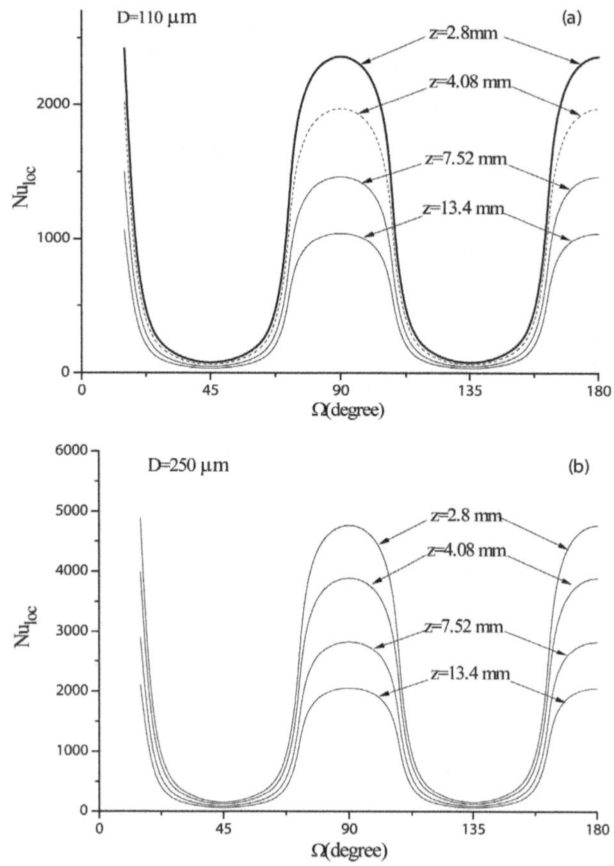

Figure 6. Evolution of the local Nusselt number around the square microchannel circumference.

To better understand the behavior of the peripheral local heat transfer coefficient, **Figures 5(a) and (b)** show predicted square channel condensate film thickness (24) plotted as a function of the peripheral coordinate Ω at four different axial locations. It can be seen from these figures that for the same z location on the microchannel, the condensate film thickness is higher for 250 μm (**Figure 5(b)**) than for 110 μm (**Figure 5(a)**). According to these representations, it is observed that for both microchannels, the symmetrical distributions of the condensate film thickness are obtained around the channel perimeter because the gravity has no effect on the condensate flow in microchannel. The condensate film thickness is very thick in the microchannel corners (Ω = 45°, 135°, 225° and 315°) under the effect of the surface tension whereas it becomes very thin at Ω = 0°, 90°, 180° and 360 where the heat transfer is the highest. These trends have been reversed for the peripheral local heat transfer coefficient (see **Figures 6(a) and (b)**) even if the symmetrical distributions are conserved.

6.3. Influence of the Cross-Section Shape on the Heat Transfer Coefficient

To indicate the influence of the channel cross-section shape, we start by the presentation of the results concerning curvature radius. In fact, **Figure 7** shows the variation of the dimensionless curvature radius along the flow direction in three cross-section shapes of microchannels. At the beginning and the end of the condensation, the curvature radius increases rapidly along the channel, while it increases slowly at the middle. We also note that for the triangular microchannel it's increasing very quickly than other geometries. The present simulation results are similar to those of the reference [21].

The annular condensation length is one of the most important parameter which influences the thermal performance of the microchannel studied here. **Figure 8** gives the dimensionless annular condensation length with respect to the boiling number Bo, for three cross-section shapes. As shown in this figure, the condensation zone decreases with Bo and it is also clear that the cross-section shape of the microchannel plays a noticeable role on the condensation.

In the heat transfer exchange point of view, the influence of the microchannel shape on the average Nusselt number is highlighted by studying condensation of steam in a square, equilateral triangular, and rectangular microchannels with the same hydraulic diameter of 250 μm. The sides of the equilateral triangular and square microchannels are 433 μm and 250 μm respectively. For rectangular microchannels having the same hydraulic diameter, the aspect ratio is about 2, 3 and 4. Three different rectangular cross sections are investigated: 375 × 187.5 μm², 500 × 166.6 μm² and 625 × 156.25 μm².

Under the same conditions, **Figure 9** compares the average Nusselt numbers with respect to the vapor Reynolds number for five microchannel cross-sections. According to this figure, it is seen that annular condensation Nusselt number is low for square microchannel cross section (aspect ratio of 1). For rectangular cross-section microchannels with aspect ratio (b/a) higher than 1, condensation Nusselt number increases by increasing aspect ratio. The highest Nusselt numbers values are obtained

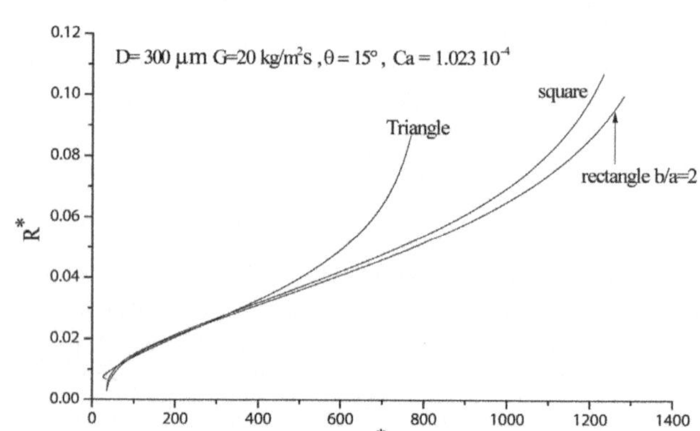

Figure 7. Dimensionless curvature radius along various microchannel shapes.

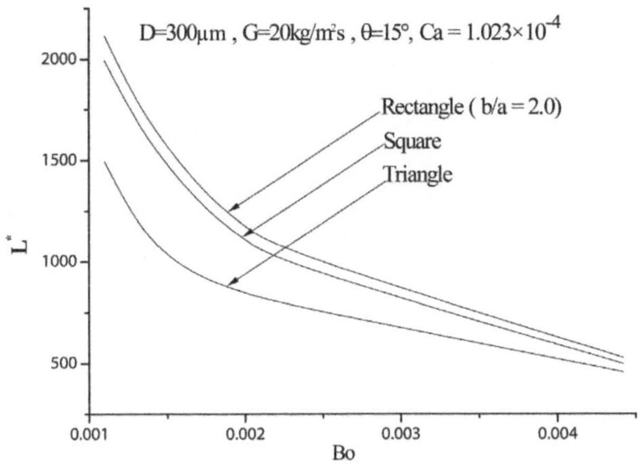

Figure 8. Dimensionless condensation length versus the boiling number.

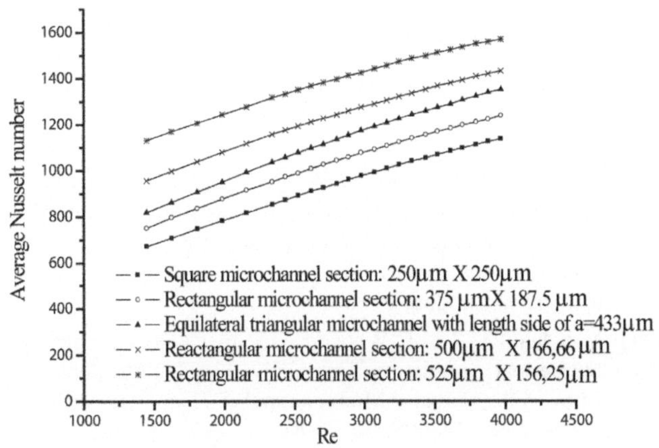

Figure 9. Comparison of the average Nusselt number for various microchannel shapes with the same hydraulic diameter (D = 250μm).

for microchannel with high aspect ratio (equal to 4) and low confinement (a = 156.25 μm). It is interesting to note that for the same hydraulic diameter, the microchannels perimeter and cross section increase with the aspect ratio leading to a reduction of the condensate film thickness. This is due to the predominant effect of the surface tension in the microchannel that increases with microchannel perimeter and thins the condensate film along the sides of the microchannel. For the same Reynolds number, average Nusselt number for equilateral triangular microchannel is between Nusselt number for rectangular microchannel with aspect ratio of 2 and that with aspect ratio of 3. Recall that the perimeter of the triangular microchannel is about 1299 μm and those of rectangular microchannels with aspect ratios of 2 and 3 are about 1125 μm and 1333 μm respectively.

7. Conclusion

The numerical model characterizing local heat and mass transfer for condensation in microchannels has been developed by including the effects of wall and liquid vapor interface shear stresses, surface tension, pressure forces, contact angle, etc. Numerical results are compared for steam condensation in square, rectangular, and equilateral triangular microchannels with the same hydraulic diameter. They are compared with the correlations available in the literature and it is shown that correlations for annular condensation in macro-channels predict heat transfer with significant deviations from those obtained for microchannels. The best predictions are obtained with the correlations of Dobson *et al.* [25] and Koyama *et al.* [24].

By using the established condensate film thickness expression [24], the influence of the microchannel hydraulic diameter is highlighted. Indeed, increasing this last geometrical factor increases the peripheral condensate film thickness and hence decreases the peripheral condensation heat transfer coefficient. Additionally, we point out that at the beginning and at the end of the condensation, the curvature radius increases rapidly along the channel, while it increases slowly at the middle. We also note that for the triangular microchannel it's increasing quicker than other geometries. As for the annular condensation length, we note that it decreases with increasing of the boiling number and its numerical values diminish in the case of equilateral triangular microchannel.

Finally, to show the major influence of cross-section shape on the condensation heat transfer, comparison of the average Nusselt number is conducted for different microchannel shapes with the same hydraulic diameter. It can be concluded that condensation average heat transfer increases with aspect ratio for rectangular microchannels. The lowest average Nusselt numbers are obtained for the square microchannel because its perimeter and cross section are lower than those of triangular and rectangular microchannels.

REFERENCES

[1] S. Kandlikar, S. Garimella, D. Li, S. Colin and M. King, "Heat Transfer and Fluid Flow in Minichannels and Microchannels," Elsevier, Amsterdam, 2006.

[2] E. Begg, D. Khrustalev and A. Faghri, "Complete Condensation of Forced Convection Two-Phase Flow in a Miniature Tube," *Journal of Heat Transfer*, Vol. 121, No. 4, 1999, pp. 904-915.

[3] H. Louahlia-Gualous and M. Asbik, "Numerical Modeling of Annular Film Condensation inside a Miniature Tube," *Numerical Heat Transfer Part A: Applications*, Vol. 52, No. 3, 2007, pp. 251-273.

[4] M. Miscevic, P. Lavieille and B. Piaud, "Numerical Study of Convective Flow with Condensation of a Pure Fluid in

Capillary Regime," *International Journal of Heat and Mass Transfer*, Vol. 52, No. 21-22, 2009, pp. 5130-5140.

[5] G. B. Ribeiro, J. R. Barbosa and A. T. Prata, "Performance of Microchannel Condensers with Metal Foams on the Air-Side: Application in Small-Scale Refrigeration Systems," *Applied Thermal Engineering*, Vol. 36, 2012, pp. 152-160.

[6] G. El Achkar, P. Lavieille and M. Miscevic, "Loop Heat Pipe and Capillary Pumped Loop Design: About Heat Transfer in the Isolated Bubbles Zone of Condensers," *Applied Thermal Engineering*, Vol. 33-34, 2012, pp. 253-257.

[7] A. Odaymet, H. Louahlia-Gualous, "Experimental Study of Slug Flow for Condensation in a Single Square Microchannel," *Experimental Thermal and Fluid Science*, Vol. 38, 2012, pp. 1-13.

[8] H. Louahlia-Gualous and B. Mecheri, "Unsteady Steam Condensation Flow Pattern inside a Miniature Tube," *Applied Thermal Engineering*, Vol. 27, 2007, pp. 1225-1235.

[9] A. Odaymet and H. Louahlia-Gualous, "Experimental Study of Slug Flow for Condensation in a Single Square Microchannel," *Experimental Thermal and Fluid Science* Vol. 38, 2012, pp. 1-13.

[10] X. J. Quan, P. Cheng and H. Y. Wu, "Transition from Annular Flow to Plug/Slug Flow in Condensation of Steam in Microchannels," *International Journal of Heat and Mass Transfer*, Vol. 51, No. 3-4, 2008, pp. 707-716.

[11] T. S. Zhao and Q. Liao, "Theoretical Analysis of Film Condensation Heat Transfer inside Vertical Mini Triangular Channels," *International Journal of Heat and Mass Transfer*, Vol. 45, 2002, pp. 2829-2842.

[12] H. Y. Wu and P. Cheng, "Condensation Flow Patterns in Silicon Microchannels," *International Journal of Heat and Mass Transfer*, Vol. 48, No. 11, 2005, pp. 2186-2197.

[13] H. S. Wang, J. W. Rose and H. Honda, "A Theoretical Model of Film Condensation in Square Section Horizontal Microchannels," *Chemical Engineering Research and Design*, Vol. 82, No. A4, 2004, pp. 430-434.

[14] H. S. Wang and J. W. Rose, "Theory of Heat Transfer during Condensation in Microchannels," *International Journal of Heat and Mass Transfer*, Vol. 54, No. 11-12, 2011, pp. 2525-2534.

[15] A. Agarwal., "Heat Transfer and Pressure Drop during Condensation of Refrigerants in Microchannels," Ph.D. Dissertation, Georgia Institute of Technology, 2006.

[16] J. Hu and C. Chao, "An Experimental Study of the Fluid Flow and Heat Transfer Characteristics in Micro-Condensers with Slug-Bubbly Flow," *International Journal of Refrigeration*, Vol. 30, 2007, pp. 1309-1318.

[17] Y. Chen, M. Shi, P. Cheng and G. Peterson, "Condensation in Microchannels," *Nanoscale and Microscale Thermophysical Engineering*, Vol. 12, 2008, pp. 1-27.

[18] Y. Chen, J. Wu, M. Shi and G. Peterson, "Numerical Simulation for Steady Annular Condensation Flow in Triangular Microchannels," *International Communications in Heat and Mass Transfer*, Vol. 35, No. 7, 2008, pp. 805-809.

[19] M. M. Shah, "A General Correlation for Heat Transfer during Film Condensation Inside Pipes," *International Journal of Heat and Mass Transfer*, Vol. 2, 1979, pp. 547-556.

[20] W. M .Kays and R. J. Moffat, "The Behavior of Transpired Turbulent Boundary Layers," In: *Studies in Convection: Theory, Measurement and Application*, Academic Press, London, 1975.

[21] Y. Chen, X. Li, J. Wu and M. Shi, "One Dimensional Numerical Simulation for Steady Annular Condensation Flow in Rectangular Microchannels," *Heat Mass Transfer*, Vol. 46, No. 1, 2009, pp. 75-82.

[22] M. K. Dobson, J. C. Chato, D. K. Hinde and S. P. Wang, "Experimental Evaluation of Internal Condensation of Refrigerants R-12 and R-134a," *Proceedings of the ASHRAE Winter Meeting*, New Orleans, 23-26 January 1994, pp. 744-754.

[23] S. Koyama, K. Kuwahara and K. Nakashit, "Condensation of Refrigerant in a Multi-Port Channel," *First International Conference on Microchannels and Minichannels*, Rochester, 24-25 April 2003, pp. 193-205.

[24] D. P. Traviss and W. M, Rohsenow, "Flow Regimes in Horizontal Two-Phase Flow with Condensation," *ASHRAE Transactions*, Vol. 79, 1973, pp. 31-39.

[25] W. W. Akers, H. A. Deans and O. K. Crosser, "Condensation Heat Transfer within Horizontal Tubes," *Chemical Engineering Progress Symposium Series*, Vol. 55, No. 29, 1959, pp. 171-176.

[26] W. W. Wang, T. D. Radcliff and R. N. A. Christensen, "Condensation Heat Transfer Correlation for Millimeter-Scale Tubing with Flow Regime Transition," *Experimental Thermal and Fluid Science.*, Vol. 26, No. 5, 2002, pp. 473-485.

Nomenclature

List of Latin symbols
A: Area (m^2)
a: Width of rectangular microchannels (m)
b: Depth of rectangular microchannels (m)
c = 1/4, 1/3: Inverse relationship of number of corners
D: Hydraulic diameter (m)
f: Friction factor
hfg: Latent heat ($J \cdot kg^{-1}$)
l: Annular condensation length (m)
l: Length of the end part in condensation (m)
M: Mass flux ($kg \cdot s^{-1}$)
P: Pressure (Pa)
Q: Heat flux density ($W \cdot m^{-2}$)
R: Curvature radius (m)
$\mathcal{P}$: Perimeter (m)
U: Velocity ($m \cdot s^{-1}$)

Greek symbols
B: Half of right angle (°)
ε = min(a,b) (m)
μ: Viscosity ($kg\ m^{-1} s^{-1}$)
θ: Contact angle (°)
ρ: Density ($kg \cdot m^{-3}$)
σ: Surface tension coefficient ($N \cdot m^{-1}$)
τ: Shear stress ($N \cdot m^{-2}$)
Ω: Peripheral angle (°)

Subscripts
L: Liquid
Lw: Liquid-wall interface
V: Vapor
Vl: Liquid-vapor interface
O: Inlet

Magnetized Water and Memory Meter

Ashraf Kotb
Department of Mechanical Power Engineering, Faculty of Engineering, Ain Shams University, Cairo, Egypt

ABSTRACT

This paper presents an experimental study to investigate the effect of using the magnetic water conditioner on the properties of water. The water flows through a closed loop, while the pH, TDS, and hardness represent its properties. For magnetic water conditioner with flux density of 170 mT, results showed that pH increased by 15.65% for 820 minutes of non-stop circulation. The increase in pH is divided to 93.5% for the first 360 minutes, and 6.5% for the last 460 minutes. TDS and Hardness of water are not affected by the magnetic water conditioner. Water remembers and keeps the impact of passing through the magnetic field for several hours, and pH decreased by 0.642 in 24 hours. While the results lead to introduce and create the magnetized water saturation curve and water memory meter.

Keywords: Magnetic; Water; pH; TDS; Hardness; Memory; Saturation; Meter

1. Introduction

Much time has passed since the discovery of the effect of magnetic water treatment by Vermeiren [1]. Having begun with a decrease in the formation of scale in steam boilers, the magnetic treatment of water and aqueous systems has found wide application in recent years in the most varied fields of practical human activity, since the overwhelming majority of energetic and industrial processes are connected with water. Magnetic treatment is presently used in combating precipitation, in the production of concrete, and in the enrichment of useful minerals, as well as for the intensification of water filtration and purification processes, etc.

Magnetic water conditioners are permanent magnets that attach to waterlines entering to "purify" or "condition" water supplies. These devices purportedly use magnetic fields to change the molecular makeup of various water constituents like calcium and iron to other more "inert" forms. The claimed result is a reduction or elimination of water contaminants.

There is virtually no valid scientific data to support any water treatment benefit from magnetic devices. Despite this, companies, sales representatives, and product brochures for the devices may make any one of the many claims [2].

Usually, printed testimonials from "hundreds of satisfied customers" support these claims. The claims put forth by manufacturers and sales representatives of these devices are without validity. They do not refer to standard physical, chemical, or biological water treatment processes. Therefore, several researchers have conducted performance evaluations of the equipment.

The researchers are divided into two sides; the first side opposes using the magnetic water conditioners, while the second side supports the benefits of the usage of magnetic water conditioners.

As early as 1977, Duffy [3] concluded that permanent magnets have no effect on the hardness of water or the formation of scales on pipes. The South Dakota School of Mines and Technology performed tests on several magnetic conditioners units. Researchers concluded that there was no change in the physical and chemical properties or the calcium ion concentration of water conditioned with the devices [4]. A third study conducted in 1985 at Purdue University tested six units placed on water supplies for water heaters and tested their effectiveness against a controlled system. Water was tested for temperature, specific conductivity, surface tension, boiling point of depression, pH, alkalinity, total hardness, calcium and scaling effect. The study concluded that no significant variation in the chemical water quality existed between the control and the "treated" systems.

The U.S. Army Corps of Engineers [5] tested three devices and the results showed that no clear advantage for any of the three devices tested versus a control for the inhibition of mineral scale formation or the corrosion of copper.

A study conducted in 2013 at Ain Shams University tested five units placed on PP-R pipe line for open loop water flow and tested their effectiveness. Water was tested for pH, hardness, and total dissolved solids. The study concluded that no significant variation in the chemical water quality existed from the units [6].

On the other side, Banejad and Abdosalehi [7] showed that the magnetic field affected on water quality and that production of magnetic water has ability on water hardness reducing until 51 percent.

Rameen *et al.* [8] reported the effect of commercial magnetic water conditioners on the total dissolved salts and pH on different solutions, the work includes laboratory evidence of water quality modifications which supplies different purposes using magnetic means.

Ghauri and Ansari [9] reported the effect of magnetic field on the water viscosity between 298 and 323 K. A higher absolute viscosity has been observed in the presence of applied magnetic field compared to no field which has been explained on the basis of stronger hydrogen bonds.

Szcze *et al.* [2] found that the magnetic field decreases the water conductivity, which is inversely proportional to the flow rate, and increases the amount of evaporated water, even after the water's distillation. The effects are due to the hydrogen bond network strengthening and the perturbation of gas/liquid interface from the air nanobubbles in the water.

Also, many testimonials are uploaded in each manufacturer site to certify the effectiveness of magnetic water conditioners

The objective of the present work is to investigate experimentally for water flows in closed loop and, the effect of magnetic water conditioner on pH, TDS, and Hardness of water. Also, the objective extends to demonstrating the effect of time on the properties of magnetically conditioned water.

2. Materials and Methods

According to [5], the test rig is constructed from nonmetallic piping, fittings, and pump to avoid increase in the turbidity of flowing water. The test rig is illustrated in **Figure 1**, it contains 4 liter glass tank, while PVC pump draws water from the tank via a plastic hose and discharges through a plastic hose followed by polypropylene PP-R pipe, the pipe is fitted with gate valve, the discharged water is recirculated into the tank through plastic hose. The test rig performs a closed loop of circulation, while the magnetic water conditioner (MWC) is placed before the valve and over the PP-R pipe. The magnetic water conditioner is investigated with the specifications listed in **Table 1** and installation as shown in **Figure 2**.

Instrumentation system is chosen to measure the pH, total dissolved solids, hardness values, and magnetic field intensity. pH is measured by high accuracy pen type pH meter with accuracy of ± 0.1 and resolution of 0.01. TDS is measured by TDS meter with range from 0 to 1999 ppm and accuracy of ± 2% f.s. Hardness is deter-

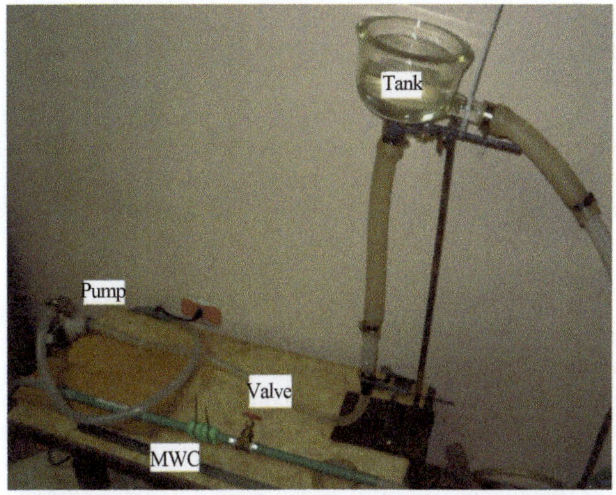

Figure 1. Components and Construction of Test Rig.

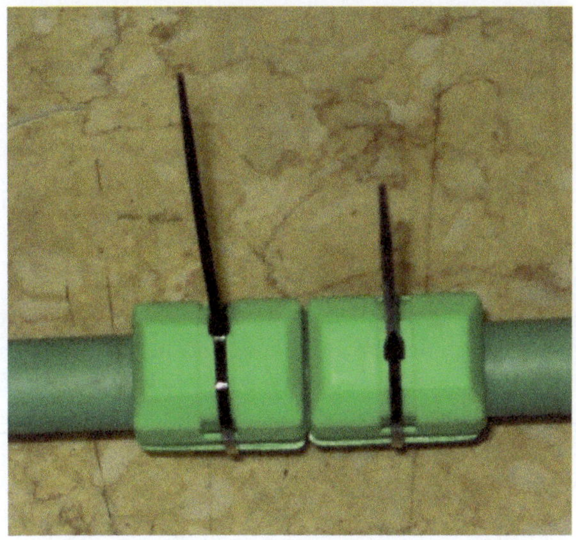

Figure 2. Installation of MWC-01 on Test Rig.

Table 1. Specifications and Installation of MWC.

Brand	Specifications and Installation
Powermag China	**Ultra Power Magnetic Water Softener UP-2** 2 pairs in one set **Description per Manufacturer**: Assembled with strong NdFeB magnet part forces +/− hydronium temporarily into cathode to exchange water molecule and mineral hydronium, makes Ca^{2+}/Mg^{2+} to unite electro, and has no cation function temporarily. Hence, even heating, it doesn't unite hydronium—for example SO_4^{2-}. The part makes water not to scale any more. Magnetic filter makes water soft, and scale will be never produced in water pipe.

mined by EDTA titration method. The magnetic flux density is measured by Gauss/Tesla meter with zero flux chamber.

Regarding to the chemical analysis to determine the hardness of samples, titration procedures provide relatively inexpensive means for the analysis of different substances. These titrations are based on chemical reactions with completion points that can be monitored by some visible change in the reaction systems. The most common system involves acids and bases with the use of an indicator that changes color as the system moves from an acidic to basic composition. Calcium and magnesium ions can be measured through reaction with a chelating agent EDTA (ethylenediaminetetraacetic acid). This molecule has four carboxylic acid (~COOH) group sites and two nitrogens, all of which have lone pairs of electrons. The EDTA molecule can form a complex with as many as six sites on a particular cation like Ca^{2+}. These EDTA complexes are generally very stable are always in 1:1 (metal:EDTA) molar ratios:

In this activity we will be titrating Ca^{2+} in water samples with EDTA. Both Ca^{2+} solutions and EDTA are colorless so an indicator is needed to signal the reaction completion. The indicator of choice is Eriochrome Black T which forms a wine-red complex with Mg^{2+}. A very small amount Mg^{2+} will be bound to the indicator through most of the titration. When all of the Ca^{2+} has reacted with EDTA, the Mg^{2+} in the indicator will react with the EDTA. The Indicator then returns to its acidic form which is a sky-blue and signals the end of the process. Below are the reactions that occur during the titration where H_3In is the general formula for the Eriochrome Black T.

During titration: $$H_2Y^{2-}(aq)+Ca^{2+}(aq)\rightarrow CaY^{2-}(aq)+2H^{+}(aq)$$

$$H_2Y^{2-}(aq)+Mg^{2+}(aq)\rightarrow MgY^{2-}(aq)+2H^{+}(aq)$$

At end point $$H_2Y^{2-}(aq)+MgIn^{-}(aq)\rightarrow MgY^{2-}(aq)+HIn^{2-}(aq)+H^{+}$$

Wine-red sky-blue

3. Results and Discussion

The manufacturer of the used magnetic water conditioner wrote the technical specifications and benefits of using the magnetic water conditioner, but there are no scientific or measured data about the construction, material, and the magnetic flux density. For this reason and before investigating the effect of magnetic water conditioner, an attempt is carried out to measure the magnetic flux density for the used magnetic water conditioner. The magnetic flux density is measured at the contact surface for each piece; it was found that the flux density is not uniformly distributed over the contact surface with the maximum of 330.0 mT. After installation the magnetic flux density inside the pipe is measured, it was found that the flux density is not uniformly distributed over the cross flow area or the covered length with the maximum of 170.0 mT.

The test rig is filled with water; initially, amount of Sodium Chloride is dissolved in the water to raise the total dissolved solids in the water, and pH, TDS, and Hardness are measured and considered to be the base line data for the water properties before recirculated through the magnetic water conditioner.

Unified method of measurements is performed throughout the work; each water sample at each time is measured and chemically analyzed 10 times, the values are statistically analyzed and the mathematical average represents as the result at the time.

Figure 3 illustrates the pH values of recirculated water versus the time of circulation, before circulation the base line data is clearly identified to be 7.343. The pH is increased by increasing the circulation time; over 820 minutes of non-stop circulation the pH reaches 8.492 with total increase of 1.149% with 15.65%.

One can explain the result as; when water flows through the magnetic field, the nuclei of its atoms are polarized, this polarization makes the atoms behave as tiny magnets with north and south poles. The net results are; increment in the pH value as a result of polarization and uniform arrangement of atoms as a result of the creating poles. Increasing the circulation time increases the residence time at which water is subjected to the magnetic field, which gives the atoms enough time for more polarization and more uniformity in the arrangement of atoms.

The pH increases 1.074 during the first initial 360 minutes of circulation, this increment represents 93.5% from the total increase achieved. While the pH increases 0.075 during the last 460 minutes of circulation, this increment represents 6.5% from the total increase achieved.

One can say that; the first 360 minutes of circulation divided the behavior of pH into two sections; the first section is from 0 to 360 minutes at which the pH increases with noticeable values, the second section is from 360 to 820 minutes at which the pH increases with very small values. This phenomenon needs to define the term (*Saturation Time*) and it may be defined as; the circulation time through the magnetic water conditioner for maximum increase in the pH value. Consequently, for the circulation time less than the saturation time one can define that the water is in the state of sub-saturation, while for circulation times greater than the saturation time the water is the saturation as illustrated in **Figure 4** which is defined as the Magnetized Water Saturation Curve.

Figure 5 illustrates the TDS values of recirculated water versus the time of circulation, before circulation the base line data is clearly identified to be 1219.9 ppm. It is deduced that; over the 820 minutes of circulation through the magnetic water conditioner, the TDS of water does not affected by recirculation through the magnetic water conditioner. The averages of measured values at each time are in the range of uncertainty.

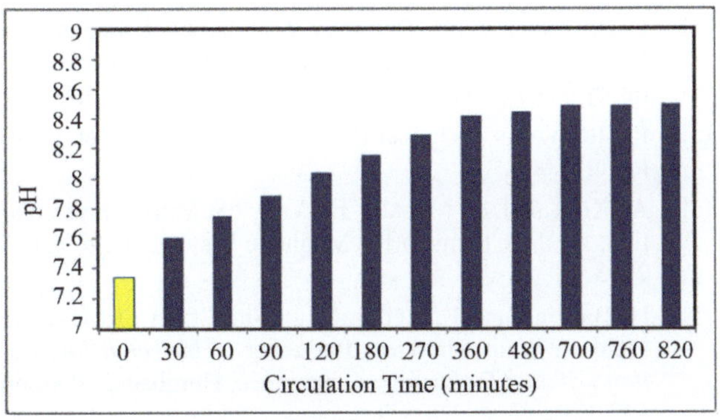

Figure 3. pH of Circulated Water.

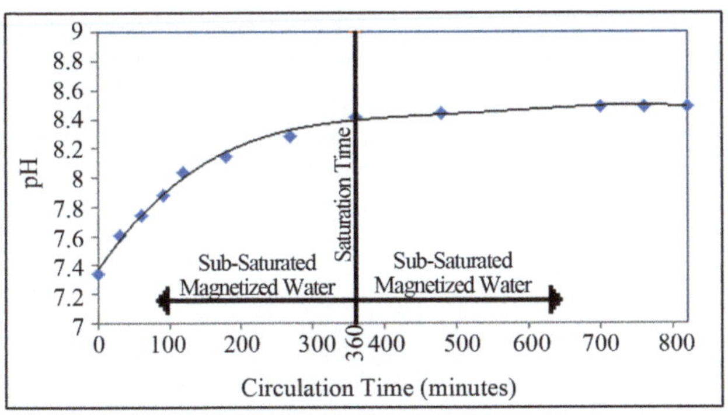

Figure 4. Magnetized Water Saturation Curve.

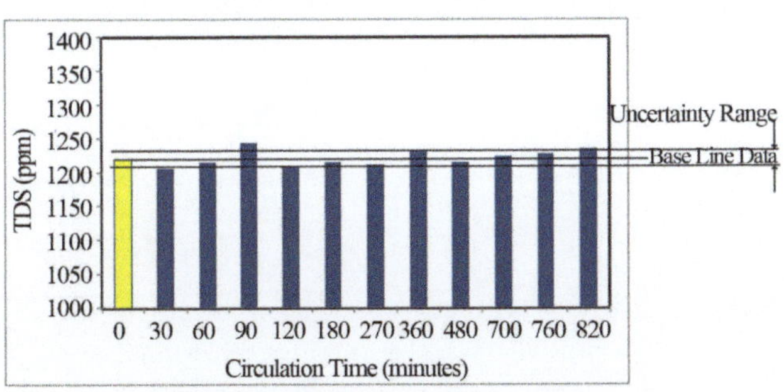

Figure 5. TDS of Circulated Water.

Figure 6 illustrates the Hardness values of recirculated water versus the time of circulation, before circulation the base line data is clearly identified to be 139.5 ppm. It is deduced that; over the 820 minutes of circulation through the magnetic water conditioner, the Hardness of water does not affected by recirculation through the magnetic water conditioner. The averages of measured values at each time are in the range of uncertainty.

Since the previous results showed that; the effect of the magnetic water conditioner is dominated on increase-ing the pH of circulating water from 7.343 to 8.492. Therefore, the question arises here, is to what time the water keep the maximum value of pH of 8.492, which is obtained from the flowing through the magnetic water conditioner for 820 minutes, after it being stored in static condition and away from the magnetic water conditioner. This question illustrates the effect of magnetic water conditioner on the memory of magnetized water.

The answer of the last question extends the present work herein; an amount of water with pH of 8.492, after 820 minutes of recirculation through the magnetic water conditioner is stored and isolated from atmospheric air to eliminate the effect of absorption of atmospheric gases by the stored water. The time counter is restarting, and after periods of time, a sample is taken from the stored water and the pH value is measured by taking as mentioned 10 measurements each time. The total storage objective time is 24 hours; **Figure 7** illustrates the pH values of stored water versus the time. As illustrated the

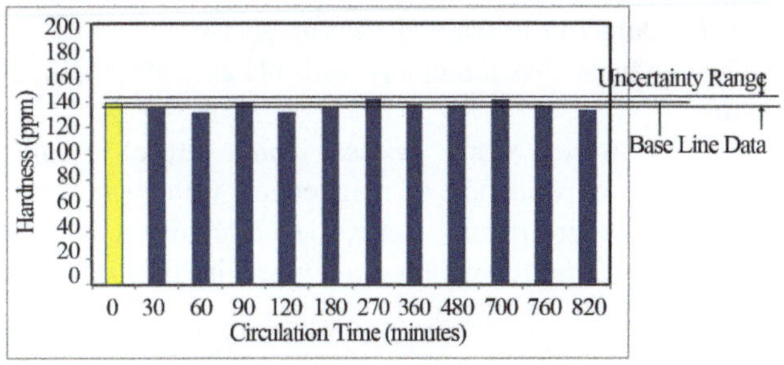

Figure 6. Hardness of Circulated Water.

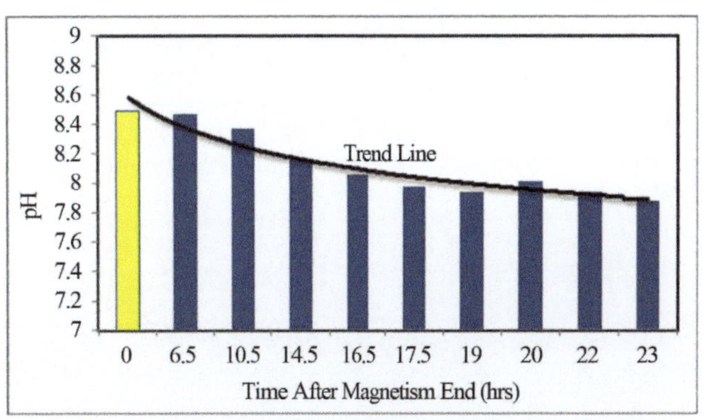

Figure 7. pH of Magnetized Water (Memory of Water).

water remember and keep the impact of passing through the magnetic field, it is declared as keeping high pH values 8.466 after 6.5 hours, and the water memory continue to forget the magnetic impact to be with pH of 7.88 over the 24 hours. One can deduce that; for 1.149 pH increase in 820 minutes, the pH decreases 0.642 in 24 hours.

This phenomenon needs to define the term (*Water Memory*) and it may be defined as; the time at which the magnetized water remember the impact of magnetic field. The pH value may be considered as the (*Water Memory Meter*).

4. Conclusion and Recommendations for Future Work

It is concluded that, for water recirculated in closed loop through the 170 mT applicable magnetic water conditioner:

1) pH is increased by 15.65% over 820 minutes of non-stop circulation.

2) pH is increased by 93.5% over the first 360 minutes, and 6.5% over the last 460 minutes.

3) TDS and Hardness of recirculated water are not affected by the circulation through the magnetic water conditioner.

4) Water remembers and keeps the impact of passing through the magnetic field for several hours, the pH decreases 0.642 in 24 hours.

Also, two phenomenons are introduced and definitions are created;

1) The magnetized water saturation curve; which governs the behavior of water during magnetism.

2) The water has a memory with pH as water memory meter.

The proposed work focuses many future research points; the dependence of magnetized water saturation curve and water memory meter, the water content of total dissolved solids, temperature of water, initial pH value, magnetic flux density and shape of the magnetic water conditioner, water flow velocity, and quantity of recirculated water. And the research points may be extended to different fluids.

REFERENCES

[1] T. Vermeiren, "Magnetic Treatment of Liquids for Scale and Corrosion Prevention," *Anti-Corrosion Methods and Materials*, Vol. 5, No. 7, 1958, pp. 215-219.

[2] A. Szcze, E. Chibowskia, L. Hołysza and P. Rafalski "Effects of Static Magnetic Field on Water at Kinetic Condition," *Process Intensification*, Vol. 50, No. 1, 2011, pp. 124-127.

[3] E. A. Duffy, "Investigation of Magnetic Water Treatment Devices," Doctorate Thesis, Clemson University, Clemson, 1977.

[4] C. E. Gruber and D. D. Carda, "Measurable Parameters in Water Conditioning Equipment as Determined in Laboratory Simulations at Rapid City, South Dakota," 1981.

[5] Public Works Technical Bulletin, "Magnetic Water Treatment," 2001.

[6] A. Kotb and A. M. Abd El Aziz, "Scientific Investigations on the Claims of the Magnetic Water Conditioners," 2013.

[7] H. Banejad and E. Abdosalehi, "The Effect of Magnetic Field on Water Hardness Reducing," *Thirteenth International Water Technology Conference*, Hurghada, 2009, pp. 117-128.

[8] R. S. Abdel Tawab, M. A. A. Younes, A. M. Ibrahim and M. M. Abdle Aziz, "Testing Commercial Water Magnetizers: A Study of TDS and pH," *Fifteenth International Water Technology Conference*, Alexandria, 2011, pp. 146-155.

[9] S. A. Ghauri and M. S. Ansari, "Increase of Water Viscosity under the Influence of Magnetic Field," *Journal of Applied Physics*, Vol. 100, No. 6, 2006, Article ID: 066101.

Relations among Main Operating Parameters of Gasifier in IGCC

Hao Xie[1], Zhongxiao Zhang[1], Zhenzhong Li[2], Yang Wang[2]
[1]University of Shanghai for Science and Technology, Shanghai, China
[2]National Plant Combustion Center, Shenyang, China

ABSTRACT

Gasification unit is one of the key subsystems in the IGCC power system; the operating parameters of gasifier directly affect syngas quality and performance of whole IGCC system. The system model of gasification unit with coal water slurry gasifier was simulated and calculated using THERMOFLEX software, and the relations of oxygen coal ratio (R_{oc}), water coal ratio (R_{sc}), gasification pressure, gasification temperature and cold gas efficiency were mostly researched. The results show that R_{oc} and R_{sc} have effect of mutual restriction on gasification temperature, cold gas efficiency and syngas composition. Gasification pressure mainly determines the capacity of the gasifier, little effects on syngas composition.

Keywords: Integrated Gasification Combined Cycle (IGCC); Gasifier; Oxygen Coal Ratio; Water Coal Ratio; Cold Gas Efficiency

1. Introduction

The gasifier is the core of gasification unit and the source of the IGCC system, gasification performance directly affects the chemical energy of syngas and cold gas efficiency, relevant to the economic operation of the whole IGCC system. However, the gasification performance mainly depends on the gasification parameters, such as oxygen coal ratio, water coal ratio, gasification pressure, gasification temperature, which is mutual influence and restrict on each other. How to configure operating parameters reasonably is the key to achieve efficient operation of the gasifier. So the research on gasification effect of operation parameters on gasification performance has important significance.

Domestic and foreign scholars have done a lot of research on gasification mechanism and models, such as JIAO Shujian [1] gave the introduction to the key technology of some IGCC, and the direction of development of IGCC system are analyzed and discussed. Li Zheng [2] established a mathematical model to predict the performance of Texaco coal gasifier, and analyzed in detail the influence of the effect of the main operating parameters on the performance of gasification furnace by the model. Wu Xuecheng [3] has established a flow gasification kinetics model, predicting and studying the influence of various gasification methods on the performance of gas bed. G.S Liu [4] established the entrained-flow gasifier model, researching the influence of coal particle size on syngas from the point of reaction kinetics. NI Qizhi [5] put forward a new simulation method of entrained-flow gasifier, analyzing the effect of O/C ratio, temperature and pressure on the gasification components and cold gas efficiency. Y.C Choi [6] studied the gasification characteristics of entrained-flow gasifier.

On the basis of the above researches, in order to seek the optimal parameters configuration mode, a gasification system model was established, the relations among mainly operating parameters and the effect on gasification performance were analyzed. The result can help to find the optimal operation conditions, to provide references for the design and operation of IGCC power plant.

2. Model and Parameters

2.1. System Model

Gasification unit comprises a gasifier, radiation syngas cooler (RSC), convection syngas cooler (CSC), air separation unit (ASU) and etc, as in **Figure 1**. Gasifier is adopted with slurry gasification and slag-tap mode. Firstly, coal water slurry and oxygen from ASU are fed into gasifier to produce raw syngas, high temperature raw syngas is cooled by RSC and CSC, and then enters syngas purification unit.

The gasifier is the use of pure oxygen gasification of coal water slurry, the composition of coal is shown as in

Table 1. The pressure losses caused by the flow of syngas, steam and water are assumed equal through any equipment in gasification unit. The main operating parameters in designed condition are shown in **Table 2**.

2.2. Operating Parameters

In coal water slurry gasification process, oxygen carbon ratio is a very important index, which usually refers to oxygen carbon molar ratio (ε), and in engineering applications, it can also refer to the ratio of the oxygen consumption and pulverized coal (R_{oc}) in gasifier, as shown in equation 1 and 2.

$$B_o = B_c \times C_{ar} \times 16/12 - B_c \times O_{ar}, \quad (1)$$

$$R_{oc} = B_o / B_c = 4/3 \times C_{ar} \times \varepsilon \times O_{ar}, \quad (2)$$

where R_{oc} denotes oxygen coal ratio in g/g, B_o and B_c respectively denote consumption of oxygen and pulverized coal, C_{ar} and O_{ar} mean carbon and oxygen elemental mass percentage of coal, ε denotes oxygen and carbon atom ratio.

Water coal ratio (R_{sc}) refers to feed water and pulverized coal mass ratio, which directly affects the flow characteristics, combustion characteristics and calorific value of coal water slurry, as shown in equation 3 and 4.

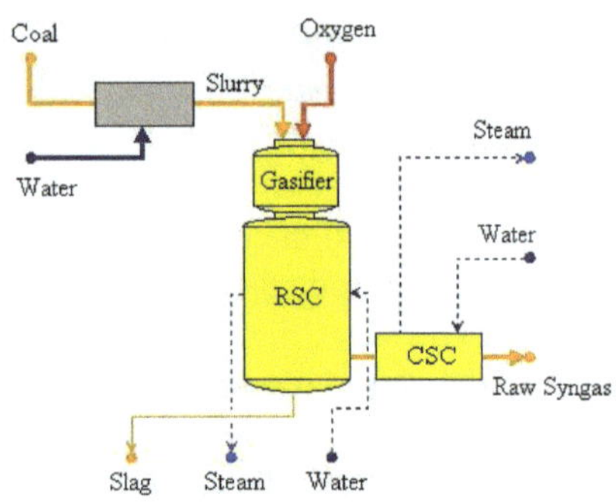

Figure 1. Gasification unit.

Table 1. Proximate and ultimate analysis of coal.

Ultimate analysis of coal /%			Mar(%)	$q_{net,coal}$ (kJ·kg^{-1})
C	H	O		
57.81	3.62	9.29	17.3	21740

Table 2. the main parameters of gansifier.

parameters	data	parameters	data
Gasifier capacity (t/d)	2 000	Oxygen coal ratio (g/g)	0.756
Gasification pressure (MPa)	3.6	Water coal ratio (g/g)	0.386
Gasification temperature(K)	1584	Carbon conversion efficiency(%)	98

$$B_w = B_c \times (1 - M_{ar}) / C_s - B_c - B_a \quad (3)$$

$$R_{sc} = B_w / B_c = (1 - M_{ar}) / C_s - 1 - \phi \quad (4)$$

where R_{sc} denotes water coal ratio in g/g, B_w, B_a respectively denote consumption of water and additive for coal water slurry, C_s denotes the concentration of coal water slurry, M_{ar} is water mass percentage of pulverized coal, Φ denotes the proportion of additive in coal water slurry (0.5%-1.5%).

Cold gas efficiency is another important index to measure the gasifier operating performance, which denotes the ratio of the chemical energy of syngas by gasification to the chemical energy of coal for gasification, as in equation 5.

$$\eta_c = (q_{H_2} + q_{CO} + q_{CH_4}) / q_{net,,coal} \quad (5)$$

The chemical energy of syngas (q_{H_2}, q_{CO} and q_{CH_4}) and coal ($q_{net,\,coal}$)are both corresponding net calorific value. Obviously to improve the cold gas efficiency of gasifier can promote transforming the chemical energy from storage in coal to syngas, improving the whole performance of IGCC.

Pressurized gasification can improve the gasification capacity effectively, and bring obvious economic benefits for transmission, distribution and subsequent chemical processing. But if the pressure is too high, the gas into the combustion engine should be reduced pressure firstly, thus leading to the waste of energy.

3. Results and Discussion

3.1. Oxygen Coal Ratio (Roc)

Figure 2 shows the effect of oxygen carbon ratio on the syngas components. With the increase of R_{oc}, the content of CO first increases and then decreases. When the R_{oc} is 0.75, the maximum mounts to 37.1%. The content of H_2 is similar to CO, the maximum of H_2 amounts to 30.43% when R_{oc} is 0.64. The content of CH_4 decreases sharply from 4.9%, and when R_{oc} is greater than 0.71, it reduces slowly, until close to 0. With the increase of R_{oc}, the content of H_2O increases, and the trend of CO_2 content is opposite to CO. The content of the effective gas composition ($CO+H_2$) firstly increases and then decreases and at R_{oc} of 0.66, the maximum reaches 66.43%.

Figure 3 shows the effect of R_{oc} on gasification temperature and cold gas efficiency. With the increase of R_{oc}, gasification temperature rises, while cold gas efficiency reduces.

Because of oxygen coal ratio increase, the oxygen content increases relatively, it is good for the combustion reaction of C and H to release much heat, which is the main cause of the rise of gasification temperature. Especially in the range of 0.67 to 0.85, R_{oc} is approximately linearly related to the gasification temperature, and gasification temperature increases about 352K when R_{oc} 0.1 per growth. Therefore, we can regulate the gasification temperature accurately and effectively by controlling R_{oc} value in the actual operation.

Based on the chemical kinetics theory, the combustion heat release of CO is basic equal to H_2 under the same pressure and volume, while the heat release of CH_4 is 3 times as much as CO. In the process of the increase of $CO+H_2$ content to maximum value, $CO+H_2$ content increases from 59.78% to 66.43%, while CH_4 decreases from 4.9% to 0.77%. Therefore, the chemical energy of syngas is reduced totally, which leads to the decrease of cold gas efficiency.

3.2. Water Coal Ratio (Rsc)

It is two situations to change water coal ratio, one assumes oxygen coal ratio is unchanged and the other assumes gasification temperature is constant. But the second is difficult to be implemented when practically operation. **Figure 4** gives the effect of R_{sc} on the components of syngas under the condition of R_{oc} is 0.75. When R_{sc} increases from 0.27 to 0.57, only the content of CO decreases, while the other main components increase. The content of CO decreases from 45.76% to 34.63%, H_2 increases from 23.97% to 25.49%, CH_4 from 0.0006% to 0.0107%, CO_2 from 9.512% to 13.06% and H_2O from 20.01% to 26.16%. **Figure 5** shows the effect of R_{sc} on gasification and cold gas efficiency. When R_{sc} increases from 0.27 to 0.57, the efficiency of cold gas decreases from 73.9% to 72.98%, , and the gasification temperature decreases from 1518K to 1838K.

The above reasons can be analyzed from the following aspects. With the increase of R_{sc}, the amount of steam also increases, while the amount of pulverized coal and oxygen relatively decreases, leading to weaken combustion reaction in the gasifier. At the same time, a large amount of heat is carried by steam, all these lead to gasification temperature decrease.

It is generally believed that methanogenesis (6) occurred in pressure and temperature lower than 1423K, and that reforming reaction of methane-steam (7) is a reversible and endothermic reaction. Therefore, with the increase of R_{sc}, gasification temperature decreases, leading to the increase of methane content.

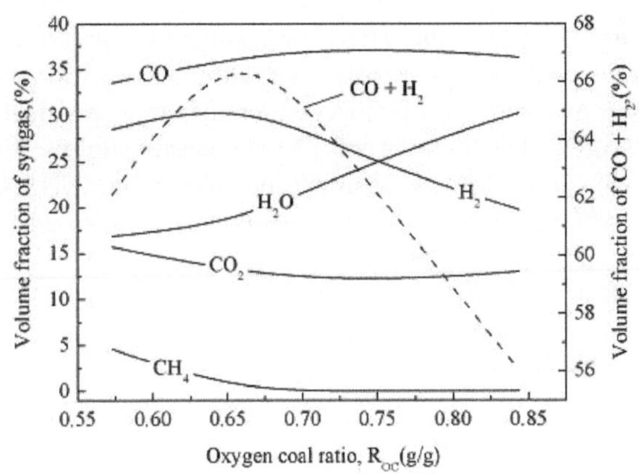

Figure 2. Effect of oxygen coal ratio on syngas components.

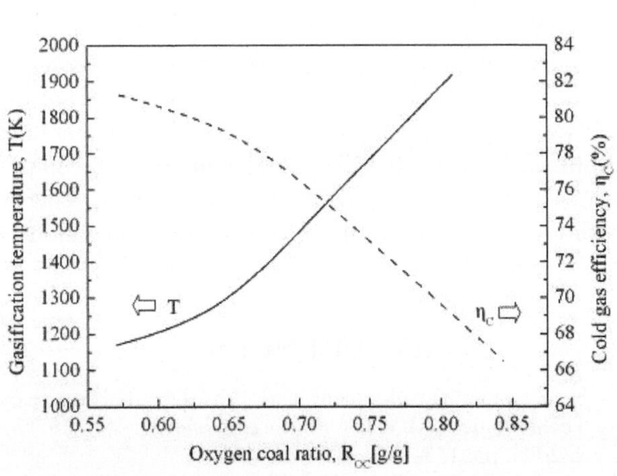

Figure 3. Effect of oxygen coal ratio on gasification temperature and efficiency.

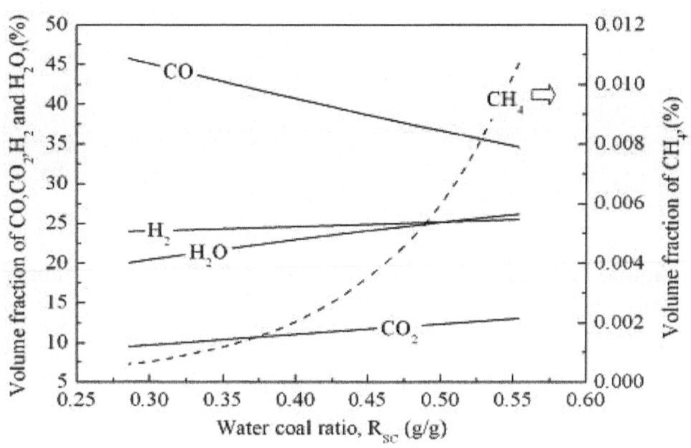

Figure 4. Effect of water coal ratio on syngas components.

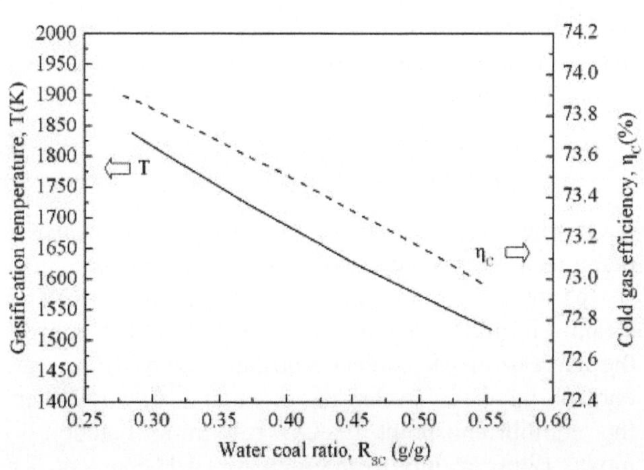

Figure 5. Effect of water coal ratio on gasification temperature and efficiency.

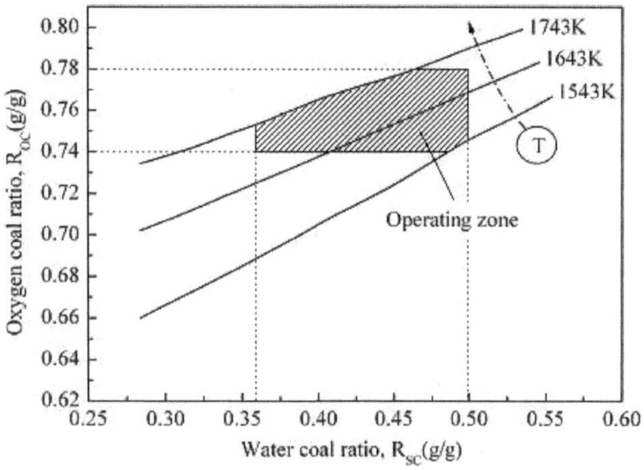

Figure 6. Double effects of water coal ratio and oxygen coal ratio on gasification temperature.

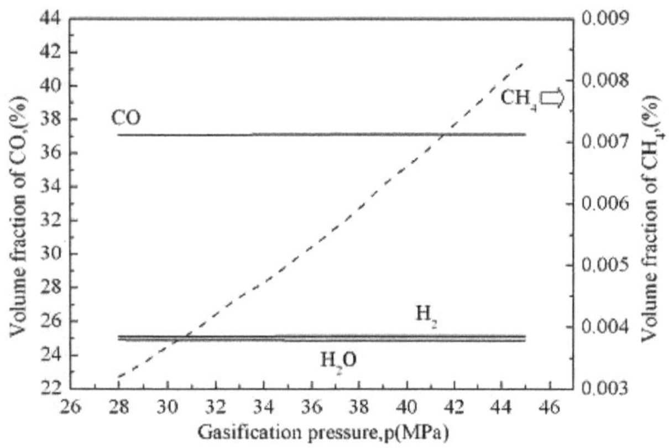

Figure 7. Effect of gasification pressure on syngas components.

$$C+2H_2 \rightarrow CH_4+74.9MJ.(kmol)^{-1} \quad (6)$$

$$CH_4+H_2O \leftrightarrow CO+3H_2-319.3MJ.(kmol)^{-1} \quad (7)$$

The decrease of gasification temperature prevents the reduction reaction (8) of CO_2, at the same time leads to the equilibrium point of CO conversion reaction (9) moves right, eventually lead to the decrease of CO content and the increase of CO_2 content.

$$C+CO_2 \rightarrow 2CO-162.4MJ.(kmol)^{-1} \quad (8)$$

$$CO+H_2O \leftrightarrow CO_2+H_2+41.2MJ.(kmol)^{-1} \quad (9)$$

On one hand, with the decrease of gasification temperature, water-gas reaction (10) weakens, methanogenesis (6) strengthens and the equilibrium point of methane-steam reforming reaction (7) moves left, all these lead to the decrease of H_2 content. On the other hand, as water content increases, water-gas reaction (10) strengthens, the equilibrium point of CO reforming reaction (9) moves right, leading to the increase of H_2 content. The combined effects of the two aspects result in H_2 content increasing slightly.

$$C+H_2O \leftrightarrow CO+H_2-131.5MJ.(kmol)^{-1} \quad (10)$$

Figure 6 shows the matching relationships between R_{oc}, R_{sc} and gasification temperature. We can see that there are double effects of R_{oc} and R_{sc} on gasification temperature. When Increasing the value of R_{oc}, gasification temperature become higher, while increasing the value of R_{sc}, gasification temperature become lower.

Generally, as far as possible to reduce R_{sc} can improve the gasification performance, at the same time, it can also be affected the flow and combustion characteristics of coal water slurry. The selection of gasification temperature depends on the characteristics of coal and the requirements of syngas. Gasification temperature is generally higher than the ash melting point 30~50 K.

According to the specific coal and IGCC system, shadow in **Figure 7** indicates the region of the optimum operating points, which R_{oc} is from 0.74 to 0.78, oxygen carbon molar ratio corresponding is 0.96 to 1.01, R_{sc} is from 0.36 to 0.5, the concentration of coal water slurry corresponding is 60% to 66%, gasification temperature is from 1543K to 1743K.

3.3. Gasification Pressure

Figure 7 shows the effect of gasification pressure on syngas components. We can see that the volume fraction of syngas has little change under different gasification pressures. The content of CO increased slightly from 37.08% to 37.12%, H_2 content increased from 25.10% to 25.14%, H_2O content decreased from 24.91% to 24.86% and CH_4 content increases slightly from 0.003% to 0.008%.

4. Conclusions

Oxygen coal ratio, water coal ratio and gasification pressure are initial operating parameters of coal gasification. With the increase of oxygen coal ratio, gasification temperature increases, while cold coal efficiency decreases. With the increase of water coal ratio, gasification temperature and cold coal efficiency both decrease. Influence of gasification pressure is mainly reflected the overall effect in the gasifier, under the high temperature conditions, pressure has little effect on syngas components. All these can provide the reference for low energy consumption, high performance gasification operation in IGCC system.

REFERENCES

[1] S. J. Jiao, "Development and Prospect of Some Key Technologies of IGCC," *Power engineering*, Vol. 26, No. 2, 2006, pp. 153-165.

[2] Z. Li, T. J. Wang and Z. M. Han, "Study of Mathematical Model for Texaco Gasifier ——Calculation and Analy-

sis," *Power engineering*, Vol. 21, No. 4, 2001, pp. 1316-1319.

[3] X. C. Wu, Q. H. Wang and Z. Y. Luo, "Kinetic Model Prediction for Various Coal Gasification Schemes in a Fluidized Bed II. Model Prediction and Analysis," *Journal of Fuel Chemistry and Technology*, Vol. 32, No. 3, 2004, pp. 292-296.

[4] G. S Liu, H. R Rezaei, J. A Lucas, *et al,* "Modelling of a Pressurized Entrained Flow Coal Gasifier: the Effect of Reaction Kinetics and Char Structure," *Fuel*, Vol. 79, No. 14, 2000, pp. 1767-1779.

[5] N. Qizhi and W. Alan. "Simulation Study on the Performance of an Entrained-flow Coal Gasifier," *Fuel*, Vol. 74, No.1, 1995, pp. 102-110.

[6] Y. C. Choi, X. Y. Li and T. J. Park "Numerical Study on the Coal Gasification Characteristics in an Entrained Flow Coal Gasifier," *Fuel*, Vol. 80, No.15, 2001, pp. 2193-2201.

Calculation of Thermal Stress and Fatigue Life of 1000 MW Steam Turbine Rotor

Shuang Bian, Wenyao Li
Energy Power and Mechanical Engineering Institute, North China Electric Power University, Beijing, China

ABSTRACT

The paper calculated the temperature and stress fields of 1000 MW ultra-supercritical steam turbine rotors which start in cold condition using the finite element calculation program (ANSYS). After getting rotor stress field, the paper use the general slope method to estimate the low cycle fatigue life loss, the rest of the conditions can be calculated in this method.

Keywords: Component; Steam Turbine; Thermal Stress; Finite Element Method; Fatigue Life

1. Introduction

When power plant stat up and shut down or change load, the internal of rotor metal will produce the temperature gradient, due to the presence of the temperature gradient, rotor will produce thermal stress in the unstable conditions. When steam turbine starts up, temperature of rotor surface gradually raise with the steam temperature raising, temperature of central part change lag behind the rotor surface, so rotor surface endure pressure stress, on the contrary rotor surface endure tension stress in the shutdown process, after a certain number of circulation, the surface of rotor may produce fatigue crack [1]. Ultra-supercritical unit woks under the high pressure and temperature, working environment of rotor is worse, so the analysis of thermal stress and fatigue damage about rotor has the important practical significance.

Using FEM, this paper analyzes the startup and shutdown process of a 1000MW steam turbine. Analysis gets the temperature field and stress field of the tutor, thus we can determine the dangerous position of the rotor, and calculate the maximum stress. According to local stress and strain method, this paper calculates the fatigue life loss of the dangerous position.

2. Thermal Stress Calculation

2.1. Calculating Object

The calculating object of this paper is the rotor of a 1000MW ultra-supercritical steam turbine. The unit is N1000-25.0/600/600 with a intermediate reheat, single shaft, four overcastting, four condenser in the plant. The unit startup mode is high casing start-up. Due to the steam enters high pressure cylinder firstly, and the steam parameters such as temperature and pressure are higher in the high pressure cylinder, the thermal stress of rotor is larger than the stress in the ip-lp cylinders. The calculated object of this paper is hp cylinder rotor. The material of rotor is improved 12Cr.mat".

2.2. Model Establishment and Grid Partition

Rotor is regarded as a column object, which has no inner heat source, and is homogeneous and is tropical. Calculating rotor temperature field belongs to solve rotational symmetry unsteady temperature function [2]. Differential equation is expressed in the following form:

$$\frac{\partial t}{\partial \tau}=\frac{\lambda}{\rho C_p}\left(\frac{\partial^2 t}{\partial z^2}+\frac{\partial^2 t}{\partial r^2}+\frac{1}{r}\frac{\partial t}{\partial r}\right) \quad (1)$$

λ —— thermal conductivity;
ρ —— density of material;
C_p —— heat capacity;

Initical condition: $t|_{\tau=0}=f(z,r)$ (2)

Boundary condition:

$$-\lambda\frac{\partial t}{\partial \vec{n}}\bigg|_{\Gamma}=a\left(t-t_f\right) \quad (3)$$

Transient analysis of the heat balance equation is expressed in the following form [3]:

$$[C]\{\dot{T}\}+[K]\{T\}=\{Q\} \quad (4)$$

F where is the conductivity matrix, whch includes thermal conductivity,convection coefficient,radiance and shape factor. is heat capacity matrix, in view of the increase of the system internal energy. is the temperature vector of node. is the time derivative of temperature. is heat flow rate vector of node.

The calculation can be completed by a integration method, which is called Crank-Nichalson/Enther θ [4].

The geometry shape of hp cylinder rotor calculated in this paper is irregular, boundary conditions has a large change, the governing stage has double flow stages, rotor has eight pressure stages. To simulate the temperature and stress field of 12Cr in the process of operation more accurately, we should build a geometric model which conforms to reality. This paper builds an ax symmetrical two-dimensional calculation model. The initial temperature of rotor distribute homogeneous, and is equal to the steam temperature. Model which be meshed is showed in the following **Figure 1**.

2.3. Initial Conditions and Boundary Conditions

The physical properties of rotor are regarded as variable physical.

Boundary conditions include the first and the third kind of boundary condition. Calculating heat transfer coefficient is a difficult point in the third kind of boundary condition. Generally, we use two kinds of formula to calculate the heat transfer coefficient, the Westinghouse formula and Nagong-Haqi formula [5]. This paper calculated the heat transfer coefficient through the Westinghouse formula.

2.4. Analysis of Temperature and Stress Field

The stress field of the rotor is influenced by the temperature field, when temperature gradient is biggest between rotor center and surface, the thermal stress is max. The time is about 24000 s, at which thermal stress is max, when unit closes to quasi steady state. **Figure 2** is the distribution of temperature field at the time that stress is max, when unit starts up according to the actual cold start-up curve. The nominal parameter of main steam in the hp cylinder is 25 MPa, 600℃.

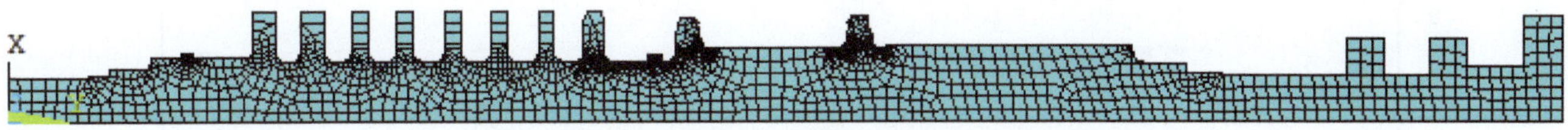

Figure 1. Mesh dividing model of steam turbine rotor.

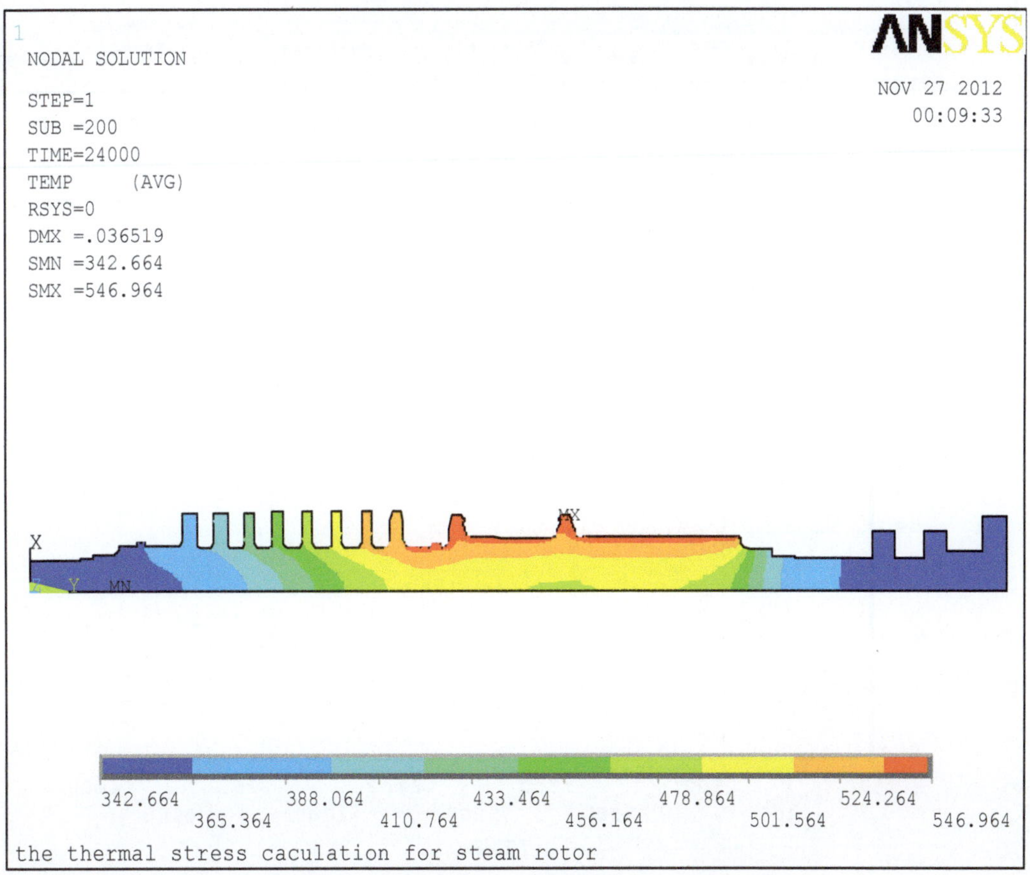

Figure 2. Temperature field of rotor at 24000s.

Figure 3 is the distribution of stress field at the time that stress is max, when unit starts up according to the actual cold start-up curve. As can be seen from the grapy, the stress concentration is large near the impeller root of governing stage, whose thermal stress is largest.

According to the analysis of the temperature and stress field of rotor, when unit starts up in the cold state, we can see the position whose thermal stress is largest is the impeller root of governing stage. Analyzing this position, this paper get the temperature-time and stress-time curves of impeller root of governing stage, shown in the following **Figure 4** and **Figure 5**.

Figure 5 shows that thermal stress has been on the rise in the startup process. When starting time reaches to 200min, the rotation rate of rotor ascend to 3000r/min, steam temperature changes faster, and stress increases speedly. Up to rated load, the stress trend tends to be gentle. When the steam temperature is kept constant, rotor temperature gradient decreases gradually, rotor stress begins to decline.

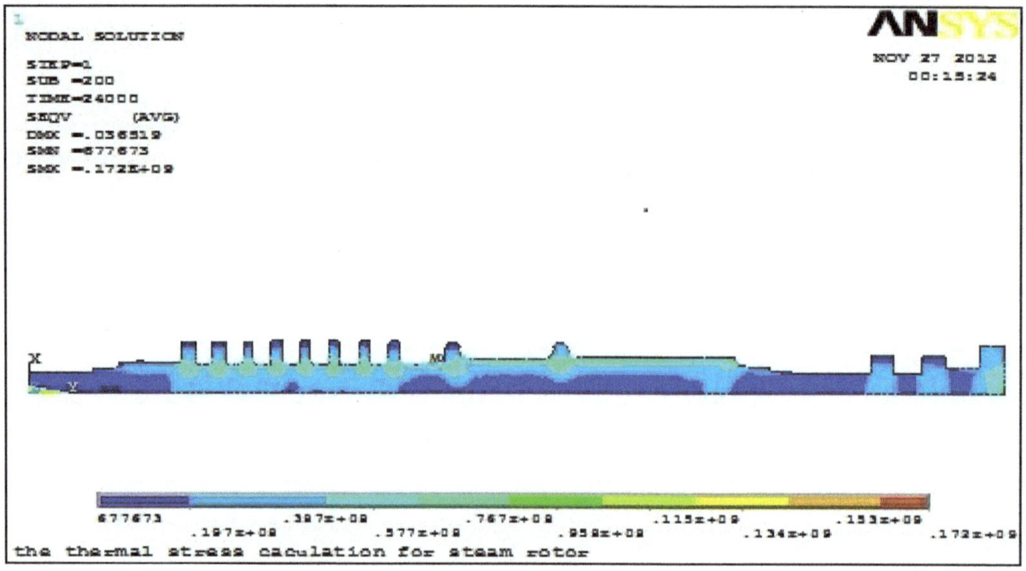

Figure 3. Stress distribution of rotor at 24000s.

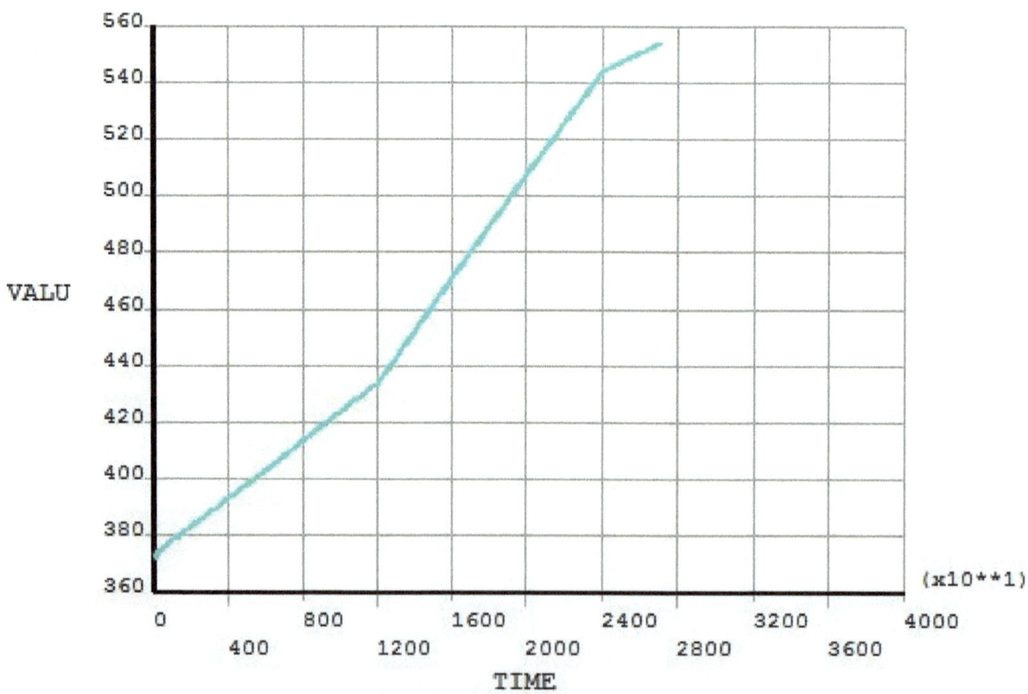

Figure 4. Temperature curve varying with time in the cold start-up condition.

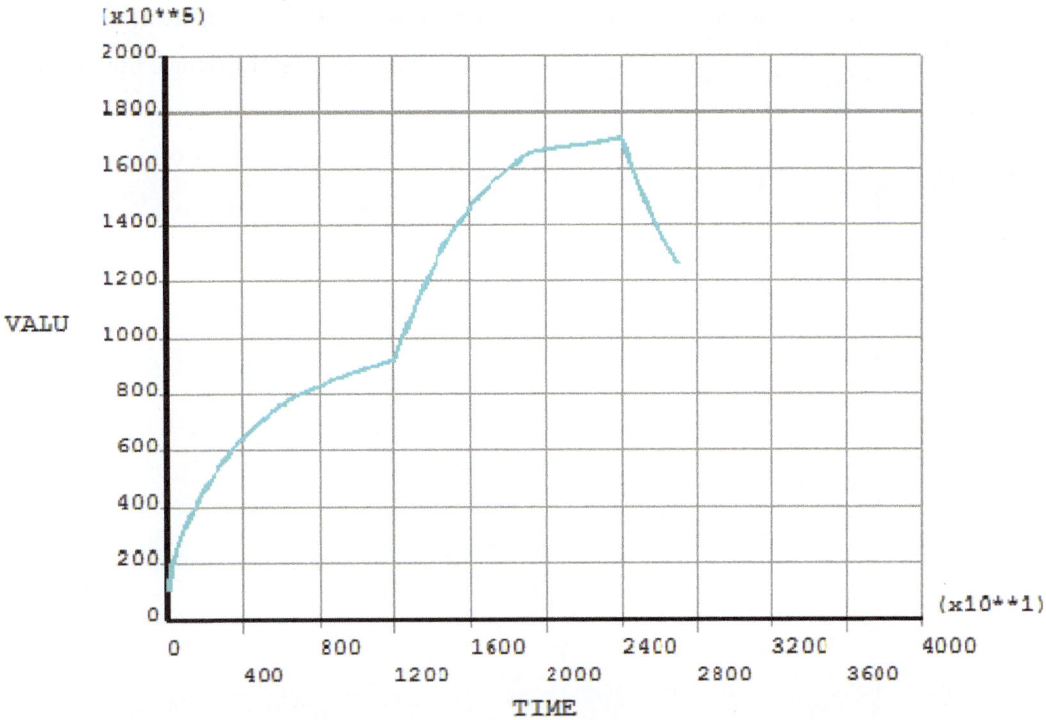

Figure 5. Stress curve varying with time in the cold start-up condition.

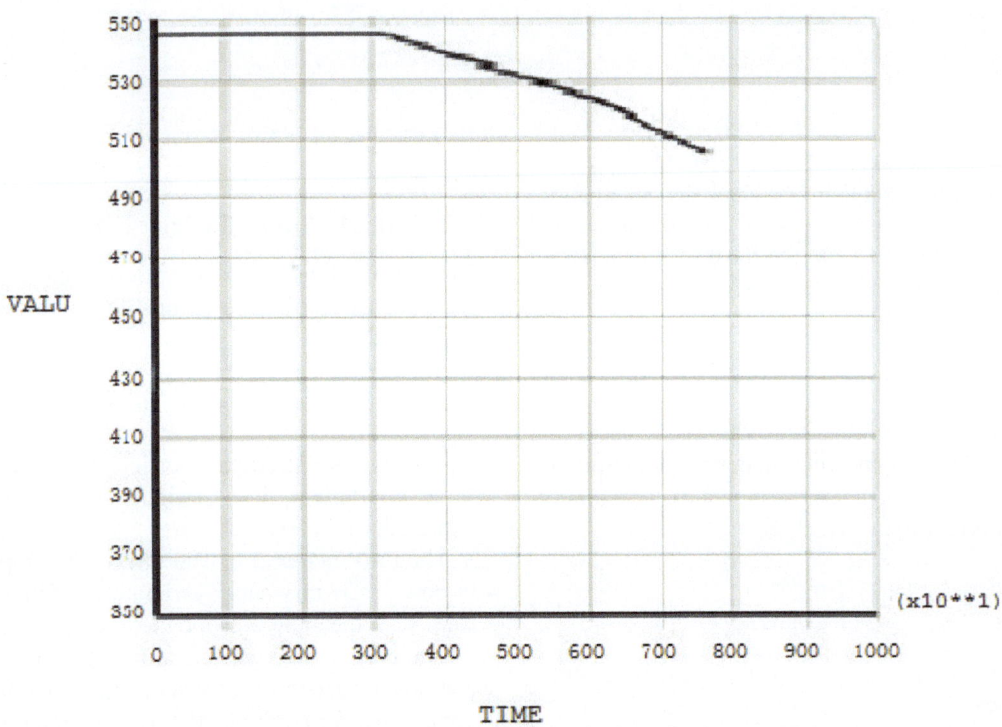

Figure 6. Temperature curve varying with time during shut-down period.

2.5. Condition of Shut-down

The pulling stress and mechanical stress of the rotor have the same direction when the unit is shutdown, thus stress is easy be large. We should control the change rate of temperature in the shutdown process.

Figure 6 is the temperature curve varying with time

during shut-down period. Figure 7 is the stress curve varying with time during shut-down period. From the **Figure 7** we can say that the maximum stress is 135MPa in the shutdown process.

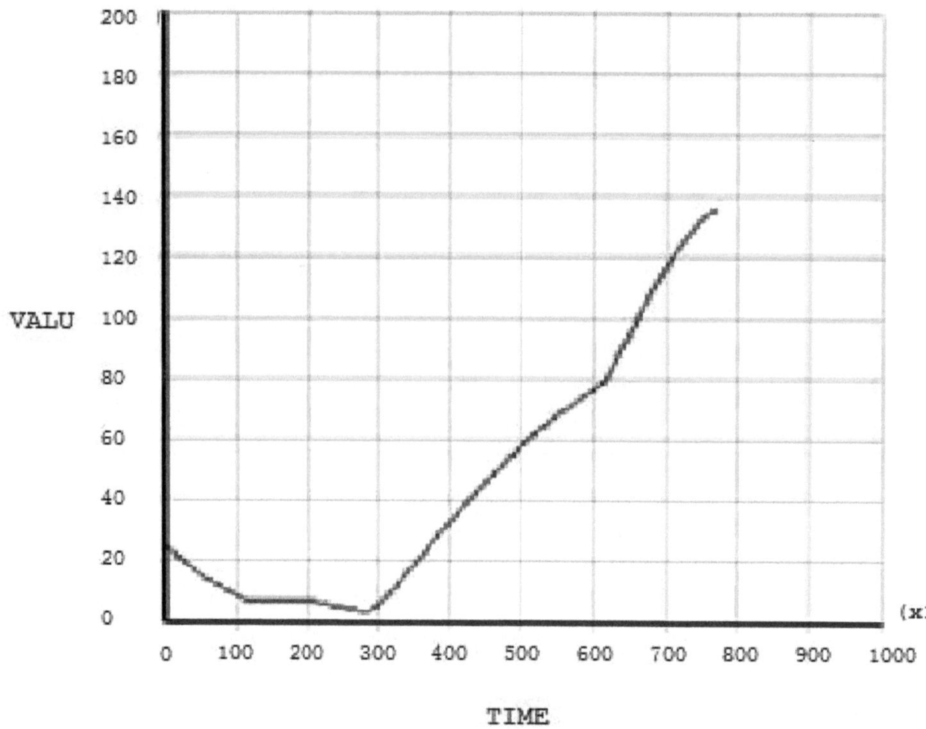

Figure 7. Stress curve varying with time during shut-down period.

3. Calculation of Fatigue Life

This paper calculates the fatigue life loss of rotor using local stress and strain method. The local stress and strain method can consider the influence of plastic strain and loading sequence. According to the local stress and strain method, we can get the fatigue life loss by the stress stain curve and stain life curve.

3.1. Calculation of Stain

There are many models which describe the stress strain curve, because the engineering material is commonly hardening material, we can use the model called Ramger-Osgood, the model can be expressed as:

$$\varepsilon = \frac{\sigma}{E} + \left(\frac{\sigma}{K}\right)^{\frac{1}{n}} \tag{5}$$

K is cycle strength factor, n is circulating strain hardening index.

Through the calculation of the stress field of the rotor, this paper get that the maximum stress of rotor is 171MPa. By referring to data, we determinate that K is 375.5[6], and the cyclic strain hardening index n is 0.086[7], when the maximum stress is 171 MPa, the strain is 0.002.

3.2. Calculation of Fatigue Life

Manson-Coffin formula is widely used in all the formulas which describe the strain-life curve, Manson-Coffin expressed as:

$$\varepsilon_a = \varepsilon_{ea} + \varepsilon_{pa} = \frac{\sigma_f}{E}(2N)^b + \varepsilon_f'(2N)^c \tag{6}$$

σ_f' is fatigue strength coefficient, ε_f' is fatigue continue coefficient, b is fatigue strength index, c is a continuation index.

The rotor material of 1000 MW ultra-supercritical unit is new 12Cr, this paper estimate the fatigue life loss of rotor with the general slope method [8] .

The best way to get the strain life curve is the fatigue test under the control of strain. But sometimes because of conditions, we can't use the experimental method, so we can approximately estimate the fatigue performance curve by some static tensile performance parameters, such as strength of extension, Young's modulus, fatigue resistance coefficient and true fracture strength. Estimation methods that we commonly use are general slope method , four correlation method an improved four correlation method, this paper select the general slope method to calculate the fatigue life.

General slope method is created by a variety of material performance date fitting. In this method, b = -0.12, c = -0.6, $\sigma_f' = 1.75\sigma_b$, $\varepsilon_f' = 0.5\varepsilon_f^{0.6}$, so Manson- Coffin can be expressed as:

$$\varepsilon_a = \varepsilon_{ea} + \varepsilon_{pa} = \frac{1.75\sigma_b}{E}(2N)^{-0.12} + 0.5\varepsilon_f^{0.6}(2N)^{-0.6} \quad (7)$$

Through the formula, this paper get that N is 5601, the fatigue life loss is 0.0095%, when unit starts up in cold condition everytime.

There are many fatigue cumulative damage models at present, in the actual operation, linear damage accumulation theory used more widely. Although linear damage accumulation theory has some shortcomings, it needs relatively few data, and is simple and easy in calculation. When we get the loss of fatigue life under different conditions, this paper suggests using the linear damage accumulation theory.

4. Conclusions

Because 1000 MW ultra-supercritical units have the high steam parameters, and rotors work in severe conditions, we should pay more attention to the fatigue life of rotor.

In this 1000 MW ultra-supercritical unit, the position which own the maximum stress of the rotor appear in the impeller root of regulation stage.

General slope method can estimate the fatigue life loss of rotor.

REFERENCES

[1] B. H. Zhang, "Management on Life Time Assessing and Running in all Case for Fiery and Electric Air Crew," Beijing: Water Conservation and Electric Power Press, 2001 (Ch).

[2] D. G. Xu, "Calculation of Temperature and Thermal Stress Field for Turbin Roter," *Chinese Journal of Applied Mechanics,* Vol. 13, No. 3, 1996, pp. 79-84.

[3] ANSYS, Ine, "ANSYS Thermal Analysis Guide Release," Third Edition. SAP, IPIne, Vol. 17, No. 2, 1997, pp. 194-210.

[4] A. Bagaviev, "Life Assessment of Turbine Components based on Deterministic and Probabilistic Procedures," *In: International Journal of Pressure Vessels and Piping,* Vol. 15, No. 2, 2004, pp. 19-21.

[5] B. H. Huang and W. T. Li, "Calculation and Analysis of Radiative Coefficient for Turbine Rotor," *Power System Engineering,* Vol. 15, No. 4, 1997, pp. 35-37(Ch).

[6] D. Z. Guo and X. L. Song, "12 cr1mov Welding of Large Diameter Boiler Steel Pipe High Temperature Strain Fatigue Characteristics," *Boiler Technology*, Vol. 10, 1995.

[7] R. Viswanothan and R. I. Jaffee, "Toughness of Cr-Mo-V Steels for Steam Turbine Rotors," *Journal of Engineering Materials and Technology,* Vol. 15, No. 2, 1983, pp. 19-21.

[8] P. F. Yuan and D. R. Sheng, "Finite Element Analyses of Fatigue Life-span about On-line Examination on Steam Turbine Rotor," *Power System Engineering*, Vol. 20, No. 2, 2004, pp. 43-48.

Load Shedding Application within a Microgrid to Assure Its Dynamic Performance during Its Transition to the Islanded Mode of Operation

Dorel Soares Ramos[1], Tesoro Elena Del Carpio-Huayllas[1*], Ricardo Leon Vasquez-Arnez[2]
[1]Department of Electric Power and Automation Engineering, University of São Paulo, São Paulo, Brazil
[2]FDTE (Foundation for the Technological Development of the Engineering Sciences), São Paulo, Brazil

ABSTRACT

This article presents the simulation results and analysis related to the response of the generators within a microgrid towards an accidental overload condition that will require some load shedding action. A microgrid overload can occur due to various reasons ranging from poor load schedule, inadequate switching of circuits within the microgrid, outage of one or more generators inside the microgrid, illegal load connections by some low voltage consumers, etc. It was observed that among the main factors that determine the survival of the microgrid during its transition from the grid connected mode to the islanded mode of operation are the size and type of the load connected (passive or dynamic load) as well as the length of time during which the unexpected load is connected. Models of a speed and voltage regulators of a diesel generator, and important for coping with the overload conditions are provided in the paper. The novelty of the work lies in the load shedding simulation and analysis of the specific generators studied herein, regarding that in many countries the microgrid technology is seen as an important alternative towards the ever increasing load demand and also to assist the system during periods of blackout.

Keywords: Distributed Generation; Islanded Operation Mode; Load Shedding; Microgrid Systems

1. Introduction

At present, the establishment of microgrid systems is regarded by the power industry as one of the alternatives to not only keep running critical loads but also provide electricity to some regular consumers during periods of prolonged interruptions (e.g. blackouts). Microgrid systems are the compelled choice in remote areas where it is difficult to provide power through the power system. The microgrid concept is not new; actually, in the early days of the power industry this was the kind of system established in the urban and industrial areas. The interconnection of systems to strengthen and form the network came years later. This was done to offer the system a high level of safety regarding possible faults; thus, cope with stability issues aside of allowing the surplus generation capacity in one area to be used elsewhere in the system.

Ironically, this kind of system networking may also lead to some major interruptions (due to cascading effect) like those that affected central, south and southeastern Brazil and all Paraguay in 2009, the northeastern part of the USA in 2003, and all Italy in 2003. Also, nearly 100 million people in Indonesia were affected by a huge blackout in 2005. The last major blackout occurred in India in 2012 leaving virtually inoperative in the Northern, Eastern and Northeast part of the country. In the aftermath of these undesired events nearly all the affected systems set up independently a common action, to seek for some alternatives capable to diminish their impact among which the establishment of microgrid systems was also pointed out.

One of the key features of a microgrid is its ability to separate itself from the distribution utility (e.g. during unscheduled periods of interruption) in order to continue feeding its own islanded portion. This is not a simple task though, especially when taking into account the compelled operational procedures and protocols to be followed.

Another outstanding characteristic of a microgrid is that, provided an agreement with the grid, it can supply its surplus generation to the utility, for example, during

*Corresponding author.

peak periods of demand or whenever the microgrid has excess capacity.

Microgrid systems, when properly designed, have the ability to separate from the distribution utility, for example, during disturbance or blackout periods experienced by the utility. This gives them the possibility to continue feeding their own islanded portion. Additionally, provided an agreement with the distribution utility, they can supply their surplus generation whenever they have excess capacity. This is particularly attractive for periods when the utility faces peak periods of demand. Nonetheless, it can occur that during the transition from the grid connected mode to the islanded mode of operation, an excessive load (larger than the microgrid can uphold) could be connected to the microgrid.

This overloading condition can occur due to various reasons, namely: poor load schedule, inadequate switching of circuits within the microgrid, an upstream tripping of the utility circuit breaker that leaves part of its load connected to the microgrid, illegal load connections by some low voltage consumers, etc. Under this condition, the most common way to save the microgrid from a complete collapse is to shed part of the load connected. This action can help the fading generator to become stable again ensuring its safe operation.

Some critics say that the application of load shedding at specific times during the 2003 major blackout in North America could have prevented the cascading outages that came after the initial tripping event.

Several interesting references addressing the load shedding issue in large systems were found. There may also be some other references having the same merit; however, due to space restrictions of this article it will not be possible to include them all.

Reference [1], for example, provides a comprehensive coverage on the load shedding issue, load restoration and generator protection schemes using underfrequency relays during abnormal frequency conditions. In [2], an underfrequency protection program developed for a certain region in North America is presented. The program reportedly optimizes the system wide load shed schedule, also checking up its coordination with a steam turbine-generator underfrequency protection scheme. In [3], a method for determining the maximum probable rates at which a power system frequency will decay, following a disturbance, is presented. Such an analysis is chiefly directed to provide underfrequency protection for nuclear power plants. Reference [4], presents a summary on System Protection and Voltage Stability prepared by the IEEE Power System Relaying Committee. It describes the risk and mechanism of voltage collapse as well as suggests some operation, system upgrades and protection solutions to avoid such a condition.

In [5] the effect of the reduced frequency upon the capacity of a power plant, with special regard to cases with deficiency of generation, is presented. It is stated that on systems with a high percentage of motor load (such as pumping), a combination of frequency and voltage reduction may secure maximum load relief during an emergency. In [6-9], some methods dealing with underfrequency load shedding, so as to avoid voltage instability and its further collapse, are presented.

In [10] an optimal load shedding algorithm based on an economic criterion is developed. In [11], a strategy to shed an optimal number of loads in an islanded distribution system, using factors like the rate of change of frequency (RoCoF) and the customers' willingness to pay during periods of outage to stabilize its frequency, is presented.

Finally, in [12] a scheme that combines frequency and voltage changes to shed loads is proposed. The premise behind this scheme is that in the last years, power systems have changed and yet no corresponding modification of the underfrequency load shedding schemes was made; thus, the load shedding procedure would still be based on the disconnection of pre-selected loads.

As can be seen, most of the above reviewed load shedding methods and strategies are directed to large systems, hence the need to develop a study on the excessive overloading effects on microgrid generators. The article presents the main control components of a gen set and the various situations the microgrid generators can face during the transition from the grid connected to the islanded mode of operation.

The microgrid current status and some of the challenges presently encountered by the microgrid technology, are presented in [13]. In [14,15] a microgrid islanding condition following a fault and its respective stability behaviour are investigated. The load shedding alternative is applied once the system frequency starts to decay from its nominal operative value (50 or 60 Hz).

According to [16,17], the two load shedding methods widely used are:

1) Traditional frequency drops with load percentage shed. Typically, the load shedding scheme can be done in 3 (and up to 6) steps [16]. For example, in a three-step method, the percentage of load shed would be:

Step	f (Hz)	(%) of load
1	59.3	10
2	58.9	15
3	58.5	as required to avoid going below 58.2 Hz

2) Use of the rate of change of frequency (RoCoF) and load percentage sheds. It evaluates the speed at which the frequency (*df/dt*) is declining. This enables characterizing the kind of contingency occurring in the system at vari-

ous instants, thus, provides the system a most adequate load shedding scheme [17]. For example, regarding the above frequency drop of 59.3 Hz, the *df/dt* could be set at:

59.3 Hz ..... *df/dt* = 0.4 Hz/s .... 10% of total load.
59.3 Hz ..... *df/dt* = 1.0 Hz/s .... 25% of total load.
59.3 Hz ..... *df/dt* = 2.0 Hz/s .... 35% of total load.

The above under-frequency control methods are commonly used by many distribution utilities. In the context of this paper, the entire load exceeding the normal power demand of the microgrid internal generation will be shed. This is done considering the simulation response of the generators in returning to the pre-overload condition. Also, the inherent differences existing between the grid and a microgrid should be considered. In the latter case, for example, the inertia of the generation sources is much smaller.

Generally, gen-sets like the one considered here have no overloading capability. Wind generators and small hydro generators may admit little overload (up to 10%). Nevertheless, a brief analysis on what would be the percentage of load to be shed, if the above methods were also used, will be included whenever appropriate.

2. Microgrid Load Shedding Application

The transition from a grid-connected to an islanded mode of operation of a microgrid can occur due to reasons like the presence of faults in the system, or due to some pre-planned conditions like the system (grid) maintenance, energy costs, etc. During such a transition the power control of the microgrid generators must act quickly as they start controlling the frequency of the islanded section.

Whenever power in the network is lost, the microgrid generators assigned to provide power to the intentionally islanded portion should be able to pick up and feed the load of the islanded system after the switch at the Point of Common Coupling (PCC) has opened. The generation sources herein referred can be any of the sources cited by the IEEE std. 1547 [18]: PV arrays, wind turbines, fuel cells, microturbines, conventional diesel, gas-fired turbines, and energy storage technologies.

The microgrid system to be analyzed is connected to the utility through a circuit-breaker (CB), in series with a 6 MVA, 13.8/2.4 kV transformer, as depicted in **Figure 1**. The energy sources in the considered microgrid are: a synchronous generator driven by a diesel engine (connected to PCC through CB-1); a wind turbine driving a synchronous generator (connected to PCC through CB-3) and another small synchronous machine which could represent a small hydro-generator (SHG) connected to the PCC through CB-2. No PV array sources (solar panels) or sources requiring energy storage elements were considered because the primary focus of this research is to analyze the dynamic behaviour of the considered sources. Also, the machine equations are not presented here as they can be found in the available literature dealing with electric machines.

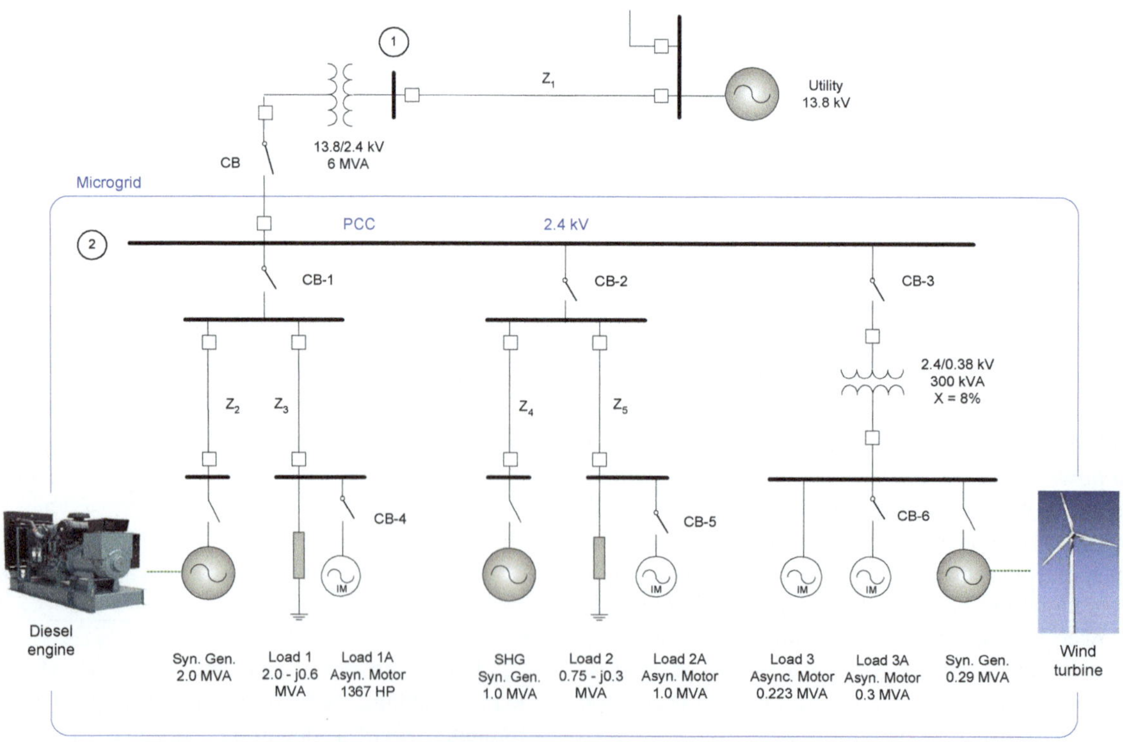

Figure 1. Microgrid system used in the simulations.

The wind generator is connected to the microgrid through a 2.4/0.38 kV transformer. In practice, due to technical and economic reasons, most of the sources used in wind power schemes are asynchronous generators. One of the main reasons has to do with its ability to operate at speeds different from the synchronous speed. However, in this study, a synchronous generator was chosen due to its less dependency from an external source for providing reactive power. A Doubly Fed Induction Generator (DFIG) was also placed instead of the wind (synchronous) generator; its response was comparable to the synchronous generator used herein.

The loads fed by each generator, as well as the extra loads (equivalent asynchronous motors) that simulate the overloading condition (*Load* 1*A*, *Load* 2*A* and *Load* 3*A*) are also shown in **Figure 1**. These extra loads are connected to their respective generator through CB-4, CB-5 and CB-6. *Load* 1 and *Load* 2 were specified as constant power loads. The normal load connected to the wind generator (*Load* 3) is a dynamic load (equivalent asynchronous motor). The complete system was implemented in the EMTDC/PSCAD® program [19]. Some components, like the electrical generators and circuit breakers were taken from the software library, while others, like the diesel generator speed and voltage regulators, etc., were independently built and defined.

The sequence of the load shedding procedure is as follows: initially, all loads are primarily being fed by the utility when, due to any reason (e.g. a three-phase fault), the circuit-breaker (CB) is open. At this moment all three generators start running and taking up their respective loads. It is assumed that along with CB the other circuit breakers (*i.e.* CB-1, CB-2 and CB-3) also trigger with the fault. This way, it will be avoided the condition of any of the generators from being carried away by another generator. This condition is critical in microgrids containing several sources; otherwise, the whole microgrid system can be jeopardized by the loss of synchronism among them. Synchronization of fully-loaded generators is commonly unachievable, hence the need to open CB-1, CB-2 and CB-3. But that is an issue that for now is put aside.

2.1. Diesel Generator Overloading

Among the control systems implemented on this generator are: a basic speed regulator, a constant mechanical power regulator (**Figure 2(a)**) and a voltage regulator (**Figure 2(b)**). As it will be shown later, a reasonably robust voltage and speed regulator may be useful in helping the machine cope with faults on the system. Most of the synchronous generator parameters like the direct and quadrature reactances as well as the transient and subtransient time constants (X_q, T''_{do}, X''_q, T''_{qo}, etc.) were estimated according to [20]. The machine starts as an ideal source (t = 0.0 s) until it reaches its steady-state condition. At t = 0.5 s the model enables the machine to pass from an ideal source to a non ideal machine, simultaneously the voltage regulator control is inserted in to the generator. No cylinder misfiring condition was simulated.

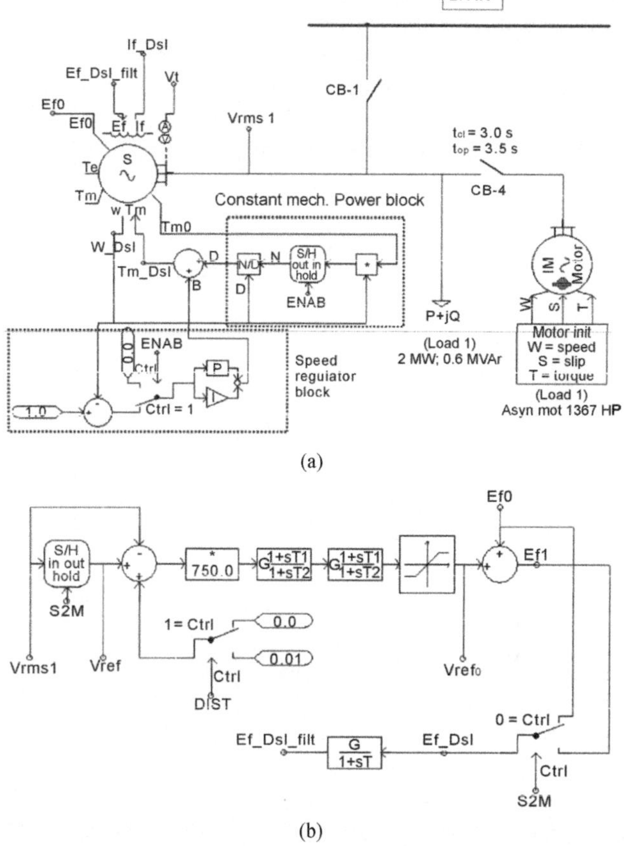

Figure 2. (a) Synchronous (diesel) generator model; (b) Voltage regulator.

At t = 2.0 s, once the initialization transient reaches a stable condition, the dynamic model of the machine is enabled. From this moment on, the machine electromechanical equations start driving the generator, enabling the variation of the speed and mechanical torque. Initially, the diesel gen-set is feeding a linear load (L_1 = 2.0 + j0.6 MVA) when at t = 3.0 s an equivalent induction motor representing 50% of excess load (L_{1A} = 1367 Hp) is connected. The extra load (*Load* 1*A*) is switched off after 500 ms. It can be seen that at first the generator intends to take up this extra load (see $P_{\_Dsl}$ in **Figure 3(a)** and also the stator current I_{L1} in **Figure 3(b)**) though failing in its attempt. The instant the load shed occurs (opening of CB-4 at t = 3.5 s) relieves the machine which quickly returns to its previous loading condition. The P-I (proportional-integral) constants define how quickly the machine will return to the pre-overloading condition.

The terminal voltage ($V_{\_Dsl\_rms}$) shown in **Figure 4(a)** drops instantly, upon which, following a few oscillations the voltage regulator carries it back to its previous operative value (1.03 pu). The gen-set frequency ($F_{\_Dsl}$) shown in **Figure 4(b)** reaches a minimum value of 58.6 Hz. If the conventional load shedding strategies were to be applied (see Section 1), the amount of load to be removed would be the highest specified.

For example, according to [16], the percentage of load to be shed, regarding the minimum value of 58.6 Hz, would be above 15% (*i.e.* as required to avoid falling further the frequency). Now, according to [17], the percentage of load to be shed for this same case would be 35%. This is because the approximate RoCoF observed in **Figure 3(b)** is about 12 Hz/s. Nonetheless, shedding the entire extra load helped the frequency to get restored in about 1.0 s. From the tests conducted, a maximum of 0.8% overload could only be upheld by the gen-set. The time taken by the frequency to get restored was about 1.5 s. The main parameters of this and the other machines used are presented in the Appendix section.

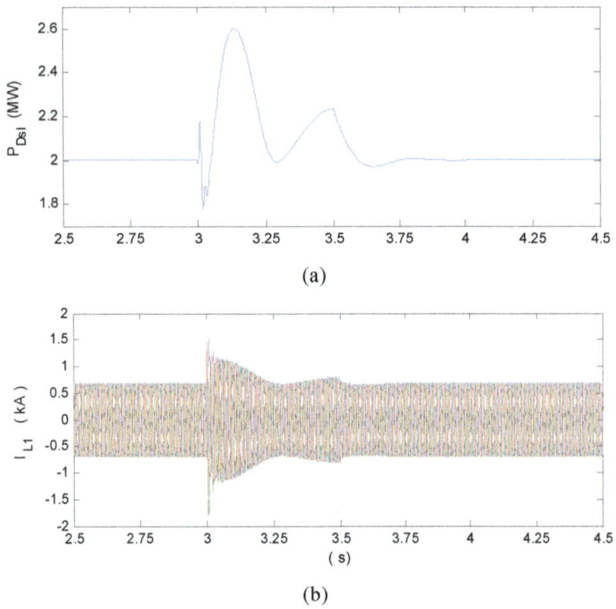

Figure 3. (a) Diesel generator power; (b) Current at the stator.

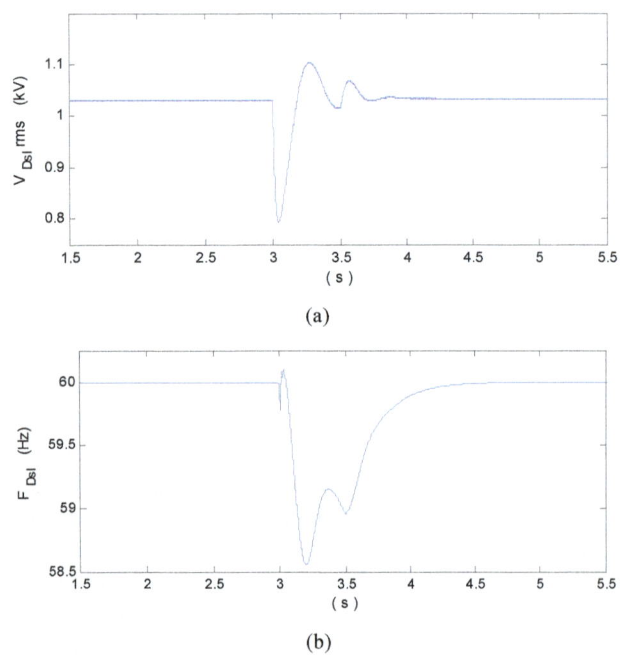

Figure 4. (a) Diesel generator terminal voltage; (b) Frequency.

2.2. Wind Generator Load Shed

The wind generator shown in **Figure 5** uses a simple PSS (Power System Stabilizer). The wind turbine model is available in the library of the program used [19]. The main link between the generator and the wind turbine is the mechanical torque (T_m) which is set to operate at a constant value. This is because the wind speed and the pitch angle of the wind turbine are also set to be constant at W_{SP} = 8 m/s and 11.5°, respectively.

At t = 5.0 s both loads (L_3 = 223.6 kVA and L_{3A} = 300 kVA) are simultaneously connected (**Figure 6(a)**). At this instant, the generator frequency ($f_{\_wnd}$) drops to about 58 Hz (**Figure 6(b)**). Notice that the overloading condition (set at t = 5.0 s) and its respective load shedding (t = 6.0 s) result in fast frequency oscillations which are effectively coped by the wind generator speed governor. The rms voltage measured at the busbar where the wind generator is connected to, showed a fair dip during the overload and a slight overvoltage after the disconnection of *Load* 3*A*.

Again, if the frequency drop and RoCoF methods were applied, the percentage of load to be shed would be that

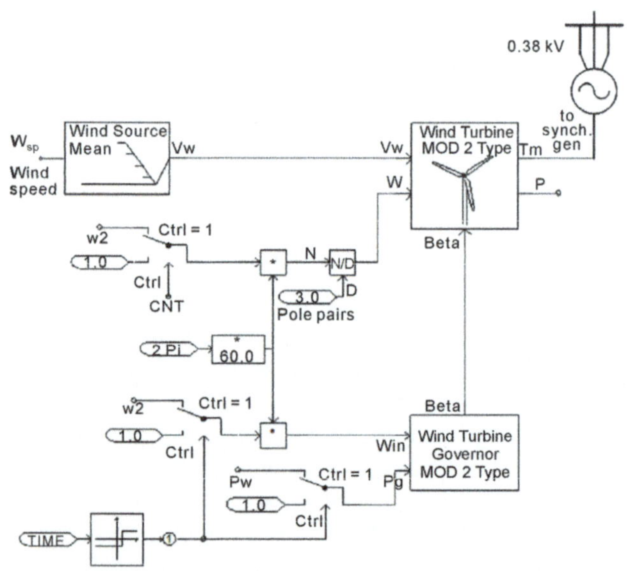

Figure 5. Synchronous wind generator model used.

exceeding the generator capacity. During this overloading period, the speed regulator damps also these oscillations. Notice that the machine power (P_{L3}), and the load current (I_{L3}), start dropping steadily (**Figure 6(a)**). At t = 6.0 s the equivalent extra load (Load 3A) is shed causing the frequency to rise up to about 60.75 Hz, though, becoming damped and stable in approximately 0.5 s. The electric torque (T_e) has an opposite behaviour compared to that of the frequency (**Figure 6(c)**). Due to the conditions specified in the model, the mechanical torque (T_m) remains constant at all times.

Wind turbines, especially small ones, are usually prevented from operating under overload conditions. To avoid this condition they are equipped with rigorous protection schemes like the stall and pitch angle control and other protection systems.

2.3. Small Hydro-Generator (SHG) Load Shed

The machine dynamics previously described in Section 2.1 (*i.e.* the diesel generator model) also apply here. The passive load and the unexpected load (equivalent motor), is shown in **Figure 7**. In this case, two consequences of the load shed procedure were considered.

2.3.1. Successful Disconnection of the Extra Load

It can be observed (**Figure 8(a)**) that at t = 2.5 s the generator is running at its nominal frequency ($F_{\_hydr}$) with an initial load (L_4 =0.75 + j0.3 MVA). Then, at t = 3.0 s, the extra load (*i.e.* L_{4A} = 1.0 MVA) is connected. The output power in the generator (P_{L4}) rises up immediately, though, failing to take up this load until at t = 3.5 s the extra load is shed. The effect of the voltage regulator ($V_{\_rms}$) in restoring the terminal voltage can be seen in **Figure 8(b)**.

Notice also how after the disconnection of the extra load the machine frequency ($F_{\_hydr}$) returns to its normal value not before facing some low frequency oscillations

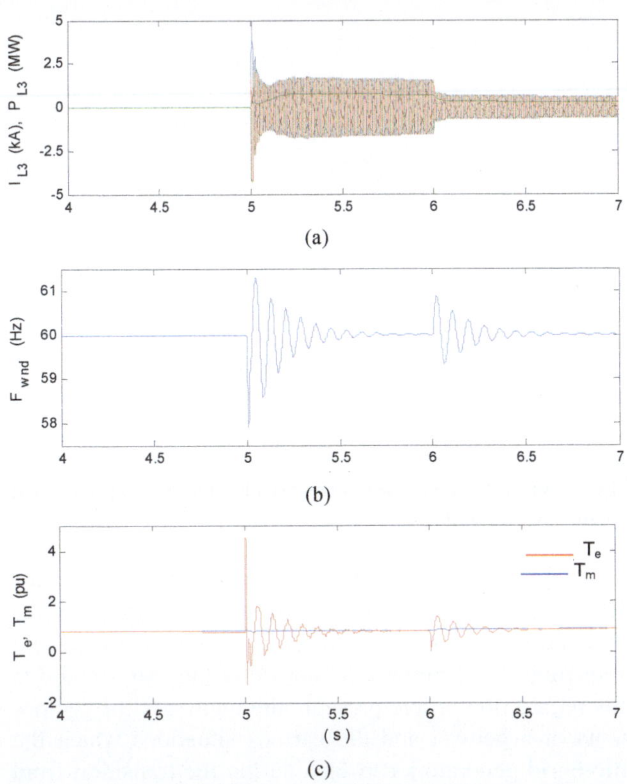

Figure 6. Wind generator: (a) Total load connected; (b) Frequency; (c) Electric and mechanical torques.

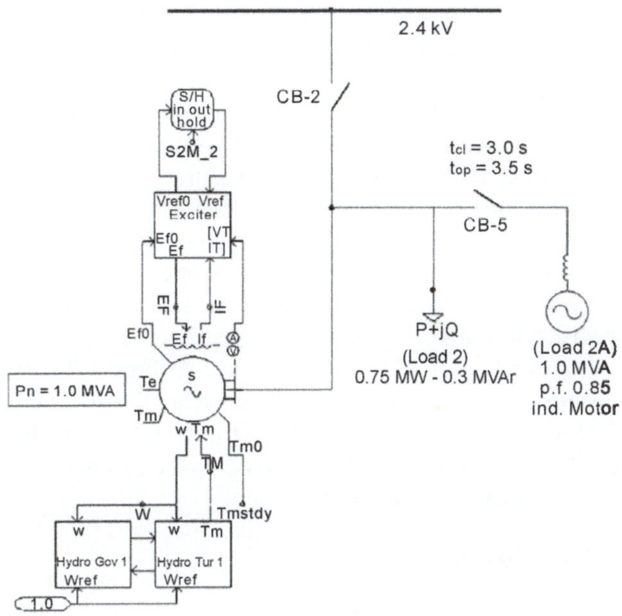

Figure 7. Small hydro-generator model used.

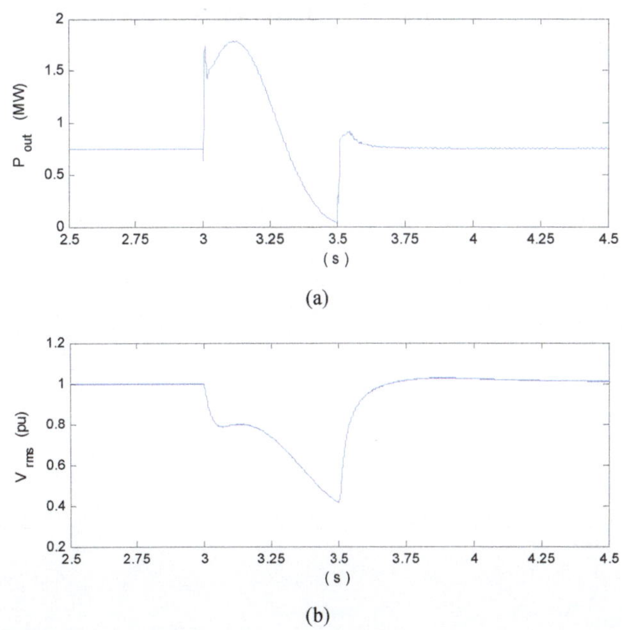

Figure 8. (a) Output power of the SHG; (b) Terminal voltage.

(**Figure 9(a)**). The electrical and mechanical torque oscillation ($T_{\_elt}$ and T_{mech}) and their subsequent damping are shown in **Figure 9(b)**.

2.3.2. Failure to Disconnect the Excessive (Extra) Load

Failure in disconnecting the extra load will inevitably lead to a steady collapse of the SHG generator. In this case, the switch CB-5 that connects the equivalent extra asynchronous motor is not opened. Similarly to the previous case, the generator initially intends to take up the extra load (P_{out} in **Figure 10(a)**); however, due to its limited capacity it quickly drops to zero. Notice that despite the machine has a speed regulator, in situations like this, there cannot be any speed regulator (or power/frequency control) able to cope with such a condition.

A similar falling pattern after the failure to disconnect the extra load can be observed in the case of the SHG terminal voltage ($V_{\_rms}$ in **Figure 10(b)**). Note that nothing was mentioned about the overcurrent protection system which would trip before the current reaches a certain specified threshold.

The steady drop of the SHG frequency (**Figure 11(a)**) towards the excess load is more evident. Finally, both electrical and mechanical torques ($T_{\_elt}$ and T_{mech} in **Figure 11(b)**) start up a continuous rise in an effort to fulfill the unexpected load demand.

3. Conclusion

The load shedding simulation results of some specific generators within a microgrid are presented in this article. A quick load shed applied to such a microgrid systems, upon which large unforeseen loads are connected, is vital for keeping the generators running normally. Provided their respective voltage and speed regulators all three generators are considered resuming their operation and frequency stabilization after sheeding the extra load. On this regard, the article presents the main control components of a gen-set and the various situations which the microgrid generators can face during the transition from the grid connected to the islanded mode of operation. The size and length of time taken to shed the extra load

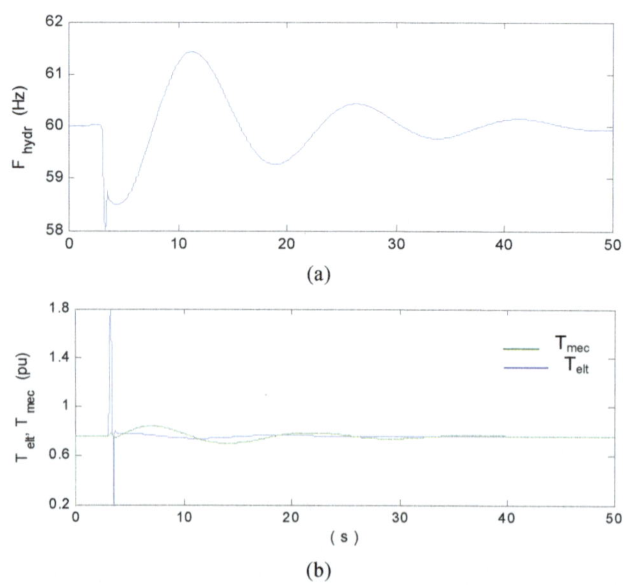

Figure 9. (a) Frequency and (b) electric and mechanical torques of the SHG.

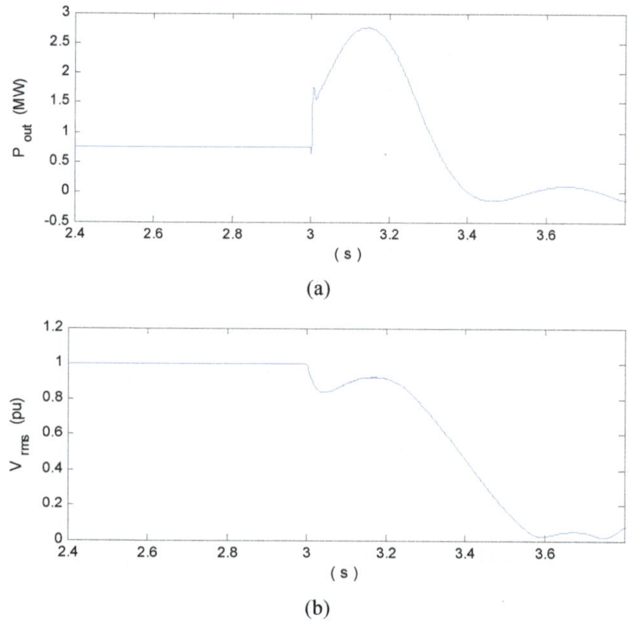

Figure 10. (a) Output power and (b) terminal voltage of the SHG.

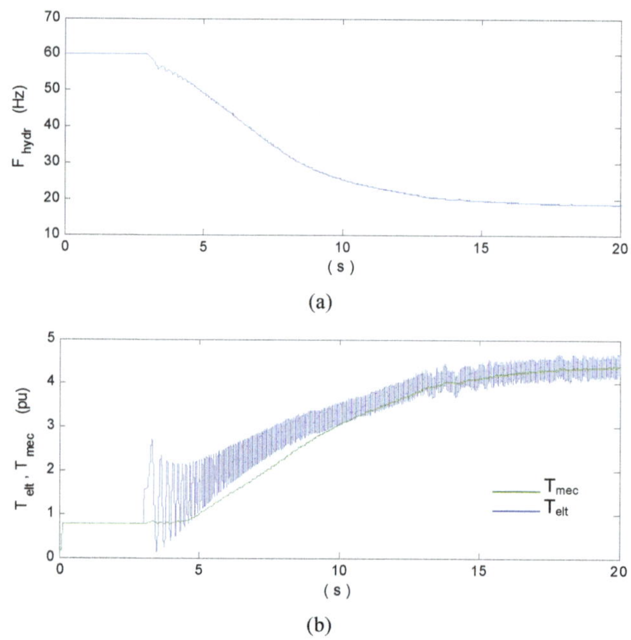

Figure 11. (a) Frequency and (b) electric and mechanical torques of the SHG.

connected are important factors that can lead to the collapse of the generator and the microgrid itself. Particularly, the size of the extra load accidentally connected will determine the frequency drop and the extent thus the speed regulator will respond.

4. Acknowledgements

The authors would like to acknowledge the help of M. T. Bassini for his aid with some of the machine control models.

REFERENCES

[1] W. C. New, J. Berdy, P. G. Brown and L. E. Goff, "Load Shedding, Load Restoration, and Generator Protection Using Solid-State and Electromechanical Underfrequency Relays," General Electric Co. Pub. GET-6449, 1974, pp. 1-32. http://store.gedigitalenergy.com/FAQ/Documents/489/GET-6449.pdf

[2] C. R. Rowland, D. W. Smaha and J. W. Pope, "Coordination of Load Conservation with Turbine-Generator Underfrequency Protection," *IEEE Transaction on Power Apparatus and Systems*, Vol. 99, No. 3, 1980, pp. 1137-1150.

[3] M. S. Baldwin and H. S. Schenkel, "Determination of Frequency Decay Rates During Periods of Generation Deficiency," *IEEE Transactions on Power Apparatus and Systems*, Vol. 95, No. 1, 1976, pp. 26-36.

[4] M. Begovic, *et al.*, "System Protection and Voltage Stability," *IEEE Transactions on Power Delivery*, Vol. 10, No. 2, 1995, pp. 631-637.

[5] H. A. Bauman, G. R. Hahn and C. N. Metcalf, "The Effect of Frequency Reduction on Plant Capacity and on System Operation," *AIEE Transactions on Power Apparatus and Systems*, Vol. 74, 1955. pp. 1632-1637.

[6] S. A. Niremberg, D. A. McInnis and K. D. Sparks, "Fast Acting Load Shedding," *IEEE Transactions on Power Systems*, Vol. 7, No. 2, 1992, pp. 873-877.

[7] T. Q. Tuan, J. Fandino, N. Hadjsaid, J. C. Sabonnadiere and H. Vu, "Emergency Load Shedding to Avoid Risks of Voltage Instability Using Indicators," *IEEE Transactions on Power Systems*, Vol. 9, No. 1, 1994, pp. 341-351.

[8] V. V. Terzija, "Adaptive Under-Frequency Load Shedding Based on the Magnitude of the Disturbance Estimation," *IEEE Transactions on Power Sys*tems, Vol. 21, No. 3, 2006. pp. 1260-1266.

[9] X. Xiong and W. Li, "A New Under-Frequency Load Shedding Scheme Considering Load Frequency Characteristics," *International Conference on Power System Technology*, Chongqing, 22-26 October 2006, pp. 1-4.

[10] D. K. Singh, R. Shekhar and P. K. Kalra, "Optimal Load Shedding: An Economic Approach," 2010 *IEEE Region 10 Conference*, Fukuoka, 21-24 November 2010, pp. 636-639.

[11] P. Mahat, C. Zhe and B. Bak-Jensen, "Underfrequency Load Shedding for an Islanded Distribution System with Distributed Generators," *IEEE Transactions on Power Delivery*, Vol. 25, No. 2, 2010. pp. 911-918.

[12] D. Prasetijo, W. R. Lachs and D. Sutanto, "A New Load Shedding Scheme for Limiting Underfrequency," *IEEE Transactions on Power Systems*, Vol. 9, No. 3, 1994. pp. 1371-1378.

[13] T. E. Del Carpio-Huayllas, D. S. Ramos and R. L. Vasquez-Arnez, "Microgrid Systems: Current Status and Challenges," 2010 *IEEE/PES Transmission and Distribution Conference and Exposition: Latin America*, São Paulo, 8-10 November 2010, pp. 7-12.

[14] A. H. Kasem Alaboudy, H. H. Zeineldin and L. Kirtley, "Microgrid Stability Characterization Subsequent to Fault-Triggered Islanding Incidents," *IEEE Transactions on Power Delivery*, Vol. 27, No. 2, 2012, pp. 658-669.

[15] T. E. Del Carpio-Huayllas, D. S Ramos and R. L. Vasquez-Arnez, "Microgrid Transition to Islanded Modes: Conceptual Background and Simulation Procedures Aimed at Assessing Its Dynamic Performance," 2012 *IEEE Power & Energy Society Transmission and Distribution Conference and Exposition*, Orlando, 7-10 May 2012, pp. 1-6.

[16] ANSI/IEEE C37.106-1987 WG, "Guide for Abnormal Frequency Protection for Power Generating Plants," New York, 1987. pp. 1-32.

[17] B. Delfino, S. Massucco, A. Morini, P. Scalera and F. Silvestro, "Implementation and Comparison of Different Under Frequency Load-Shedding Schemes," *Power Engineering Society Summer Meeting*, Vol. 1, 2001, pp. 308-312.

[18] IEEE Std 1547.2-2008 WG, "Standard for Interconnecting Distributed Resources with Electric Power Systems," New York, 2008. pp. 1-207.

[19] PSCAD/EMTDC® Program, Manitoba HVDC Research Center, v.4.3.1.0 (x4), 2010. https://hvdc.ca/pscad/

[20] M. S. Sarma, "Electric Machines: Steady-State Theory and Dynamic Performance," 2nd Edition, PWS Pub. Co., Boston, 1994.

APPENDIX

The following are the parameters and constants used in the simulations:

$Z_1 = (0.397 + j0.089)\ \Omega/km,\ L_1 = 1.0\ km$
$Z_2 = Z_3 = (0.460 + j0.104)\ \Omega/km,\ L_2 = L_3 = 0.8\ km$
$Z_4 = Z_5 = (0.55 + j0.130)\ \Omega/km,\ L_4 = L_5 = 2.0\ km$

Diesel generator:

- Rotor inertia constant (H) : 0.55 s
- Proportional (P)* : 5.0
- Integral (I)* : 40.0 s
- Gain (G) : 1.0
- Time constant (T) : 0.005 s
- Lead time constant (T_1) : 1.0 s
- Lag time constant (T_2) : 1.0 s

*Offer the gen set a stabilizing and damping torque characteristic.

Wind generator:

- Rotor radius : 20 m
- Air density : 1.229 Kg/m^3
- Gear box efficiency : 0.97
- Angular mech. Speed : 16.667 Hz
- Wind power coefficient (Cp) : 0.28

Small Hydro Generator:

- Rotor inertia constant : 3.5 s
- Mechanical friction & windage : 0.0 pu

Reciprocating Engine Step Load Response in Islanded Power Generation

Yun Tian, Rudie Crous
Power Group, WorleyParsons Ltd., Perth, Australia

ABSTRACT

The step load response of reciprocating engines is one of the key characteristics when considering its application in medium to large scale stationary power generation especially with islanded generation. This paper discusses the impacts of power frequency deviation on the generators and electrical equipment in the network and presents the relationship between step load capabilities and generator operating parameters. For a power plant consisting of a number of generators both step load and power output requirements must be satisfied. An analysis method is proposed to facilitate the development of an operation strategy which can meet both step load and power demand requirements in the full load range. Typical reciprocating engine step load curves are used to demonstrate the analysis method and the results are further optimised for lower operational cost. This analysis method provides a general approach to operation strategy of large reciprocating engines used in islanded power generation.

Keywords: Step Load; Islanded Power Generation; Reciprocating Engine; Power Frequency Control

1. Introduction

Reciprocating engines have been widely used as power sources or mechanical drives in both transportation (including on-road and sea-borne) and stationary power generation (including base load generation, emergency or backup power generation). In recent years, reciprocating engines have seen its applications expanded to medium and large scale stationary power generation markets, including both islanded generation and commercial grid connected power supply.

Reciprocating engine technology evolved at a fast pace in the last decade, and this enabled larger capacity, better efficiency, dual fuel capability, and higher availability engines to be available to the market. Large reciprocating engines used for stationary power generation are typically in the 6MW to 18MW capacity range per unit. These engine generators typically operate at medium speed (500 rpm to 1000 rpm, depending on capacity and grid frequency) due to trade-offs between performance and operational life / maintenance interval requirements for base load or continuous operation. Lower speed engines (<500 rpm), feature low dynamic performance and massive physical size and weight, due to low power intensity (output per unit of volume/displacement) related to the low speed, are less attractive than medium speed engines in stationary power generation applications.

These advantages have made large reciprocating engines especially attractive for islanded power generation where power quality requirements are not as stringent as those for commercial power grid or utility power generation. However, reciprocating engines still have to meet project requirements and in some cases, be suitable for interconnection to a commercial power grid. These requirements include emissions, availability, frequency control, heat rate, etc. Among these requirements, frequency control or engine's step load response is often neglected in manufacturer's performance guarantee but could prove to be a major issue for some applications.

Frequency response to load changes for gas turbines and steam turbines are well known thanks to their dominance in large scale power generation for more than a century. With a short history in medium to large scale stationary power generation, reciprocating engine's performance on frequency control is a relatively new area for power engineers, planners and plant owner, although it is an important issue for both grid connected and islanded power generations.

Frequency control is implemented through generator governor control, assisted inherently by the power system's moment of inertia, frequency-dependent load response and ancillary service. For commercial grid connected application, any load/supply imbalance and subsequent frequency control requirements could be shared by the large generators connected to the network, especially gas turbines, together with contracted ancillary service [1]. The large capacity and inertia of a power grid can often absorb most imbalances easily.

Compared to gas and steam turbines, reciprocating engines has lower inertia to help overcome the initial frequency/speed change, mainly due to its low speed. For reciprocating engines operating in power island mode where no gas/steam turbine or ancillary service is available, frequency control depends totally on the reciprocating engine's capability thus network stability becomes crucial to islanded generations that have large load variations.

This paper first discussed frequency requirements from both power network and consumer equipment perspectives, followed by the relationship between operation parameters and step load capability under design frequency change limits for large capacity reciprocating engines. The focus of this paper is to provide an analysis method to identify reciprocating engine power plant operating conditions based on individual engine generator step load capability, fuel type, number of operating generators and power demand to satisfy frequency control requirements. Typical reciprocating engine step load curves have been used in the analysis and the results are discussed. The analysis method also includes optimisation of operating conditions to minimise fuel cost.

2. Frequency Requirements for Power Network and Island Power Generation

Frequency control is very important for the safe operation of a power network, mainly because of the strong relationship between frequency change, supply/demand imbalance and generator protection. Frequency changes have the following impacts to the equipment in the network:

- Generators connected to the power grid: When a power demand/supply imbalance occurs, grid frequency and generator speed will change as a response to this imbalance because of time lag of mechanical power input. Generators are designed to operating within a certain speed range with automatic protection to disconnect them from the grid when frequency change is more than a predetermined value.
- Rotating equipment driven by electric motors which are essentially reverse of generators: There is a fixed relationship between frequency, number of poles in the induction motor field and the rotation speed. Changes in grid frequency will change the rotating equipment speed therefore its work load.
- Some electronic equipment or electric timers depend on the grid frequency as a time reference.
- Harmonic filters. Component overstress may happen when impedance change significantly due to frequency deviation.

Commercial power grid frequency deviation is usually limited to a very small range by regulations in many countries in the world. For example, in Australia, the 'normal operating frequency band' as provided by the National Electricity Rules is set at 49.85 Hz to 50.15 Hz, or 50±0.15Hz [2][3][4]. This frequency band was set despite some studies on grid frequency finding that the actual tolerance for frequency change are in general much wider than the regulated limits from the perspectives of generators, grid stability, timer control and consumer equipment [5].

For islanded power generation, the frequency deviation depends on contractual requirements, technology selection and the possibility of future interconnection with a commercial power grid. Considerations should be given to generator speed control range for stability of the power grid and motor maximum allowed speed for the protection of mechanical driven equipment (motor, gear, belt, etc.).

Motors built in accordance to National Electrical Manufacturers' Association (NEMA, US) standards are designed to operate successfully at rated load and at rated voltage with a variation in the frequency of up to 5% above or below the rated frequency (MG1-12.42) [6]. Large motors used in industrial process are often controlled by Variable Speed Drives (VSDs) which are insensitive to frequency deviation.

Frequency deviation will affect AC generator and transformer performance. To avoid the overheating of AC generator core laminations due to flux density, International Electrotechnical Commission (IEC) standard IEC 60034-1 requires industrial AC generators to be designed to operate within a maximum ratio between voltage and frequency (the volts-per-hertz ratio). The frequency deviation limits for temporary operation with reduced performance (Zone B) is between +3% and -5% [7].

AC Generators are diving by prime movers, e.g., turbines or reciprocating engines, whose speed controls the power line frequency. The speed of the prime movers must avoid resonance and is controlled by the speed governor. The maximum frequency change corresponding to the governor response by changing the air or fuel inlet vane/gate, or the governor droop, is typically set to 5% by manufacturers. Generator protection system will disconnect the generator from network when frequency deviation is outside of the allowed range, this may cause instability of the power network as the power imbalance may deteriorate. This is especially the case for an under frequency situation.

A tighter range of frequency control limits can protect equipment in the network and reduce the risk of system collapse. It is also recognized that for island power generation, wider limits for voltage and frequency changes are appropriate to accommodate the effects of a weak distribution network. The purpose of this paper is to establish an analysis method which can be used for step load increase and step load decrease at any frequency change limits. To facilitate the analysis, we assume ±4% as the targeted frequency change limit.

3. Reciprocating Engine Step Load Response

Step loads are instantaneous load changes in a power network. For an island power network, there are basically three types of power imbalance that can occur instantaneously:

- Failure of one generator which causes insufficient power supply;
- Failure of an electrical component in the electrical network which causes either oversupply or undersupply of power depending on the location of the equipment;
- Large consumer load start-up or trip which causes either sudden increase or decrease in power demand.

Generators respond to the imbalance of power demand/supply by increasing or decreasing power output, and in extreme situations, disconnecting from the network. The response will not be instantaneous mainly due to time lag in the engine or turbine's mechanical components which control the fuel and air input. During this transit time, the prime mover's speed will increase or decrease as the first response to the load imbalance and the maximum speed or frequency change is a function of its starting point and system inertia, i.e., mass and speed of the rotor of engine generators.

As discussed above, reciprocating engines have lower inertia compared to gas or steam turbines. The speed/frequency change of reciprocating engines depends on engine design, load imbalance, engine load factor (ratio between initial load and generator net capacity) and fuel type (gas/diesel). The maximum step load allowed for a specific reciprocating engine generator without exceeding frequency limits can be determined by its load factor and fuel type. Reciprocating engine generators step load response has the following characteristics:

- Allowed step loads (increase or decrease) for a certain frequency limit vary with load factor and fuel type;
- Dual fuel reciprocating engines operating in diesel mode has a higher step load capability than in gas mode;
- Reciprocating engines have a different capability for step load increase and decrease. Engine generators typically allow equal or larger step load decrease than step load increase for the same speed change.
- For a step load increase, an engine's step load capability increases with the increase of load factor until it reaches its maximum value, then starts to decrease when load factor continues increase.

The analysis method proposed in this paper focuses on step load increase, because step load increase causes more stress to the engine generator rotor and requires a narrower operation range than step load decrease. To meet the step load increase requirement, operating generators must maintain some level of spinning reserve above actual power demand. A typical performance data for step load increase with maximum speed change of 4% is presented in the **Table 1** and illustrated in **Figure 1**. This step load increase curve, not associated with any specific engine generator model, is used as a start point to demonstrate the analysis approach described in the following sections. The same approach can also be used for step load decrease analysis.

4. Analysis Approach for Power Plant Operation Under Frequency Change Limits

The step load curves provided individual engine generator's step load capability under given speed/frequency change limit. Maximum step loads can be conveniently established for different power demand for each individual generator using the curves. For a power plant consisting of a number of reciprocating engine generators, the operation scenarios become more complicated than a single generator. For a given power demand, the operating engine generators must satisfy the following two contradicting requirements:

- The net power output of all operating generators must meet the power demand. The number of generators in operation will be determined based on the power demand and should be operating at a load factor as high as possible to obtain better efficiencies for the same fuel type;
- Power plant must have sufficient spinning reserve to cope with a step load increase. The spinning reserve of all the operating generators shall be capable of coping with a step load increase and maintain frequency change under the limit.

Table 1. Typical reciprocating engine step load increase capability.

Diesel Mode										
Load Factor	%	0%	50%	50%	70%	100%				
Step Load	%	15%	15%	30%	30%	0%				
Gas Mode										
Load Factor	%	0%	20%	20%	40%	40%	60%	60%	80%	100%
Step Load	%	0%	0%	10%	10%	15%	15%	5%	5%	0%

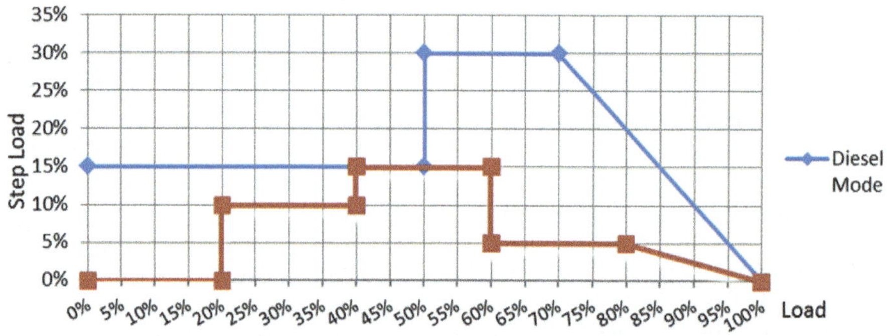

Figure 1. Typical reciprocating engine step load increase curve.

During the engineering phase, for example, the frequency limit and maximum step load, instantaneous power demand from starting of the largest motor in a process plant are typically defined in the Basis of Design document. Power plant configuration will be determined based on a number of criteria including both technical and commercial requirements. The technical factors are mainly contracted maximum power demand, plant availability, emission limits and step load capability.

4. Analysis Approach for Power Plant Operation Under Frequency Change Limits

The step load curves provided individual engine generator's step load capability under given speed/frequency change limit. Maximum step loads can be conveniently established for different power demand for each individual generator using the curves. For a power plant consisting of a number of reciprocating engine generators, the operation scenarios become more complicated than a single generator. For a given power demand, the operating engine generators must satisfy the following two contradicting requirements:

The net power output of all operating generators must meet the power demand. The number of generators in operation will be determined based on the power demand and should be operating at a load factor as high as possible to obtain better efficiencies for the same fuel type;

Power plant must have sufficient spinning reserve to cope with a step load increase. The spinning reserve of all the operating generators shall be capable of coping with a step load increase and maintain frequency change under the limit.

During the engineering phase, for example, the frequency limit and maximum step load, instantaneous power demand from starting of the largest motor in a process plant are typically defined in the Basis of Design document. Power plant configuration will be determined based on a number of criteria including both technical and commercial requirements. The technical factors are mainly contracted maximum power demand, plant availability, emission limits and step load capability.

Once the power plant configuration is determined, power plant operation scenarios can be established for the whole range of power demand to satisfy the above technical requirements. The maximum step load allowed for a power plant is a function of power demand, generator step load curve for giving frequency limit, fuel type and generator net capacity. Among these parameters, the power demand, number of operating generators and fuel type are variables once plant configuration is fixed. To determine the operating conditions, a two-step analysis method is proposed in this paper. The first step is to establish all possible operating conditions for the whole range of power demand, which can be optimized in the second step based on fuel cost. To simplify the analysis, all generators are assumed to be identical, but the method can also be applied to plants consist of generators in different capacity or technology.

In the first step, the relationship between power demand, step load and load factors need to be established. Load factor is the link between power demand and step load capability in step load curves and can be determined by:

$$LF = \frac{PD}{n \times CP} \tag{1}$$

where

LF - Load factor
n - Number of operating generators
PD – Power demand
CP – Individual generator's net power output

Equation (1) can also be written as

$$PD = (n \times CP) \times LF = b \times LF \tag{2}$$

For a specific reciprocating engine, the relationship between individual generator's load factor and step load capacity for giving speed/frequency limit can be expressed as

$$SLG = f(LF, FT) \tag{3}$$

where

SLG - step load capability of individual generator, as a

percentage of the generator's design net power output, and

FL - fuel type

Generator net power output using different fuels may be slightly different even at the same load factor; therefore net full load power output for the specific fuel should be used in the conversion to absolute values in MW. To simplify the analysis, the slight difference in net power output and heat rate for different fuel types at same load factor is ignored.

For the whole power plant, plant step load capability for the frequency limit can be calculated by

$$SLP = SLC \times (n \times CP) = k \times SLC \quad (4)$$

For a giving step load, or design step load, *SLD*, the power plant step load capability under this operating condition must satisfy:

$$SLP \geq SLD \quad (5)$$

Combining equations from (1) to (5), leads to the expression for the operating conditions to satisfy step load requirement for a certain power demand and frequency limit:

$$k \times f\left(\frac{PD}{k}, Fl\right) \geq SLD \quad (6)$$

or

$$(n \times CP) \times f\left(\frac{PD}{n \times CP}, FL\right) \geq SLD \quad (7)$$

where *n* is restrained by:

$$n \leq N \quad (8)$$

where *N* is the number of installed generators in the power plant.

The power output of operating generators shall also satisfy power demand and maximum generator capacity requirements:

$$PD \leq (n \times CP) \quad (9)$$

for step load decrease, or

$$(PD + SLD) \leq (n \times CP) \quad (10)$$

for step load increase.

For step load increase, because the instantaneous step load is larger than the actual demand increase after stabilisation, the power output will meet power demand if equation 10 is satisfied.

Equation (7) shows that for a given power plant configuration and design step load, the only variables are fuel types Fl, number of operating generators n and power demand PD. The step load curves of generator are fixed when generators are selected, and n is constrained by equations (8), (9) or (10), depending on step load decrease or increase. This means power plant step load capability could either meet design step load for a certain range of power demands, or may not be able to satisfy the whole range of power demand, depending on the generator's step load curve characteristics and fuel type.

For a power plant consisting of 5 x10MW generators and based on the typical generator step load curves shown in **Figure 1**, the relationship between step load capability and power demand can be illustrated in the **Figure 2**, assuming design step load of 6.0MW and all 5 generators are in operation (n = 5).

Figure 2 shows the initial load (or initial power demand) range that satisfies the design step load under frequency change limit is from 20MW to 30MW in gas mode, or from 0MW to 44MW in diesel mode.

The relationship shown in **Figure 2** gives the possible operational conditions for 5 operating generators, but for a certain power demand, this may not be the best operation mode. The load factors, hence generator efficiencies, may be lower than operating in modes, such as using 4 operating generators, which may also be able to cope with the same design step load.

To establish an optimal power generation solution for the whole range of power demand, it is necessary to determine the heat rate or efficiency of all the operating conditions that can satisfy both step load requirements and power demand requirement. In order to do this, all the power plant's initial load ranges that can meet these requirements at different operating scenarios are explored and the relationship between heat rate and power demand established.

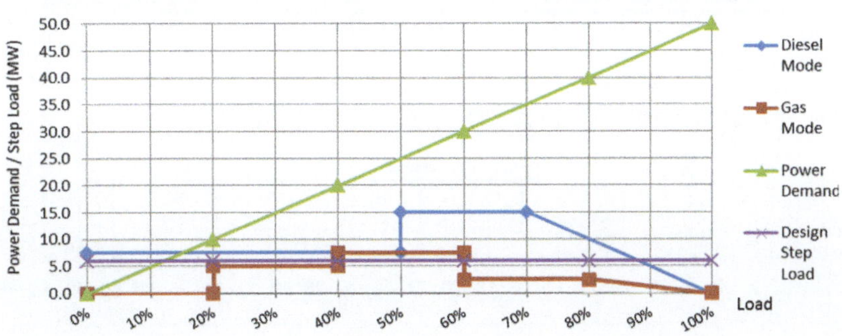

Figure 2. Relationship between step load, load factor and power demand.

From results of the first step, initial loads satisfying step load requirements under certain number of generators are obtained. Repeating the first step for all the other operating scenarios with different number of operating generators and fuel types, the relationship between power demand and initial loads are determined. **Table 2** below summarizes the operating conditions that meet step load requirement using different numbers of operating generators and fuel types. The operating conditions that cannot meet the step load requirement for the whole load range are not shown in the **Table 2** including 1 generator in either gas or diesel mode, and 2 or 3 generators in gas mode.

The initial loads are plotted against power demand in **Figure 3**.

Figure 3 shows that for each power demand, there are often more than one operation mode meeting all the requirements, therefore optimisation is required to determine the best operation mode for each power demand.

Step 2 is to optimise the operation mode based on heat rate and fuel price. A reciprocating engine generator's heat rate is a function of fuel type and load factors. Fuel cost ($/MWH) is the product of heat rate and fuel price ($/GJ). Based on the results from step 1, using typical heat rate curves for reciprocating engines at 10MW capacity range, and assuming fuel price of 6 $/GJ (gas) and 20 $/GJ (diesel), the heat rate and fuel cost can be established against the power demand, as shown in **Figure 4**.

Figure 4 shows that although reciprocating engine plant in diesel mode dominates the suitable operation conditions, the fuel costs for power plant operating in gas mode are much lower than that in diesel mode, albeit at lower load factors than in diesel mode, for example, a power plant operating in gas mode with 4 or 5 generators has lower fuel costs that it operating in diesel mode with 3 or 4 generators. However, optimal operating modes should also consider other factors, e.g., maintenance cost, which is related to the total engine firing hours and fuel types.

5. Results and Discussion

The results show that suitable reciprocating engine power plant operating modes for a giving step load depends on a number of factors. While increasing the number of operating generators may increase the initial load range, this will also reduce load factors which will increase the heat rate, i.e., reduce efficiency. Reciprocating engine power plant operating in gas mode is limited to a small range of initial loads with a higher number of operating generators; while in diesel mode, the power plant has a

Table 2. Generator operating scenarios for step load.

Operating Conditions	Each Engine Step Load Increase (MW)	Engine Load Change (%)	Engine Mode and Initial Load Factor Requirement for maximum 4% Speed Change	Required Plant Initial Load for maximum 4% Speed Change (Min/Max, MW)	
2 Engines	3.00	30.00%	Diesel Mode: 50.0% - 70.0% load factor	10.0	14.0
3 Engines	2.00	20.00%	Diesel Mode: 50.0% - 80.0% load factor	15.0	24.0
4 Engines	1.50	15.00%	Diesel Mode: 0% - 85% load factor	0.0	34.0
			Gas Mode: 40% - 60% load factor	16.0	24.0
5 Engines	1.20	12.00%	Diesel Mode: 0% - 88% load factor	0.0	44.0
			Gas Mode: 40% - 60% load factor	20.0	30.0

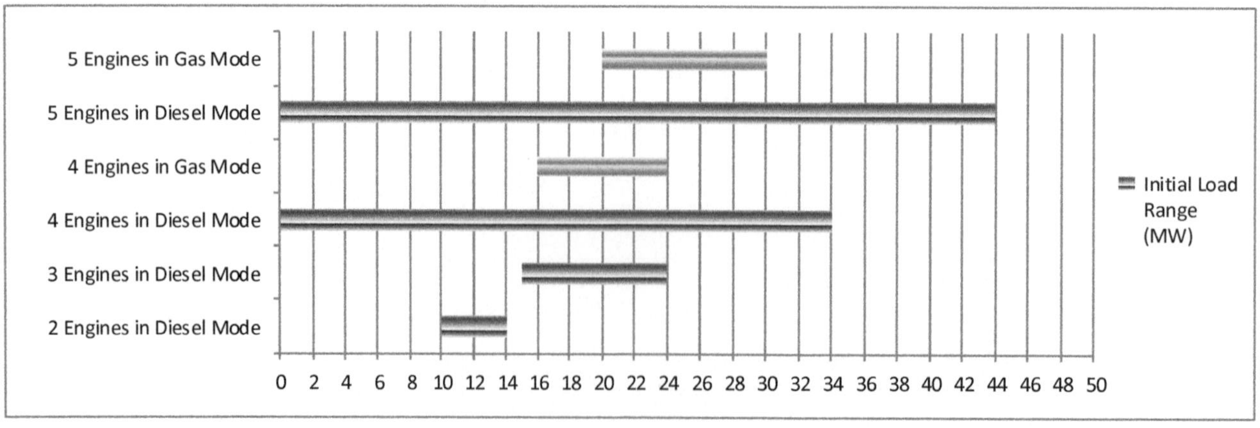

Figure 3. Initial load range at different operating scenarios.

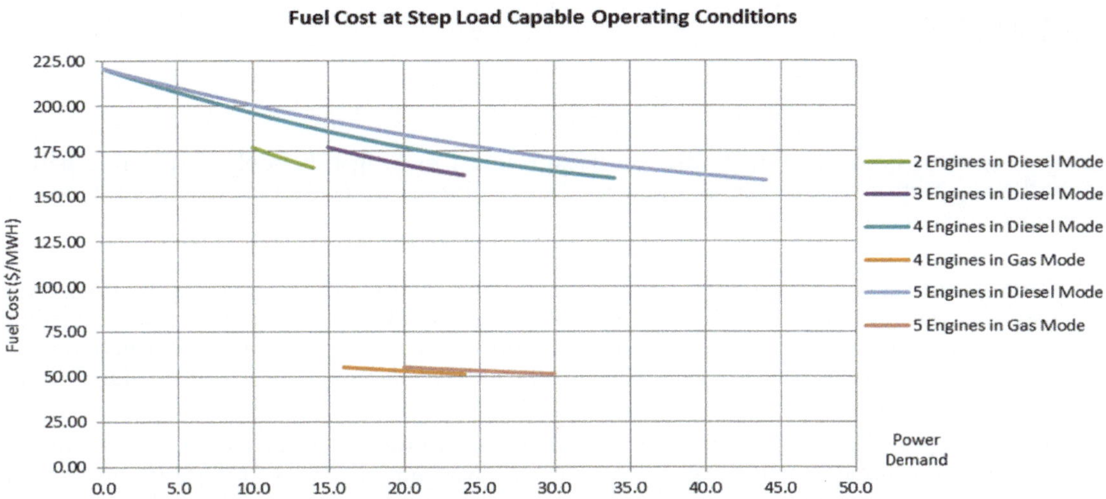

Figure 4. Fuel cost at step load capable operation conditions.

larger initial load range with less operating generators. For a certain power demand, less operating generators can improve load factors, reduce emissions and maintenance levels therefore diesel mode is more favourable from step load response and frequency control perspective. The analysis results also show that when both fuels are suitable for a certain power demand, the fuel costs of power plant in diesel mode are much higher than in gas mode based on the above fuel price assumptions.

6. Conclusions

Step load capability and frequency change represents one of the key issues for reciprocating engine's application in medium to large scale stationary power generation. This paper reviews the frequency impacts to generators and consumer equipment in the network and the influence factors to step load response are discussed.

An analysis method is proposed to establish the operation conditions for a given step load and to allow for optimisation of the operating conditions. The proposed power plant step load capability analysis method includes two steps. The first step is to identify all the operating modes that can satisfy both step load and power output requirements for the whole range of power demand. The second step investigates the fuel cost for all the suitable operating modes and allows for optimisation of the operation the whole range of power demand.

The analysis method can be applied to any reciprocating engine step load curves and will help create operation strategies for the whole range of power demand to meet both step load frequency control and power supply requirements. The optimisation of the operation conditions can reduce the fuel cost while maintaining network stability. The results can greatly reduce the complexity of frequency control in manual control procedure prior to the occurrence of step loads, or could be programmed into generator automatic control settings with an early speed change detection system to control the frequency changes in design limits.

REFERENCES

[1] M. A. Hanley, "Frequency Instability Problems in North American Interconnections," National Energy Technology Laboratory (United States), 2011.

[2] Power Quality Manager, "Supply Standards: Electricity Supply Standard," Essential Energy, 2011.

[3] National Electricity Rules (Australia), Version 55 (2013).

[4] Power System Frequency and Time Deviation Monitoring Report – Reference Guide, Australia Energy Market Operator, 2011.

[5] National Electricity Code Administrator, "Reliability Panel Frequency Standards Consulting Paper," 1998.

[6] B. J. Kirby, etc., "Frequency Control Concerns In The North American Electric Power System," Oak Ridge National Laboratory (United States), 2002.

[7] International Electrotechnical Commission (IEC) Standard 60034-1, "Rotating Electrical Machines - Part 1: Rating and Performance".

Permissions

The contributors of this book come from diverse backgrounds, making this book a truly international effort. This book will bring forth new frontiers with its revolutionizing research information and detailed analysis of the nascent developments around the world.

We would like to thank all the contributing authors for lending their expertise to make the book truly unique. They have played a crucial role in the development of this book. Without their invaluable contributions this book wouldn't have been possible. They have made vital efforts to compile up to date information on the varied aspects of this subject to make this book a valuable addition to the collection of many professionals and students.

This book was conceptualized with the vision of imparting up-to-date information and advanced data in this field. To ensure the same, a matchless editorial board was set up. Every individual on the board went through rigorous rounds of assessment to prove their worth. After which they invested a large part of their time researching and compiling the most relevant data for our readers. Conferences and sessions were held from time to time between the editorial board and the contributing authors to present the data in the most comprehensible form. The editorial team has worked tirelessly to provide valuable and valid information to help people across the globe.

Every chapter published in this book has been scrutinized by our experts. Their significance has been extensively debated. The topics covered herein carry significant findings which will fuel the growth of the discipline. They may even be implemented as practical applications or may be referred to as a beginning point for another development. Chapters in this book were first published by Scientific Research Publishing Inc.; hereby published with permission under the Creative Commons Attribution License or equivalent.

The editorial board has been involved in producing this book since its inception. They have spent rigorous hours researching and exploring the diverse topics which have resulted in the successful publishing of this book. They have passed on their knowledge of decades through this book. To expedite this challenging task, the publisher supported the team at every step. A small team of assistant editors was also appointed to further simplify the editing procedure and attain best results for the readers.

Our editorial team has been hand-picked from every corner of the world. Their multi-ethnicity adds dynamic inputs to the discussions which result in innovative outcomes. These outcomes are then further discussed with the researchers and contributors who give their valuable feedback and opinion regarding the same. The feedback is then collaborated with the researches and they are edited in a comprehensive manner to aid the understanding of the subject.

Apart from the editorial board, the designing team has also invested a significant amount of their time in understanding the subject and creating the most relevant covers. They scrutinized every image to scout for the most suitable representation of the subject and create an appropriate cover for the book.

The publishing team has been involved in this book since its early stages. They were actively engaged in every process, be it collecting the data, connecting with the contributors or procuring relevant information. The team has been an ardent support to the editorial, designing and production team. Their endless efforts to recruit the best for this project, has resulted in the accomplishment of this book. They are a veteran in the field of academics and their pool of knowledge is as vast as their experience in printing. Their expertise and guidance has proved useful at every step. Their uncompromising quality standards have made this book an exceptional effort. Their encouragement from time to time has been an inspiration for everyone.

The publisher and the editorial board hope that this book will prove to be a valuable piece of knowledge for researchers, students, practitioners and scholars across the globe.

List of Contributors

Dequan Gao and Yiying Zhang
Beijing University of Posts and Telecommunications, Beijing 100876, China
State Grid Information & Telecommunication Company Ltd., Beijing 100761, China

Jinping Cao
China Electric Power Research Institute, Beijing 100192, China

Xiuli Wang
Information Communication Branch of Beijing Electric Power Company, Beijing 100031, China

Mojgan Hojabri
Faculty of Electrical and Electronics Engineering, University Malaysia Pahang (UMP), Pekan, Malaysia

Arash Toudeshki
Department of Electrical and Electronic Engineering, University Putra Malaysia (UPM), Serdang, Malaysia

Han-Chang Liu, Chung-Teng Huang, Wen-Kuei Lee and Mei-Hsiu Lin
Green Energy and Environment Research Laboratories, Industrial Technology Research Institute, Hsinchu, Taiwan

Medhat Elkelawy and Hagar Alm-Eldin Bastawissi
Department of Mechanical Power Engineering, Tanta University, Tanta, Egypt

Olayinka S. Ohunakin, Oluwafemi R. Leramo, Olatunde A. Abidakun, Moradeyo K. Odunfa and Oluwafemi B. Bafuwa
Mechanical Engineering Department, Covenant University, Ota, Nigeria

Bouchra Abouelkhayrat, Hamid Hamza, Jawad Lahjomri and Abdelaziz Oubarra
Laboratory of Mechanics, Faculty of Sciences Ain Chock, Hassan II University, Casablanca, Morocco

Jianbo Yang and Faris Al-Naemi
Materials and Engineering Research Institute (MERI), Sheffield Hallam University (SHU), Sheffield, UK

Weiping Zhang and Xiaoping Chen
Lab of Green Power & Energy System (GPES), North China University of Technology (NCUT), Beijing, China

Ayaz Ahmad
Singhania University, Pacheri Bari, India

Rajaji Loganathan
P.B. College of Engineering, Chennai, India

Song Nie
Beijing Key Laboratory of High Voltage & EMC, North China Electric Power University, Beijing, China

Yang-chun Cheng
Beijing Key Laboratory of High Voltage & EMC, North China Electric Power University, Beijing, China
State Key Laboratory of Alternate Electrical Power System with Renewable Energy Sources, North China Electric Power University, Beijing, China

Yuan Dai
Electric Power Research Institute of Guangdong Power Grid Corporation, Guangzhou, China

O. R. AL-Hamdan and A. A. Saker
Water Resources Dept., High Institute of Energy, Public Authority of Applied Education & Training, Kuwait

Abdel-Karim Daud
Electrical and Computer Engineering Department, Palestine Polytechnic University (PPU), Hebron, Palestine

Basim Alsayid
Electrical Engineering Department, Palestine Technical University (PTU), Tulkarm, Palestine

Jieting Wei
Changchun Institute of Technology, Changchun, China

Yang Liu and Yu Gu
Northeast Dianli University, Jilin, China

Yi Wang, Yulan Hu and Ranran An
Electric Power Research Institute, Guangdong Power Grid Corporation, Guangzhou, China

Jiaman Li and Zexiang Cai
School of Electric Power, South China University of Technology, Guangzhou, China

Ruiwen He
School of Automation, Guangdong University of Technology, Guangzhou, China

John Karatzaferis and Emmanuel Tatakis
Laboratory of Electromechanical Energy Conversion (LEMEC), Department of Electrical and Computer Engineering, University of Patras, Patras, Greece

Nick Papanikolaou and Michael Loupis
Department of Electrical Engineering, TEI of Lamia, Lamia, Greece

John Spanoudakis
MILTECH HELLAS S.A., Koropi, Greece

Kazuma Hanada and Takataro Hamajima
Department of Electrical and Electronics Systems, Hachinohe Institute of Technology, Aomori, Japan

Makoto Tsuda and Daisuke Miyagi
Department of Electrical Engineering, Tohoku University, Miyagi, Japan

Takakazu Shintomi
Advanced Research Institute for the Sciences and Humanities, Nihon University, Tokyo, Japan

Tomoaki Takao
Department of Engineering and Applied Sciences, Sophia University, Tokyo, Japan

Yasuhiro Makida
High Energy Accelerator Research Organization, Ibaraki, Japan

Masataka Kajiwara
Iwatani Corporation, Tokyo, Japan

Yuan-Kang Wu and Ming-Yan Zou
Department of Electrical Engineering, National Chung-Cheng University

Shao-Hong Tsai
Department of Electrical Engineering, Hwa Hsia Institute of Technology

Albana Ilo
Siemens AG Austria, Vienna, Austria

Xing Wu, Xiaogang Yin, Quan Wei, Yifan Jia and Jing Wang
Research and Development Center, Xi'an High Voltage Apparatus Research Institute Co., Ltd, Xi'an, China